Neuroscience

Neuroscience

A Historical Introduction

Mitchell Glickstein

The MIT Press
Cambridge, Massachusetts
London. England

First MIT Press paperback edition, 2017

This book was set in Sabon by Toppan Best-set Premedia Limited, Hong Kong.

Library of Congress Cataloging-in-Publication Data

Glickstein, Mitchell.
Neuroscience : a historical introduction / Mitchell Glickstein.
pages cm
Includes bibliographical references and index.
ISBN 978-0-262-02680-2 (hardcover : alk. paper), 978-0-262-53461-1 (pb.)
1. Neurosciences—History. 2. Neurology—History. 3. Brain—Physiology.
I. Title.
RC338.G55 2014
612.8'233—dc23
2013021935

To Ben and Hannah

Contents

Acknowledgments

I am very grateful to the people who helped me to write this book.

Ron Bailey edited the text as I wrote it; Sue Bailey read it and helped me to recognize and correct obscure passages. Susan Buckley was a patient and helpful editor for MIT Press.

Jane Pendjiky was outstanding in helping to find and to convert nearly two hundred figures to a usable format.

Dale Purves and Bill Hall gave me enthusiastic permission to use several figures from their book.

The following colleagues as well as present and former students read one or more chapters in the book, correcting errors and reminding me of additional content: Joan Baizer, Jim Baker, Linda Bartoshuk, Giovanni Berlucchi, Martyn Bracewell, Stuart Cull-Candy, Ford Ebner, Pauline Field, Lea Firmin, Mike Gazzaniga, Alan Gibson, Guy Goodwin, Germund Hesslow, Mark Hollins, Bob Leaton, Mike Loop, Marjorie Lorch, Jim McIlwain, Josef Miller, Ann Mudge, Larry Squire, and Peter Thier, Jan Voogd.

1

Introduction

We are our brains: we see, we feel, we hear by way of messages sent from the sense organs to the brain. We move by commands from the brain to our muscles, relayed by way of the spinal cord. Over two thousand years ago the Greek physician Hippocrates wrote:

> Men ought to know that from nothing else but thence [from the brain] come joys, delights, laughter and sports, and sorrows, grief, despondency, and lamentations. And by this, in a special manner, we acquire wisdom and knowledge, and see and hear, and know what are foul and what are fair, what are bad and what are good, what are sweet and what unsavory. Some we discriminate by habit, and some we perceive by their utility. By this we distinguish objects of relish and disrelish, according to the seasons; and the same things do not always please us. And by the same organ we become mad and delirious, and fears and terrors assail us, some by night, and some by day, and dreams and untimely wanderings, and cares that are not suitable, and ignorance of present circumstances, desuetude, and unskilfulness. All these things we endure from the brain, when it is not healthy. . . .[1]

This book is about the brain and the spinal cord and their connections to the sense organs and the muscles of the body. The aim of this book is to discuss how we know about these processes. What were the observations and experiments that help to understand the structure and function of the brain and spinal cord? One unifying theme is that of localization. Are there parts of the brain and spinal cord that do different things? What do the different parts of the brain and spinal cord look like? How are they constructed and interconnected? If we consider any specific region of the brain or spinal cord, we can ask: What is known about what that particular region does?

Our knowledge of neuroscience comes from many sources. Some of the earliest clues came from observing the effect of injuries to the brain and spinal cord in humans and animals. An ancient observation relates to the role of the spinal cord in movement and posture. In the sixth

Figure 1.1
From the British Museum Assyrian collection ca. 600 BC. The figure is from a wall panel showing the effect of a spinal cord lesion. The lioness has been pierced by a spear, interrupting the spinal cord at a midthoracic level. Note the flaccid paralysis of the lower limbs and the normal posture in the upper limbs.

century BC, Assyria occupied a region of the Middle East between the Tigris and Euphrates rivers, now mostly inside Iraq. At the time of the Assyrian empire there were wild lions in the region that posed a danger to humans and livestock. One of the functions of the Assyrian kings was to protect their people and their flocks by hunting these Mesopotamian lions. Figure 1.1 is from a sculpture on the walls of a sixth century Assyrian temple showing a lion whose spinal cord had been severed.

The hind paws appear to be paralyzed; they no longer support the animal's hindquarters. Cutting the spinal cord produced paralysis in the parts of the body below the level of the cut. Paralysis as well as lack of sensation below the injury occurs in people with similar injury to the spinal cord.

Some of what we know about the brain and its functions comes from such chance observations and clinical descriptions of the effects of lesions in humans. But most of what we know came from experiments.

The field of neuroscience arose from many parent disciplines. The aim of this text is to introduce neuroscience from a historical perspective. I believe that a science can often be best understood by studying the work of its pioneers, those people whose work has led to our current understanding.

Here are some of the key questions that have been asked in arriving at our present knowledge:

- How were neurons—the nerve cells that are the basic building blocks of the nervous system—and their processes first identified?
- How do axons—the long fiber processes that carry sensory information and motor commands—interconnect nerve cells and activate muscles?
- How are pain, temperature, smell, and taste received and coded by the brain?
- How does a person or animal recognize objects by seeing them, hearing them, or touching them?
- What is the mechanism whereby an image that is formed by the eye is analyzed to yield the percept of an object?
- How is movement organized by the brain?
- How do we manage to avoid bumping into other people when we walk on a crowded street?
- How is it that we reach accurately for a glass of water or catch a moving ball?
- Are improvements in skills and perception as we grow from infancy to adulthood determined innately, or must they be learned?
- What sorts of complex behavior are innate, requiring no learning, and how is such behavior elicited?
- What is sleep; how is a relaxed state different from an alert state of mind?
- Why does a person eat or drink; what makes us hungry or thirsty?
- At a given moment why do we select one particular goal over all of the other goals that we might choose?
- If we have had no food or water for a day, which are we likely to seek first? What are the factors that determine such a choice?
- What is thought; what happens in the brain when we think?
- How do motives and feelings arise from activity of the brain?

There are many more questions we could ask, but this list gives an idea of the range of the questions we can ask. Some remain unanswered, and some have been answered only in part. Neuroscience attempts to answer these questions from the biological point of view: How are these functions controlled by the brain and spinal cord?

Study of the brain has an ancient history and derives from several branches of medicine and science. Neuroanatomy is one of these core sciences. Anatomy's task is to describe biological structure from its grossest features such as the large muscles and bones of the leg down to the level of a single cell and its smallest constituent parts. Neuroanatomy became recognizable as a distinct branch of anatomy several hundred years ago. From the neuroanatomists of previous centuries we learned about the basic parts of the brain; about the cell groups and fiber tracts that make it up. The nineteenth-century neuroanatomists first described the nature of nerve cells and how they connect with one another. The neuroanatomists of the twentieth century continued to teach us about the structure of nerve cells, their constituent parts, and how they are interconnected.

Just as neuroanatomy became distinct from general anatomy because of the special nature of brain, for the same reasons neurophysiology represents a highly specialized area of physiology. Neurophysiologists have demonstrated that nerves function electrically, and from their work we have understood the physical basis of that electrical activity. Neurophysiology has helped to understand the mechanism of reflexes—stereotyped, predictable, and unlearned responses to certain sensory stimuli. Absent or altered reflexes can be related to damage to the brain or spinal cord.

Allied to neurophysiology are neurochemistry and neuropharmacology. Nerve cells activate or inhibit one another by releasing a small amount of a chemical substance at nerve terminals that is called a neurotransmitter. How many such substances are there, and how do they work? Neurochemistry and neuropharmacology are directed toward study of how nerve cells communicate with one another by way of these chemical transmitter substances. Their work forms the basis for the analysis of how drugs affect the brain.

An important source of input into our thinking about the brain and its functions comes from the work of physicians who have attempted to correlate the site or nature of a brain injury with the symptoms that are caused by such injury. We know something about brain localization for speech and language from the writings of nineteenth-century physicians who studied the location of brain injuries in patients who lost the power to produce speech or understand language following damage to the brain. We know the location and organization of the visual area of the human brain by studying cases of blindness or partial blindness produced by brain injury.

Neuroscientists have used techniques for stimulating, recording, or ablating parts of the brain of experimental animals. By experiments of this kind, we learned about the role of certain areas of the brain in the learning and storage of visual memories.

Science often makes progress from intensive analysis of simple systems. Much of what we now know came from the study of animals with simple nervous systems. A giant axon in the squid conducts impulses in the same way as a mammalian axon, and yet it has a cross sectional area that is about a hundred times as big as that of the largest mammalian nerve fiber. Because squid axons are so big, they can be impaled with recording electrodes that allow us to analyze the physical and chemical basis of their electrical activity.

The mollusc *Aplysia*, or sea-hare, has a wide range of behavior. Like all animals, it must feed, escape from predators, and reproduce. But whereas mammals typically have millions or billions of nerve cells in their brains, *Aplysia* has only an estimated 20,000 neurons. Some of *Aplysia*'s neurons are very large—as much as 1,000 times the volume of a large vertebrate nerve cell—and are present in the exact same location and have the same connections and function in every member of the species. These so-called identified neurons often mediate simple behavior. The presence of large identified cells, which are relatively few in number, has allowed detailed analysis of the interconnection and functions of the nervous system of *Aplysia*.

The behavior of invertebrates often seems so well understood that it is as if the animal were a bit of clockwork—a passive captive of its environment—behaving in highly predictable ways to specified stimuli. Study of invertebrates poses a question: How much of *Aplysia*'s behavior can be understood and predicted on the basis of the stimuli presented to it and the interconnections of its nervous system? If *Aplysia* can be understood in this way, how much of the behavior of a fish, or bird, or a cat, or a person can similarly be understood and predicted? Study of brain function should give us an intuitive beginning toward answering such questions.

On the Structure of This Book

It should be clear that the subject of neuroscience is both broad and deep. No single introductory book can deal with all that we know. This book is aimed at explaining not only *what* we know but *how* we know. What were the experiments and clinical observations that gave us the knowledge that we have. I hope the reader can feel the same admiration, awe, and gratitude that I have for the people whose work has given us the field of neuroscience. Not only have they contributed to what we know now, but their work also has given us guides for thinking about how to solve the puzzles that remain.

2

Overview of the Nervous System: Structures and Functions

The structure of the nervous system can be described in two ways. One way is at the level of its major subdivisions—the groups of nerve cells and the tracts of nerve fibers that link them together. The other way is in terms of its microscopic elements—the individual cells that form the larger structures. In this chapter we focus on the former view of the nervous system. In the chapter that follows we consider the nerve cells and their supporting tissue.

Subdivisions of the Nervous System

The nervous system has two major subdivisions. The *central nervous system* consists of the nerve cells and fibers within the brain and spinal cord. The *peripheral nervous system* refers to the groups of nerve cells and fiber tracts that are distributed throughout the body outside the brain or spinal cord. Both systems are necessary for normal function. Nerves that control voluntary movement have their cell bodies in the spinal cord or brain. Nerves for sensory input have their cell bodies outside the central nervous system and relay sensory information to the spinal cord or brain by peripheral nerves.

The terminology that is used to describe the brain and its parts developed haphazardly over hundreds of years of anatomical study. To a new student the terminology presents an array of unfamiliar terms, often baffling in their number and difficulty. But we need such labels to understand what is known about the functions of each part of the nervous system. Much of the terminology comes from Greek and Latin roots. Structures were typically named for their similarity in appearance to a common object. One structure, shaped like an almond, was named *amygdala*—Greek for almond. Similarly, the *hippocampus* got its name from the Greek for seahorse, which it seemed to resemble—at least, in the anatomist's fancy (figure 2.1). An alternate name for the hippocampus, *Ammon's horn*, was proposed by the French anatomist Duvernoi, who dissected the structure along with the fiber tract that connects it to other structures in the brain. He noted that the resultant structure looked rather like a ram's horn.

Translations might not help to understand function, but they take some of the mystery out of the terminology and perhaps provide aids in remembering the many parts of the nervous system.

In addition to carrying sensory messages and motor commands, the peripheral nervous system regulates the functioning of internal organs such as the heart, the stomach, and the gut. These systems function relatively automatically; hence, they are collectively known as the *autonomic nervous system*. The autonomic nervous system is further subdivided into a *sympathetic* and a *parasympathetic* division. The two systems usually have opposite effects on the organs that they innervate. For example, the parasympathetic nerves decrease the heart rate, whereas the sympathetic nerves to the heart speed it up. In general, the sympathetic system mobilizes bodily resources for emergencies; the parasympathetic system tends to conserve resources.

This chapter presents a broad overview of the central nervous system, labeling its principal parts and, where possible, associating known functions with some of those labeled structures. Many of these structures are discussed in greater detail in later chapters.

The Central Nervous System

The brain and spinal cord make up the central nervous system (CNS). Both are encased in bone and are the best-protected parts of the body. The brain is contained within the skull. The spinal cord is contained within a tube formed by the spine's 32 donut-shaped bones. These bones, known as vertebrae, are held together by ligaments. At the upper end of the neck, the level of the highest of the vertebrae, the spinal cord is continuous with the brain through the *foramen magnum* (Latin for large opening) in the base of the skull.

In the sixteenth century, anatomy based on the dissection of human cadavers became a part of education for surgeons and physicians. The greatest of the Renaissance anatomists was Andreas Vesalius. Vesalius represented a new generation of scholars who were not content to repeat the thousand-year-old anatomical descriptions but, instead, relied on their own observations. Figure 2.2 is from Vesalius's *Fabrica*, published in 1543.

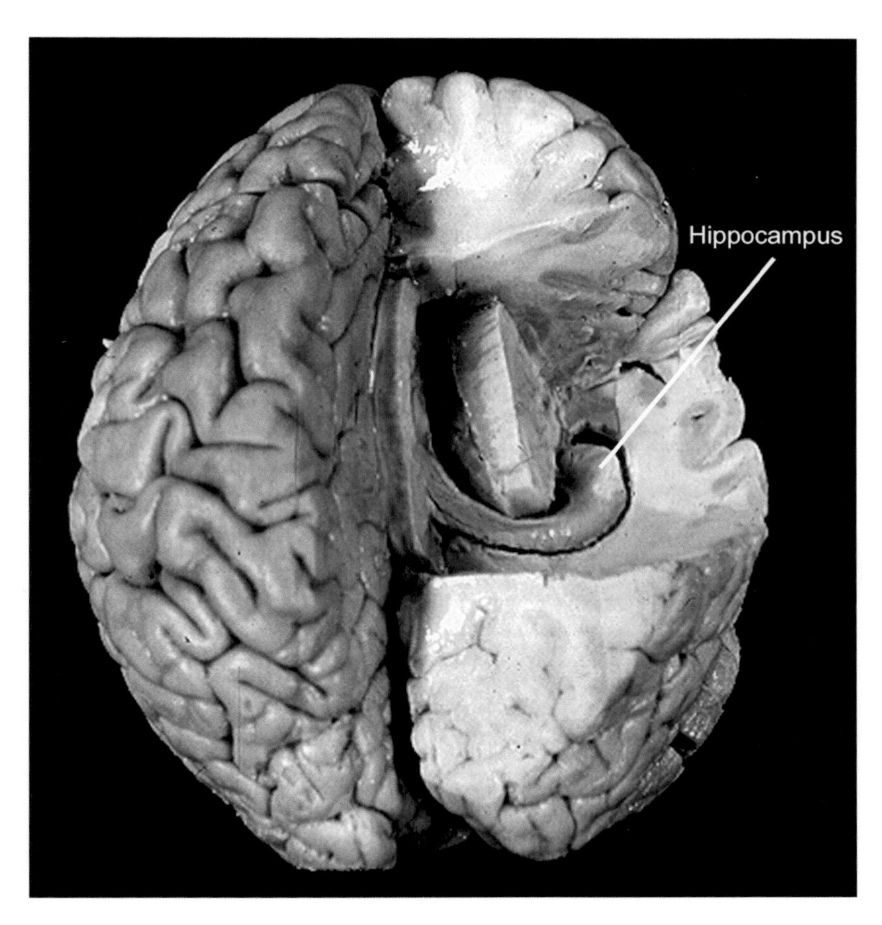

Figure 2.1
A dissection of the human brain. The hippocampus is shown along with its major output fiber tract, the fornix. The shape resembles that of a ram's horn, which led to the alternate name; *Cornu ammonis* or Ammon's horn. Courtesy University of Washington.

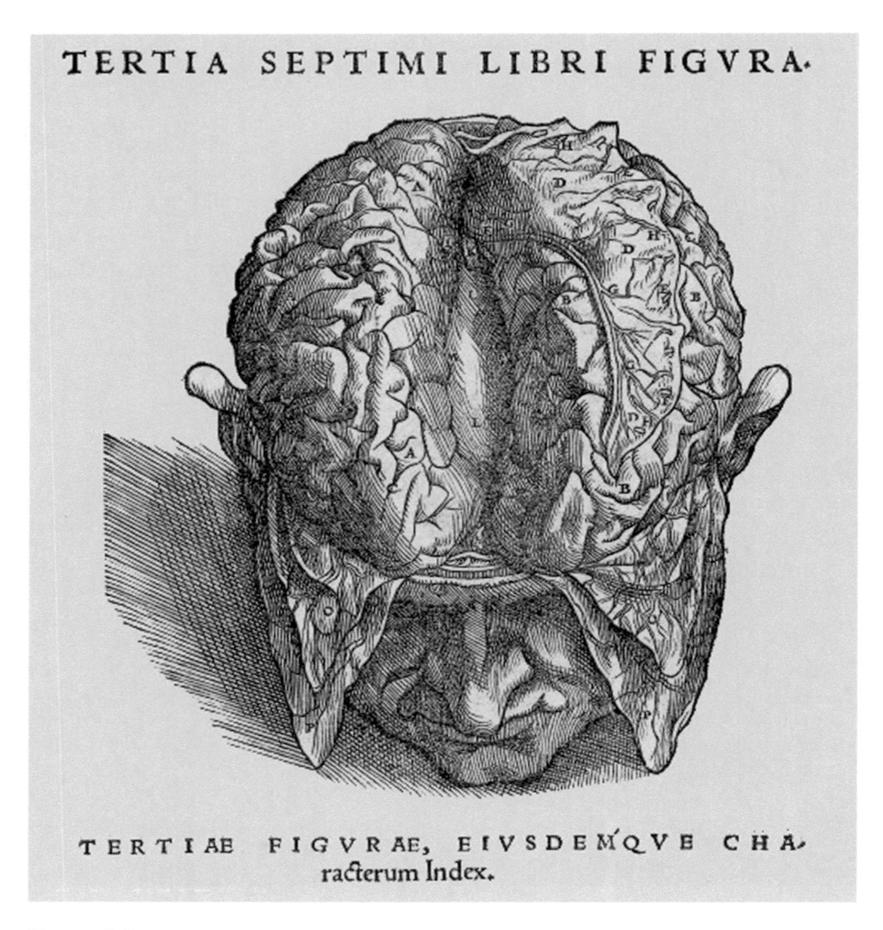

Figure 2.2
Andreas Vesalius (1514–1564) was an early pioneer in anatomical dissection. The cerebral hemispheres are gently retracted to show the corpus callosum (L), the great band of white matter between the two hemispheres. Courtesy Wellcome Library, London.

Vesalius's figure illustrates a typical dissection that he would have done. Scalp and bone form the first layers of protection for the brain and spinal cord. If the skull is removed with bone-cutting tools, there is a heavy, skin-like membrane, the *dura mater,* which is the outermost of the *meninges.* Dura mater, or "hard mother," surrounds the entire brain and spinal cord. In the brain of a mouse or rat, the dura is thin and delicate. In the human brain and spinal cord it is tough and leathery, reminding early anatomists of an elephant's skin, hence the dura mater's alternative name, *pachymeninges.*

The dura is the outermost of the three meninges, the layers of connective tissue that surround the CNS. Not obvious in the Vesalius dissection are the *arachnoid* and *pia mater.* Pia, the innermost of the three meninges, is a delicate membrane—the "tender mother"—that is fused to the surface of the brain. Pia is very soft and thin and surrounds the entire brain quite closely, extending in a continuous sheet into the folds and furrows in the brain. Between the pia, the innermost covering, and the dura, the outermost, is the arachnoid—so named because its delicate, filamentous structure resembles the filaments of a spider's web. Inflammation of the meninges through bacterial or viral infection is called meningitis, a disease that may produce severe neurological symptoms because of the nearness of the meninges to the brain and spinal cord.

The brain and spinal cord are surrounded by a clear cerebrospinal fluid, abbreviated CSF. In the living brain the space between the arachnoid and the pia mater is filled with CSF. This clear fluid, which is similar to blood plasma but without the blood cells, also fills the four hollows, or *ventricles,* within the brain (figure 2.3), and it is continuous with the CSF surrounding the spinal cord.

Although early anatomists knew that the ventricles contained a fluid, the connection with the fluid surrounding the spinal cord was not recognized until the eighteenth century. Cotugno, an Italian physician whose home was in Naples, wrote: "It seems astonishing that eminent men who have very carefully examined the fluid in the cavities of the human body have overlooked the very remarkable and, so to speak, chief and abundant collection of fluid in the spine."[1]

Ancient theorists thought that the ventricles were reservoirs for "vital spirits" that controlled muscle movement and mental activities. We now know that the CSF in the ventricles supplements the brain's circulatory system by supplying nutrients and stabilizing its chemical environment. CSF also serves as a buoyant shock absorber, not only cushioning the brain against the impact of blows to the head but also protecting it from

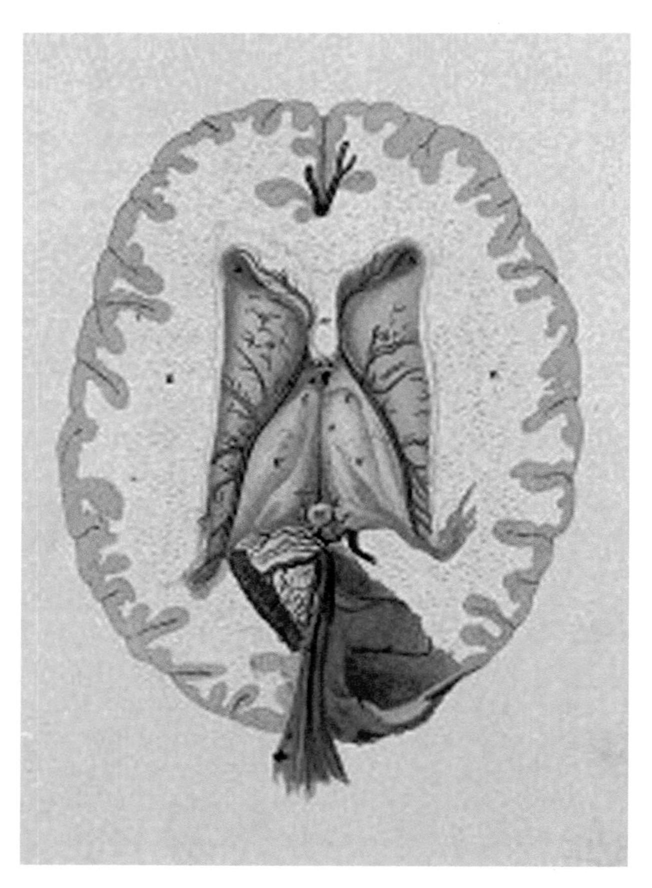

Figure 2.3
A horizontal cut through the top of a human brain. The open spaces on either side reveal the two lateral ventricles. These two lateral ventricles are connected to the third ventricle in the midline. In life, the ventricles are filled with cerebrospinal fluid (CSF). Courtesy Wellcome Library, London.

the pull of gravity. Brain tissue is soft, with a gelatinous consistency. If the brain were placed unsupported on a hard surface such as a table, gravity alone would distort it.

The brain is continuous with the spinal cord. The human spinal cord is a flattened cylinder about as thick as a thumb and roughly 18 inches long. In lower vertebrates such as fish, the spinal cord and brain are aligned along the straight axis of the animal's head and body. In humans and monkeys the continuity between brain and spinal cord is less obvious; the brain is much larger relative to the spinal cord, and the axis of the nervous system is bent almost 90 degrees. The brain is not on a direct line with the spinal cord but seems to sit on top of it, like an apple on a stick.

The spinal cord contains local involuntary reflex connections and the major pathways that link them to the brain and body. A cross section of the spinal cord shows a butterfly-shaped gray mass in the center, principally made up of nerve cells. Outside the gray center are white columns—bundles of myelin-covered nerve fibers that interconnect structures in the nervous system.

Substructures of the Brain

The human brain is very roughly spherical. Viewed from the top, the exposed brain is pinkish gray and wrinkled like a walnut. The infoldings are called fissures or sulci. This fissured surface, which covers much of the brain's exterior, is the cerebral cortex—cortex from the Latin for bark or shell. If you were to cut through a fresh brain you would see a color difference between the cortex, which is gray, and the white matter that lies below it. In the brain and the rest of the nervous system, gray matter is made up of masses of nerve cells and their supporting structures. White matter is made up principally of nerve fibers, many of which are covered with a white, fat-like coating called the myelin sheath. Myelin covers many of the nerve fibers, which are grouped into tracts that connect structures in the brain and spinal cord with one another.

Many of those tracts are named simply for the structures that they link—beginning with the area in which the tract originates and ending with the area to which it connects. Thus, for example, the spinocerebellar tracts arise in the spinal cord and terminate in the cerebellum. The spinocerebellar tracts arise from nerve cells whose cell bodies are in gray matter of the spinal cord and whose fibers travel upward in the white matter of the spinal cord to the cerebellum. Similarly, the spinothalamic tract has its cell bodies in the gray matter of the spinal cord, and many of its fibers travel in the white matter of the spinal cord to reach the thalamus. Cells in the thalamus give rise to fibers that project to the cerebral cortex.

Although there are variations in size and fissure patterns, all human brains are fundamentally alike. Popular lore would have it that the bigger the brain the more intelligent the creature, but there appears to be no correlation between size and function in the human brain. The exceptions are those rare instances where the brain remains abnormally small and never produces normal function.

The brains of notable historical figures—among them, Einstein, Lenin, and Napoleon—have been weighed and measured with great interest at postmortem but without any appreciable gain to the understanding of brain function. On average, the human brain weighs about 1,500 g. The brain of the Russian novelist Ivan Turgenev weighed more than 2,000 g, but that of another gifted writer, the American poet Walt Whitman, weighed only 1,282 g. Variation in brain size is more closely related to body size than to intellect or talent. The weight of a woman's brain, for example, averages about 10 percent less than that of a man—approximately the same difference shown in average body weight. The largest mammalian brain belongs to the whale.

In addition to its massive cerebral cortex the brain has many subcortical nerve cells. Like the cortex these subcortical masses of nerve cells are typically interconnected locally and also send out tracts of nerve fibers that make connections between brain structures.

In this brief survey of structures we begin at the base of the brain where some of the most basic functions are localized and move upward to the so-called higher centers. We also briefly describe a few of the structures' functions to go along with the colorful Latin and Greek names.

At the base of the skull, rising stalk-like from the spinal cord into the base of the brain, is the brainstem (figure 2.4). The brainstem is essentially a continuation of the spinal cord, and it is composed of both gray and white matter. The brainstem is composed of three major parts: the medulla oblongata, the pons, and the midbrain. The pons and the medulla have centers that control respiration by way of nerve fibers that descend to the spinal cord where they connect with motor nerve cells. These motor cells in turn connect with the muscles that expand the chest for breathing. A broken spinal cord at the level of the neck causes death by severing those nerve connections. The hangman's noose brings an end to breathing.

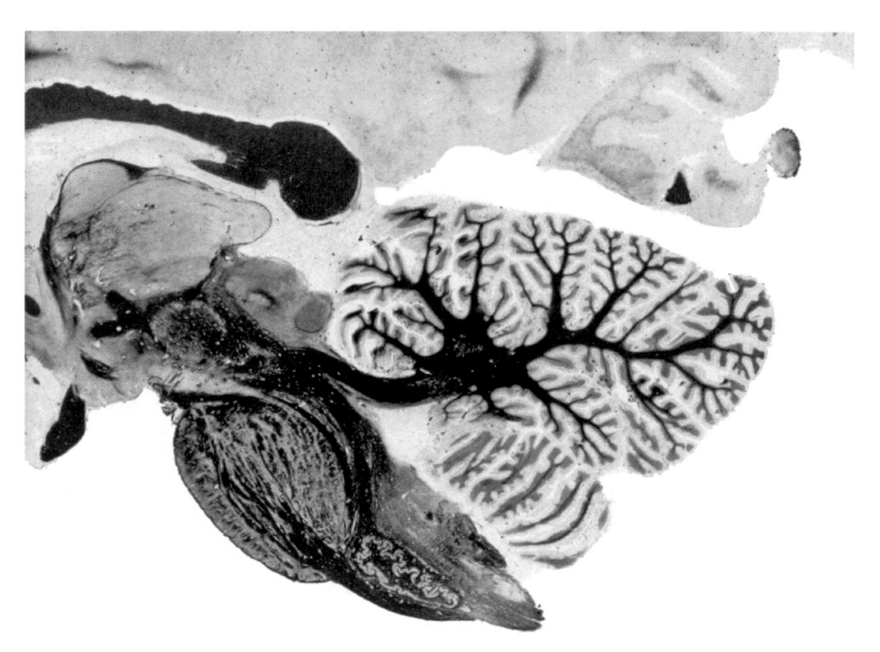

Figure 2.4
A parasagittal cut through the human brainstem and cerebellum (fiber stained). The large dark-stained tract between the cerebellum and the brainstem is the superior cerebellar peduncle, the major fiber tract leaving the cerebellum. Note also the pontine nuclei, the prominent egg-shaped structure below the cerebellum.

If you view the brainstem from its underside (figure 2.5), you will see an arching sweep of fibers connecting the pons to a large structure at the rear of the brainstem. To the sixteenth-century Italian anatomist Constanzo Varolius, this region looked like a bridge over a Venetian canal; hence, he named it the pons—from the Latin for bridge.

The *cerebellum* is a large structure, about the size of a peach; it sits just above the brainstem at the level of the pons. Named the cerebellum, or little brain, it looks like a smaller version of the cerebrum, the massive front part of the brain that overshadows it. Like the cerebrum, the cerebellum is densely furrowed (figure 2.6). If you look at the surface of the human cerebral cortex, you see only about a third of it, with the rest being in the walls and depths of the fissures. If you look at the human cerebellum you see only about one-tenth. The human cerebellum consists of a narrow midline structure—the *vermis* (Latin for worm)—and two massive cerebellar hemispheres.

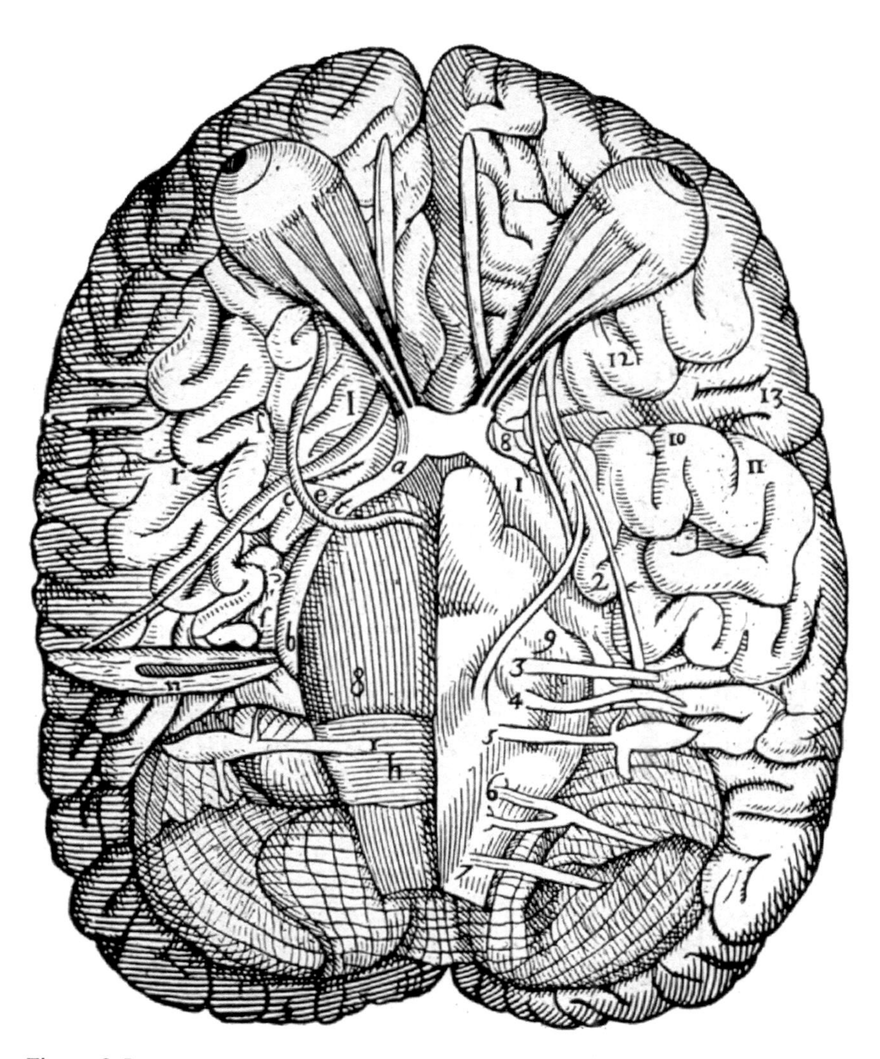

Figure 2.5
Varolius (Constanzo Varolio, 1543–1575) was among the first anatomists to dissect the human brain from below. Because of that approach he could see and describe, for the first time, many of the important structures at the base of the brain. The prominent structure at the base of the brain seemed to him to resemble a bridge; hence, Varolius called it the pons.

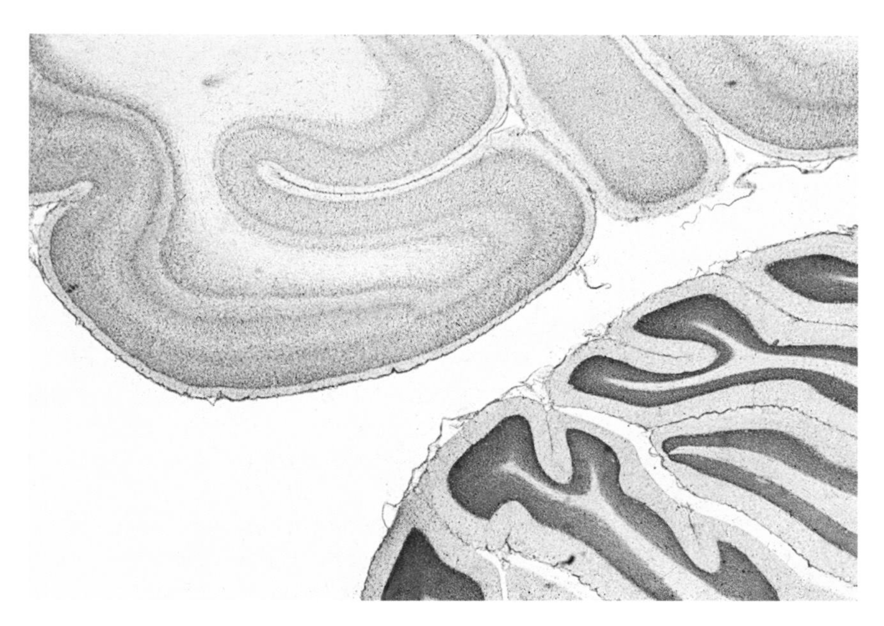

Figure 2.6
Parasagittal section through the monkey cerebral cortex (upper left) and cerebellum (lower right), Nissl stained. The cerebellar cortex is about half the thickness of the cerebral cortex, and it is more deeply folded.

The cerebellum is joined to the brainstem at the level of the pons by three paired *peduncles*, or stalks; the middle one is continuous with the "bridge" of pontocerebellar fibers that gave the pons its name. The cerebellum maintains the body's equilibrium and coordinates muscles during fine movements such as threading a needle. As we shall see, the cerebellum is also involved in certain forms of motor learning, and some have argued that it plays a role in the more complex functions of language and thought.

The midbrain consists of several prominent cell groups and large collections of nerve fibers that pass through the midbrain to interconnect structures above and below it. On the top are four small hillocks; the two in front are related to sight, and the two in back to hearing. Attached to the midbrain is a little gland, the *pineal*, which, because of its central location, was thought by the seventeenth-century philosopher René Descartes to be "the seat of the soul." In some amphibians the pineal's light-sensitive cells function as a kind of third eye that adjusts day-night activity levels. In humans the pineal may also serve as a built-in timekeeper. At the base of the midbrain there is a massive bundle of fibers known as the cerebral peduncles or, more properly, the *basis pedunculi.*

The great majority of the fibers of the basis pedunculi go from the cerebral cortex to the pons, medulla, and spinal cord. Those fibers that descend past the medulla to the spinal cord form one of the motor tracts of the brain, the corticospinal or pyramidal tract—so called because the fibers have a pyramidal shape as they run along the base of the brain at the level of the medulla (figure 2.7). Just above the basis pedunculi in the midbrain there is a group of cells called the *substantia nigra* (Latin for black substance), so called because of the dark pigment granules contained within the cells of one of its subdivisions. Loss of nigral cells is associated with the severe motor defects of Parkinson disease. We learn more about this system in a later chapter.

At the very top of the midbrain is the diencephalon. The diencephalon is composed principally of the *thalamus*—a relay station for incoming pathways to sensory and motor areas of the cerebral cortex. There is not complete agreement on the origin of the term thalamus. One interpretation is that it was the designation of the antechamber just inside the entrance to a temple. The *hypothalamus*, which lies below the thalamus, governs the autonomic nervous system. The hypothalamus is involved in the regulation of body temperature, appetite, thirst, and sexual activities. Hanging down by a stalk from the hypothalamus is the pituitary gland, which secretes hormones that influence a wide range of functions—from body growth to reproduction.

The brainstem and its associated structures are largely hidden by the two massive bodies that arch above and around them, the *cerebral hemispheres*. It is the relative size of the cerebral hemispheres and especially the extent of their folded gray surface, or cortex, that most distinguishes the human brain from that of other organisms. If we cut the two sides of the brain apart, we can see a large band of nerve fibers joining the two hemispheres. This band of fibers—the *corpus callosum* (Latin for hard body)—is about one-quarter of an inch thick and three and one-half inches long.

The *limbic system* is a set of structures forming a ring-like collection of interconnected centers, some of which are related to emotion. Initially the hippocampus (Ammon's horn; see figure 2.1) was considered a vital link in that circuit, but its functions are now recognized to be principally related to memory. On the other hand, the amygdala, not initially included, is definitely part of the group of structures related to

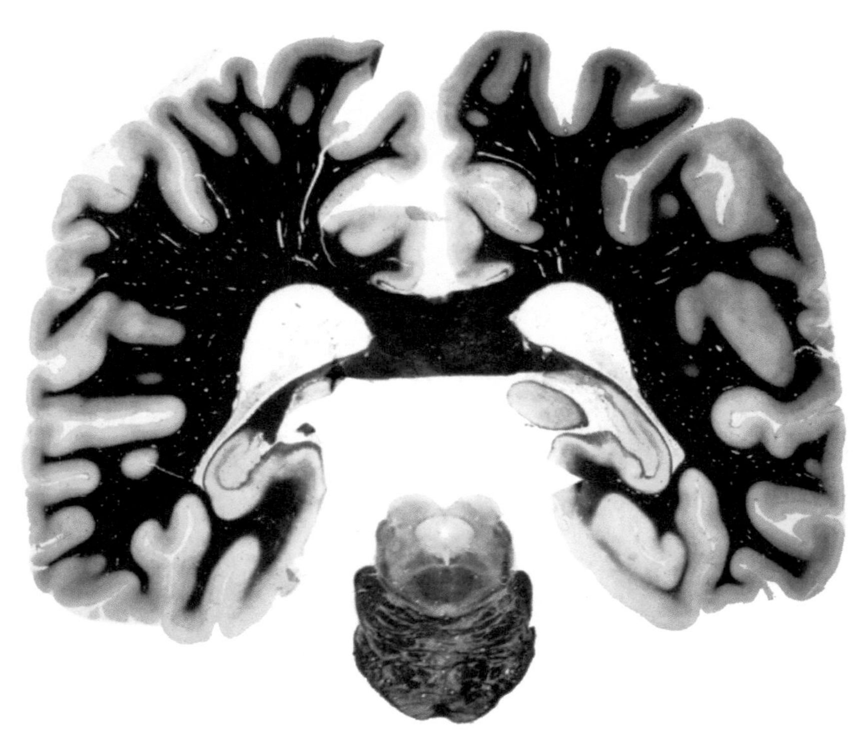

Figure 2.7
A fiber-stained section through the human cerebral cortex and brainstem. The midbrain and pons are below. The spherical structure at the roof of the midbrain is the superior colliculus; the large structure filling the base of the midbrain is the pons.

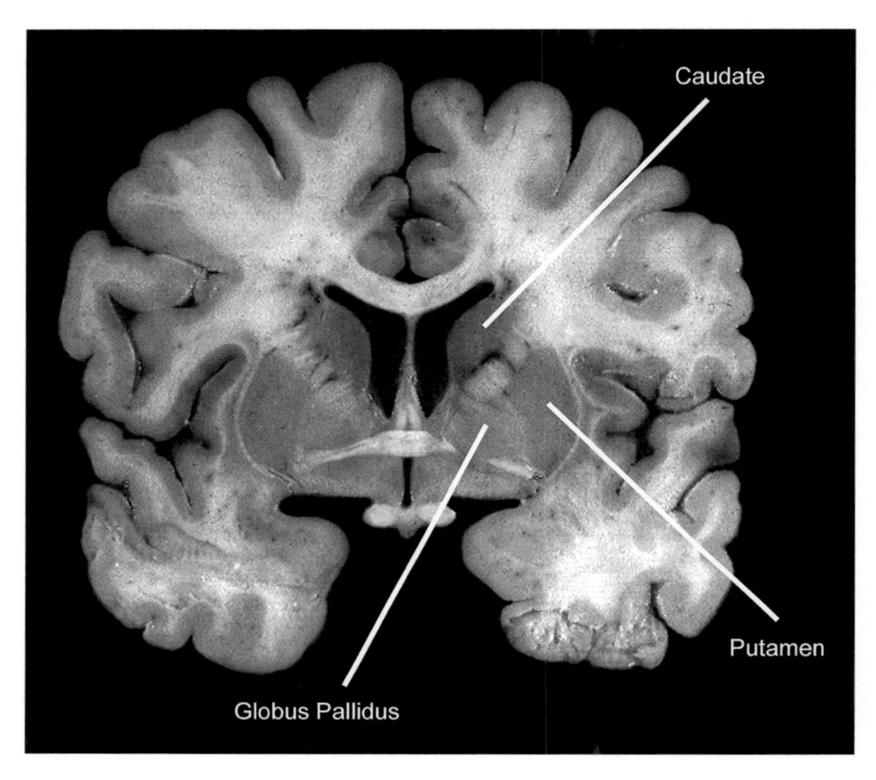

Figure 2.8
MRI image of the human brain in cross section. The position of the three massive basal ganglia structures—caudate, putamen, and globus pallidus—are labeled. Courtesy University of Washington.

controlling emotion. These centers, together with the hypothalamus and a part of the frontal lobe, are involved in controlling emotions such as anger, fear, and pleasure. Individual structures and their functions are described in more detail in later chapters.

Below the cortex and in front of the thalamus are the massive *basal ganglia* (figure 2.8). The basal ganglia consist of the *caudate* ("tailed") *nucleus*, whose tail seems to wrap around the thalamus, the *putamen* (Latin, a shell), and the *globus pallidus* (the pale globe). Collectively these nuclei are involved in the regulation of movement. Diseases of the basal ganglia produce profound deficits in movement.

If we wanted to locate objects on a two-dimensional flat surface, we would need only two dimensions; North–South and East–West would

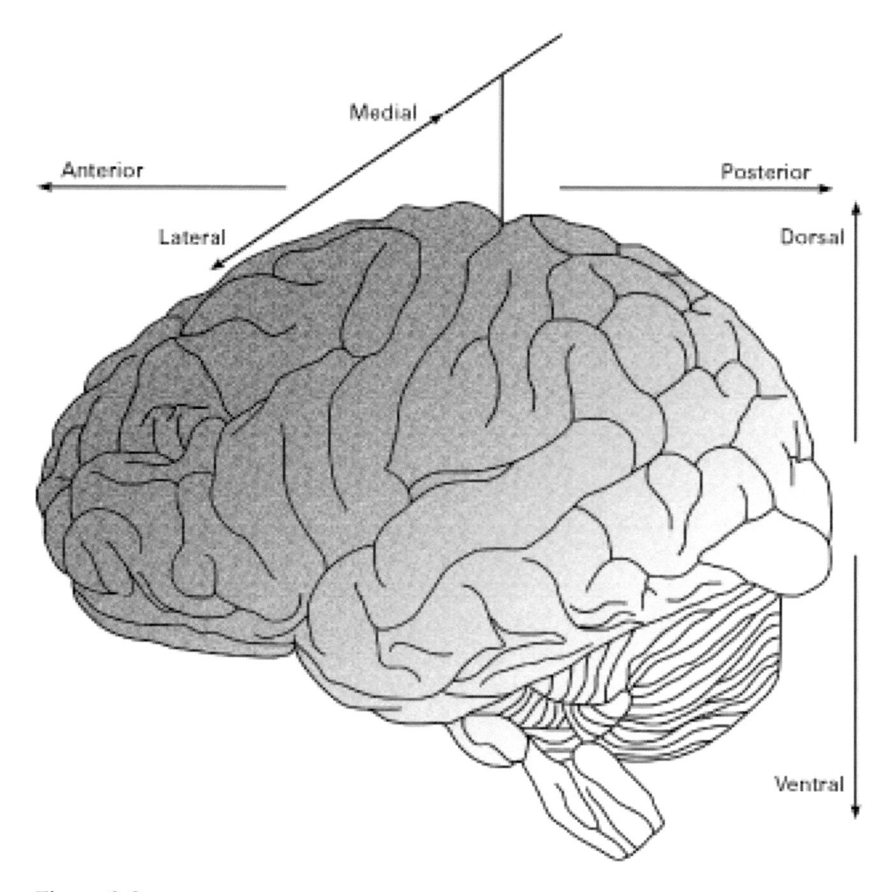

Figure 2.9
The brain is a three-dimensional object, so we need three dimensions to specify locations within it.

do. To map the location of structures in the brain we require three directions (figure 2.9). Anatomists describe the location of structures with a system of coordinates related to the main axis of the brainstem. Structures close to the head end of the nervous system are called *rostral* or superior. Structures close to the tail end of the nervous system are called *caudal* or inferior. This is one coordinate.

For the second coordinate, think about a plane passing through the very center of the body (between the eyes and extending from front to back). Structures closer to the midline are called *medial* structures; those placed laterally are called *lateral*.

We need a third coordinate, for the brain has three dimensions. The generally used descriptions for this third dimension are *dorsal* (toward the back) and *ventral* (toward the belly). In the case of the human spinal cord, *anterior* is sometimes used instead of ventral, and *posterior* is sometimes used in preference to dorsal.

We can practice these directional terms with some structures whose names we have already learned. The midbrain is rostral to the pons. The spinal cord is caudal to the medulla. The cerebellum is dorsal to the pons and medulla.

The Cerebral Cortex

The uniqueness of the human brain resides principally in the folds and fissures of the cerebral cortex. This crumpled covering of the gray matter of nerve cells is a total of about 2 to 4 mm deep and conventionally divided into six layers. Fish and amphibians have no true cortex, and in reptiles the cortex is much simpler in its construction than it is in mammals. Small mammals like the rat have a six-layered cortex, but the hemisphere is smooth, without fissures. Human cortex, by contrast, folds in upon itself so as to triple the surface area that can fit within the confines of the skull.

Prominent grooves between the folds create landmarks that serve as convenient reference points on the cortical terrain. These grooves are called *fissures* or *sulci* (Latin for furrows); the domes of cerebral tissue between them are *gyri* (Latin for rolls). Unusually deep sulci are known as fissures. Two fissures serve as major landmarks on the surface of each cerebral hemisphere. Both were first named for the early anatomist who described them: the central sulcus (of Rolando), which runs from medial to lateral across the surface, and the lateral fissure (of Sylvius), which emerges from the bottom of the hemisphere and curves upward and backward along the side. These two fissures segregate the hemispheres into major regions, or *lobes* (figure 2.10). Each lobe has functions that will help us associate the name and region with particular function of everyday behavior such as speech, movement, hearing, and sight.

Mapping of cerebral functions is a relatively recent development in the brain sciences. In naming and describing the regions of the cortical terrain, the earliest anatomists did not envisage the possibility that a particular region might be identified with a specific function of the brain. A keystone of modern neural science was the discovery of localization of functions in the cerebral cortex. The prevalent notion, up to the early

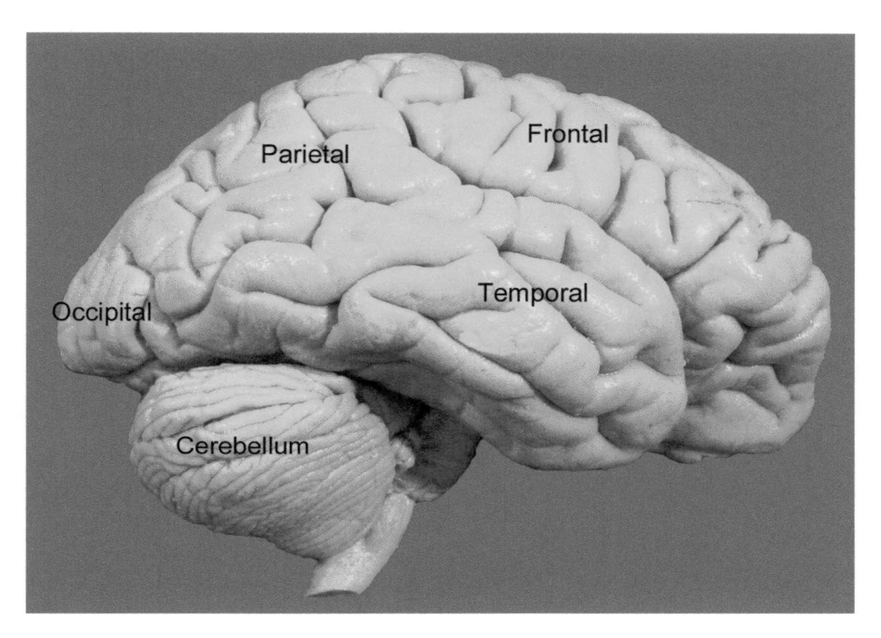

Figure 2.10
A photograph of the right side of a human brain with the four major lobes indicated. The cerebellum is seen to the left and below.

nineteenth century, was that the cerebral cortex acted as a whole to control behavior.

One of the early exponents of localization was the Viennese physician Franz Joseph Gall (figure 2.11, right). He taught that the capacity for memory was centered in the front of the cerebrum, just behind and above the eyes, because he felt that people with good memories had bulging eyes. Gall and his disciples expanded this approach to localization into the pseudoscience known as cranioscopy, later called phrenology, from the Greek for study of the mind. They pinpointed a large number of different areas of the cortex and related each to behavioral traits—from mathematical talent to cruelty and greed. Development of a particular trait was assumed to expand that portion of the cortex and produce a bulge in the skull (figure 2.11, left). Followers of phrenology believed that a person's character and mental abilities could be assessed by measuring the bumps on his head.

Phrenology was preposterous. Although it gave localization a bad name, functional specialization of different cortical areas was soon

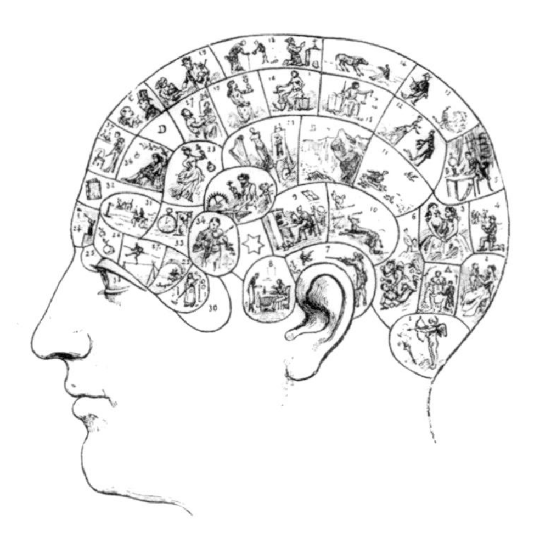

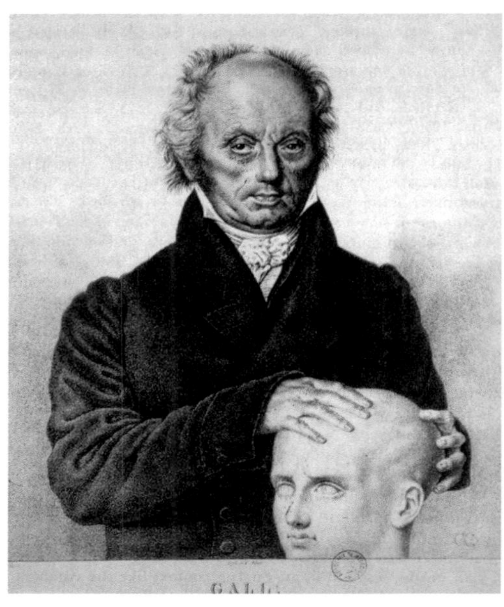

Figure 2.11
Franz Josef Gall (1757–1828) is seen on the right; a phrenological diagram is on the left. The phrenologists believed that traits are localized in the human cortex and that those traits can be revealed by palpating bumps in the head. Portrait of Gall courtesy Bibliothèque de l'Académie nationale de Médecine (Paris).

established on a scientific foundation. Postmortem study of brain damage in people who had impaired language gave increasing evidence that language, movement, and vision are controlled by different areas of cortex. These observations were reinforced by the results of electrical stimulation or surgical lesions in the cortex of animals. The relationship between cortical areas and their functions is described in subsequent chapters. For now, we briefly sketch some of the principal functions known to be associated with the four lobes of each cerebral hemisphere. The lobes, which take their respective names from prominent nearby bones in the skull, are:

• *Occipital lobe*, at the back of the head, contains the visual cortex, which receives fibers relaying visual sensations originating in the eyes.

• *Parietal lobe*, in front of the occipital lobe, further processes visual signals and also monitors touch and awareness of the location of the body in space.

• *Temporal lobe*, next to the temple, has areas concerned with hearing and the understanding of language and the processing as well as storage of visual memory for objects and faces.

• *Frontal lobe*, receives the sensations of smell; in its posterior portion are the areas that initiate voluntary movements.

The localization of functions noted here does not cover the entire cerebral cortex. The rest is a series of regions that used to be called silent areas because electrical stimulation of these areas did not elicit either movement or sensation. They also were presumed to integrate or associate information arriving at the many different cortical centers for sensation. Only in the past few decades have the functions of many regions of so-called *association cortex* become clear. Much of the current research in the neural sciences is directed toward learning more about the functions of these areas that control the most subtle aspects of human behavior.

Symmetry of the Brain

One of the striking characteristics of brain anatomy was only touched on in our tour of the various structures. Like most parts of the body these structures come in pairs and are symmetrically arranged. Just as we have two lungs, two arms, and two legs, we have two cerebral and two cerebellar hemispheres, each a rough mirror image of the other. Even single structures, such as those in the brainstem, are divided so that there is one on the left and one on the right side of the brainstem. Descartes located his "seat of the soul" in the pineal gland, an apparently unpaired structure, because he reasoned that there could be only one soul. This arrangement by which structures are paired or divided into similar halves is called *bilateral symmetry*.

Most of the nervous system is bilaterally symmetrical. Both body and brain can be visualized as duplicated by a vertical line down the middle. In this oversimplified image half of the brain relates to half of the body—but not the same half. Most nerve fiber tracts in the CNS cross over to the opposite side when they connect the body and the brain. The effect of this crossover is that one side of the brain presides over the opposite half of the body. Thus, voluntary movement of the right hand is principally controlled by the left side of the brain. Similarly, a sensory input from the left leg makes it presence known principally by its connections with the right side of the brain.

Although the cerebral hemispheres appear to be anatomical twins, they are not at all identical in function. Each hemisphere has uniquely developed capabilities in carrying out certain tasks. The most evident instance of this specialization is the almost universal tendency of human beings to favor one hand over the other for tasks such as writing or throwing a ball, which require finely tuned coordination. Most of us are right-handed, meaning that highly developed motor control is exerted by the left hemisphere, and thus, the left hemisphere is said to be dominant. Less obvious is the fact that in practically all human brains, the capacity for speech and language also resides in a single hemisphere; in 97 out of 100 persons, it is the left hemisphere. For this reason, a stroke that paralyzes the right size of the body (controlled by the left hemisphere) often also impairs the patient's ability to speak or comprehend written language.

The specialization of one or the other hemisphere for an important function is known as *cerebral dominance* or *lateralization of function.* Recent research has shown that cerebral dominance is not limited to language or hand use. In most people, for example, the left hemisphere also appears to control mental activities of a logical nature such as mathematics. The right hemisphere, by contrast, may be involved in other activities such as the perception of objects in space and music. The effect of this specialization, remarkably, is that each of us has in effect two brains. Ordinarily, these two connected brains work together. But when the principal connection between the cerebral hemispheres, the corpus callosum, is surgically severed for medical reasons, the existence of two separate minds becomes dramatically evident—as detailed in chapter 16.

Tracing the Pathways

Although the nervous system consists of discrete structures, its parts are complexly interdependent, intricately linked by tracts of nerve fibers. The fibers and their connections are usually so intertwined, in fact, that they can be sorted out and identified only with the aid of specialized techniques. In some cases, however, particularly large fiber tracts stand out distinctly and can be followed for considerable distances by simple examination. By simply tracing these nerve pathways with the naked eye, we often can learn something about the functions of a given part of the brain.

For example, we can see that a large fiber tract—the optic nerve—begins at the back of each eye and proceeds toward the brain. At the base of the brain the optic nerves from each eye appear to unite in an X shape, the *optic chiasm* (Greek for x-shaped). Physiological study shows that each optic nerve fiber—about 1,000,000 in the monkey and 100,000 in cats—carries visual information. If we follow the course of the fibers beyond the optic chiasm, we can see that they go to a subdivision of the thalamus called the *lateral geniculate body* (Latin for *genu*, a knee; thus geniculate, "resembling a little knee"). All of the cells in the geniculate are involved in the transmission of visual information from the eye to the cerebral cortex.

The visual pathway from the eye to the cerebral cortex constitutes one of the major sensory pathways of the brain. There are similar fiber systems that convey other senses and project upward in the brain, connecting at one or more relay stations along the way. Sensory information reaches the cerebral cortex by a series of relays that end in separate and identifiable cortical areas. Thus, the lateral geniculate body projects to the cortex on the banks and depths of a major fissure located on the medial side of the hemisphere, the *calcarine fissure*—so called because the shape reminded an ancient anatomist of the spur of a chicken.

Just as there are many sensory pathways from peripheral structures to the cortex, there are also motor pathways from the cerebral cortex to the brainstem and spinal cord. These originate in the brain and descend (figure 2.12) to connect to the nerve cells that control muscles to initiate movements. The corticospinal or pyramidal tract arises principally in the frontal lobe of the cerebral cortex and projects all the way down to the end of the spinal cord. Some of its fibers end directly on spinal motor nerve cells that connect to muscles.

The discussion of sensory and motor areas of the brain raises an important question that will concern us in future chapters: How are sensory and motor areas linked? The remarkable coordination of vision and movement is revealed not only in such delicate tasks as threading a needle or playing the violin but also in whole-body skills such as catching a fly ball or in the simple act of walking down a crowded street without bumping into people. As we also learn in subsequent chapters, a start has been made toward understanding how such links are made. Learning to ask the right questions is the first step toward answering them.

Does Size Matter?

In addition to identifying "aberrant" individuals, the nineteenth-century scientists were keen to establish differences among the races discernible

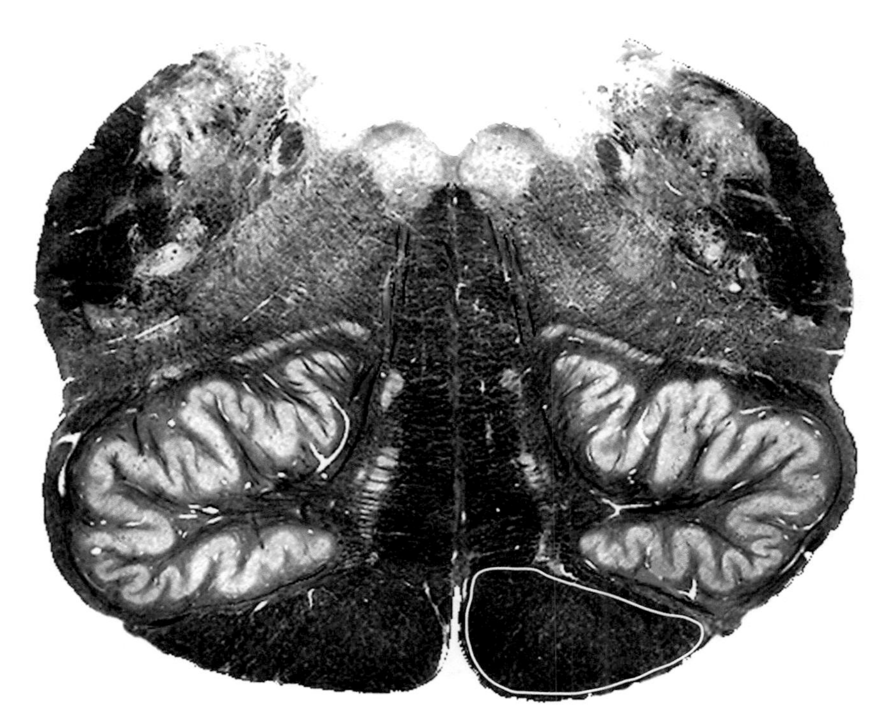

Figure 2.12
Cross section of medulla to show location of the pyramidal tract (outlined) and inferior olive. A fiber-stained section through the human medulla, with the pyramidal (corticospinal) tract indicated. The tract is one of the major descending fiber pathways controlling movement. The inferior olive, seen above, resembles a leather purse. The olive provides an important input to the cerebellum. Courtesy University of Washington.

in the human brain. One technique was to collect skulls and determine their volume. In that way, they argued, there could be an objective study of intelligence by comparative study of the cranial capacity of different peoples of the earth. Emil Huschke, a German anatomist working at Jena in the first half of the nineteenth century, asserted he had located intellect in the frontal lobes and temperament in the parietal lobes. Given that assumption, some suggested that one could achieve an indirect measurement of the relative size of the frontal and parietal lobes by studying the area of the corpus callosum—the fiber tract that connects the two sides of the brain. But despite an initial claim for racial differences, analysis of the same material without knowledge of its source showed no differences.

In addition to the comparison of brain areas and volume of different races, similar attempts have been made to identify human genius by postmortem evaluation of brain size or fissural pattern. That attempt did not work too well either. The brain of the great nineteenth-century German mathematician Gauss was average in size. Cuvier, the great French naturalist had a big brain, but Anatole France the French novelist had a small one. Maybe novelists have small brains? No, because the brain of Turgenev, the Russian novelist, weighed 2,012 g; that of Anatole France weighed about half that. Gall's own brain weighed only 1,198 g.

Yet enthusiasm for this approach has never really died. Vladimir Lenin, the leader of the Russian Revolution and the Soviet state died in 1924 after a series of strokes. The Soviet government invited the German anatomist Oskar Vogt to study Lenin's brain. He characterized it as that of an "association athlete" based on some larger pyramidal cells in the cortex. Lenin's brain weighed less than average. If size did not predict much, what about the fissural pattern? This idea is still with us. Claims are still made that Einstein had fewer fissures in a region of the temporal lobe, whereas the mathematician Gauss had more fissures in the same area. Such crude attempts have not helped. It is clear, however, that study of the functions of the brain and its parts can teach us about the nature of human sensation, movement, and thought.

3

The Structure of Nerve Cells and Their Supporting Tissues

In the last chapter we looked at major subdivisions of the nervous system and some of the fiber tracts that interconnect them. Here we look in more detail at its basic structural elements.

One of the fundamental insights in biology was the recognition by the German scientist Theodor Schwann in 1839 that all tissue is made up of cells—individual elements, typically microscopic in size, each of which usually contains a nucleus and is surrounded by a continuous membrane. The fundamental principle of neuroanatomy, called the neuron doctrine, is simply an extension of Schwann's cell theory to the nervous system. The neuron doctrine is the recognition that the main functional units of the nervous system are nerve cells, or neurons. Nerve cells interact by contacting one another, but they remain separate cells.

Our understanding of the structure of nerve cells began early in the nineteeth century. It was helped along by advances in techniques that allowed people to prepare brain tissue for study and by the development of better microscopes to study brain tissue. One remarkable early description of a characteristic type of nerve cell was made by the Czech anatomist Jan Evangelista Purkinje (figure 3.1). Two years before Schwann's cell theory was published, Purkinje described a cell type found in the cerebellum of all vertebrates. He called it a "globule," but it is now known as the Purkinje cell (figure 3.2).

The Purkinje cell is one type among many. There are billions of nerve cells in the human central nervous system. They come in a wide variety of shapes—stars, pyramids, rhomboids—and in many different sizes, but there are some properties that are common to all of these cells. If we understand the structure of a single nerve cell in some detail, we will have a good basis for understanding all nerve cells.

Figure 3.1
The Czech anatomist and physiologist Jan Evangelista Purkinje (1787–1869). Among his many contributions was a clear description of the cell type in the cerebellum that bears his name. Courtesy Wellcome Library, London.

Looking at a Neuron

The techniques available to Purkinje allowed him to see the nerve cell body of Purkinje cells as well as of much smaller cells called *granule cells* of the cerebellum. With those techniques, however, Purkinje could not see that these cells had prominent extensions or processes. Visualization of these extensions was begun by anatomists in the middle of the nineteenth century who learned to harden brain and spinal cord and then to dissect out individual nerve cells along with many of their processes. The nineteenth century anatomist Otto Deiters managed to dissect individual nerve cells from fixed tissue, a heroic task (figure 3.3).

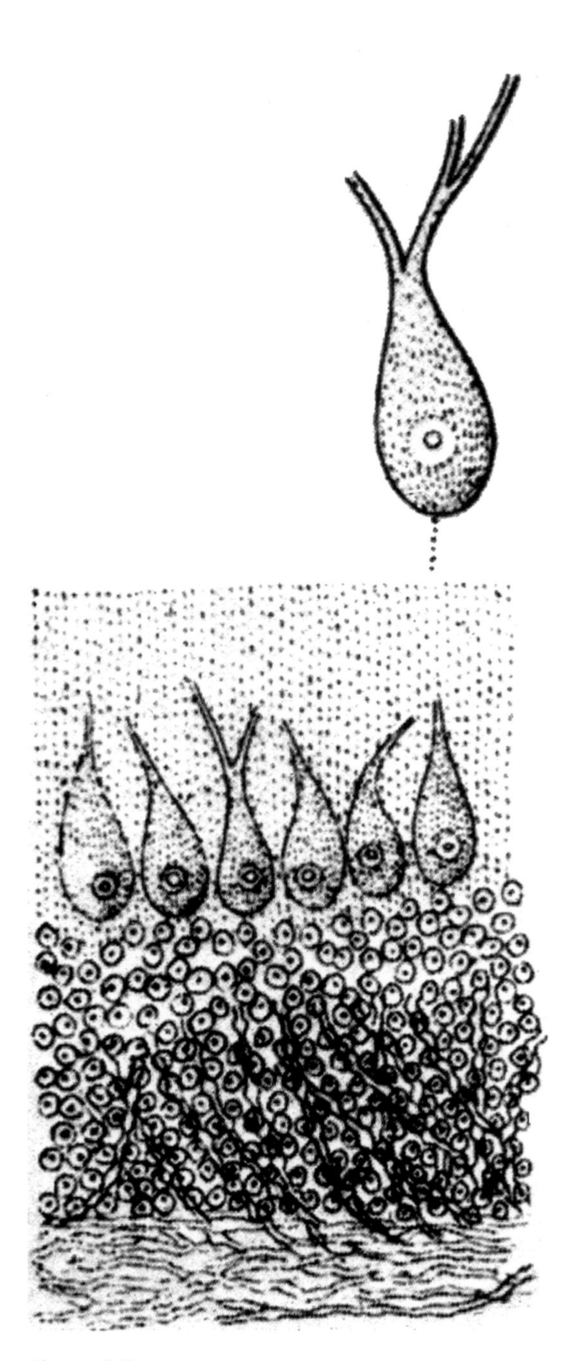

Figure 3.2
From Purkinje's 1837 paper. The figure shows the characteristic shape of the Purkinje cell body and the much smaller granule cells below. Purkinje cells form a single layer in the cerebellar cortex, and their axons are the only output from the cerebellar cortex. The elaborate dendritic tree of the Purkinje cells was not recognized until the development of the Golgi stain some years later.

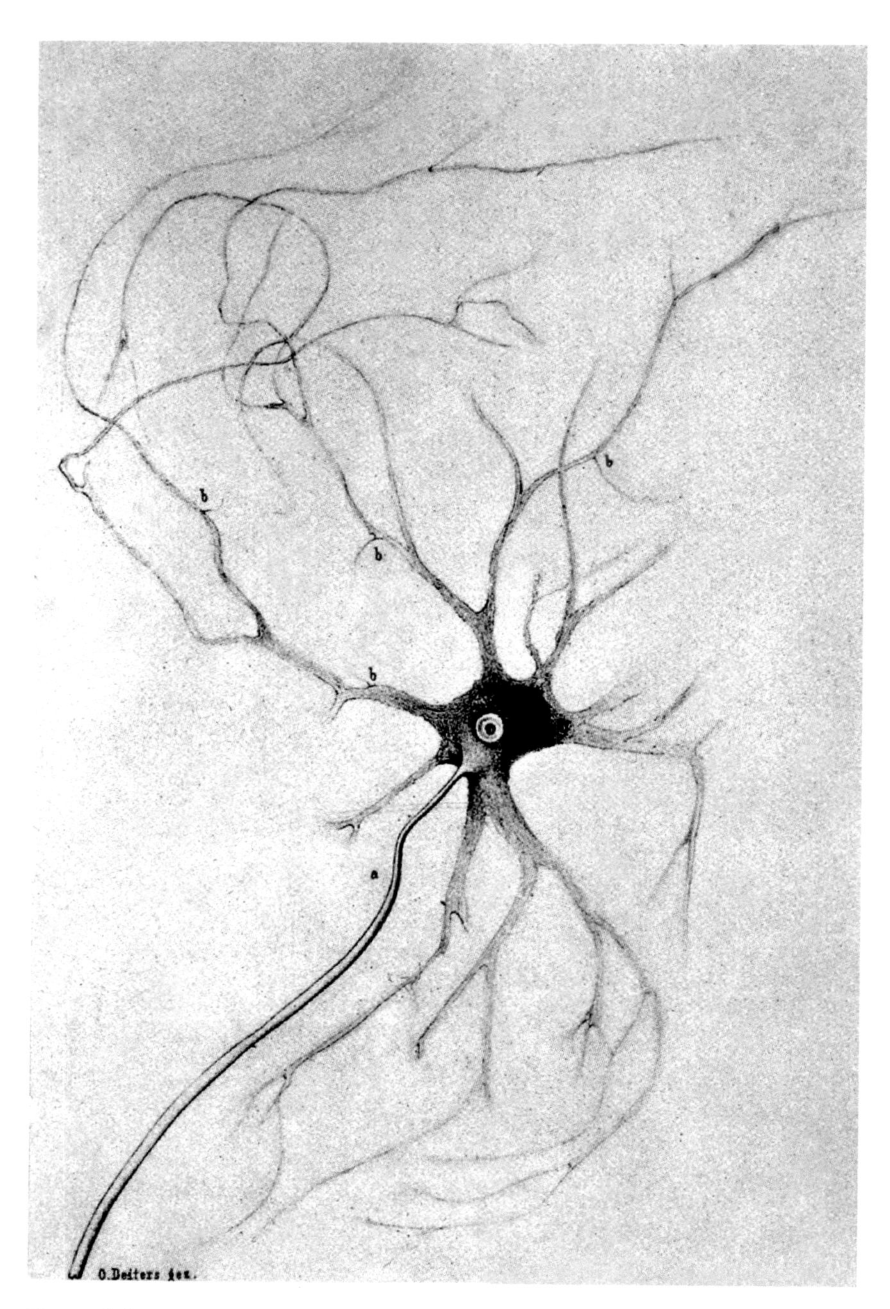

Figure 3.3
Otto Deiters (1834–1863) hardened spinal lobe tissue and painstakingly dissected out a single motor neuron, for the first time showing the great extent of the dendritic tree of those cells. A few years later Camillo Golgi devised a method whereby nerve cells of all types could be visualized along with their full dendritic trees.

Jutting out from the cell body are two types of appendages or processes—*axons* and *dendrites*. Structural characteristics help us differentiate between them. Dendrites (Greek for tree-like) are thickest at their base, near the cell body. They taper gradually and extend at most for a millimeter or so. They often branch repeatedly, forming an elegant tree-like expansion of the nerve cell.

Neurons typically have many dendrites but only one of the other type of extension, the axon. Unlike dendrites, axons do not taper. Axons are thinner, and they often extend for much greater distances. For example, the axon arising from a neuron in the spinal cord that controls a muscle in the foot of a tall person may be over a meter in length. Axons may branch, but when they do, they usually do so at right angles to the parent shaft. Dendrites usually branch at a more gradual angle.

The structure of neurons and their extensions can only be seen with the light microscope. To obtain this view, the anatomist must first properly prepare the tissue. It must first be fixed, usually with formalin in order to kill the cells and preserve their structural features. The tissue must then be hardened by a method such as freezing or embedding in a medium such as paraffin. Next the tissue is sliced with a microtome, an instrument for making thin slices. After thin sections are cut they are treated with one of a variety of special stains. These stains are chemical agents with special affinities for certain aspects of the cell, which enables the observer to distinguish particular features in the tissue. One class of stains, for example, is taken up only by particular structures within the cell body, causing those structures to stand out clearly from the surrounding tissue.

The neuron shown in figure 3.4 has a relatively large cell body, or soma, about 50 μm in diameter. (The fundamental unit used in studying the brain with the light microscope is the micrometer, abbreviated μm. One micrometer is one-millionth of a meter, or about 1/25,000 of an inch.) The nerve cell body typically contains a prominent nucleus. The region of the cell body outside of the nucleus is sometimes known as the *perikaryon* (from the Greek, peri, around; karyon, nucleus or nut).

Until the late nineteenth century the only way in which the entire nerve cell could be seen was by the tedious method of dissection. In 1873 the Italian histologist Camillo Golgi (figure 3.5) discovered a method that was to revolutionize the study of the structure of the nervous system. The Golgi method, which uses a solution of silver salts, stains all of certain neurons including the axon and dendrites. For reasons that are still obscure, the Golgi method stains only a few of the neurons in a

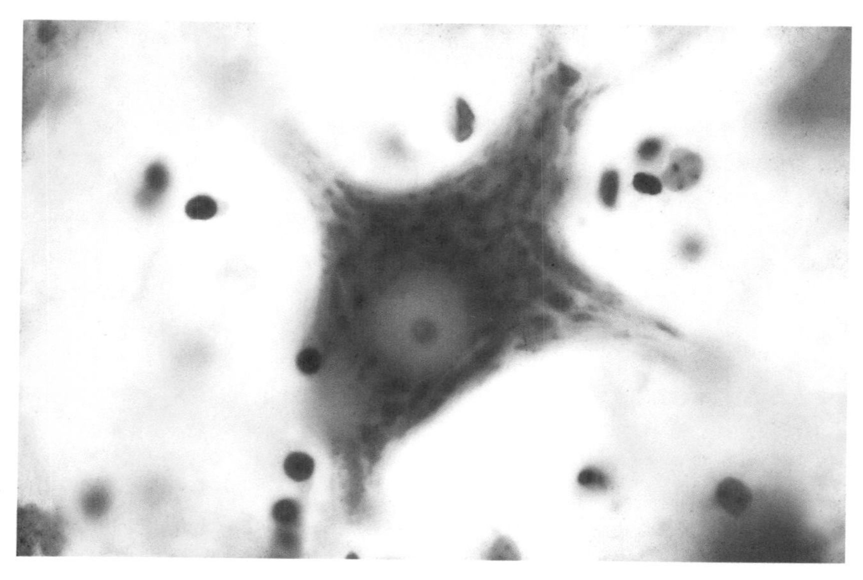

Figure 3.4
A spinal motor neuron stained with cresyl violet, a Nissl stain. The stain shows the content of the cell body, including the so-called Nissl bodies (rough endoplasmic reticulum). The pale-staining nucleus is seen inside the cell body, and the small, dark nucleolus is shown inside the nucleus.

block of tissue even though the section of tissue may be densely packed with nerve and glial cells. These few neurons stand out with stunning, three-dimensional clarity under the light microscope (figure 3.6).

Although the Golgi stain was a major advance, the resolution of the neuron's structure is limited by the built-in constraints of the light microscope, which magnifies at most with a power of about 1,000 times. Using the light microscope and appropriate staining methods, anatomists could see and describe structure down to the level of a single micrometer or so in size. And although special types of optics and illumination with short wavelengths of light enabled them to increase the possible magnification slightly, the wavelength of light, about one-half of a micrometer, sets an ultimate limit to the fineness of detail that can be resolved and precludes seeing anything that is much smaller. This level of resolution is too gross for seeing some crucial details of the structure of nerve cells and their interconnections.

The development of the electron microscope and its application to the study of nerve cells and fibers helped to reveal structural details that

Figure 3.5
Camillo Golgi (1843–1926) was one of the early histologists to study the cellular structure of the nervous system. His staining method was critical for the later work of Santiago Ramon y Cajal and other anatomists. Courtesy Wellcome Library, London.

could not be approached by light microscopy. The electron microscope is 1,000 times more powerful than the light microscope, and so we can now add to the description of the neuron significant details of structure that can only be provided by this high magnification.

The nerve cell is surrounded by a very thin membrane—on the order of 8 nm (1 nanometer = 1/1,000 of a micrometer or 10^{-9} m in thickness). All nerve cells are surrounded by such a membrane, which extends continuously around its dendrites, cell body, and axon. Although there are regional variations at different parts of the nerve cell, membranes are composed of layers of protein and lipid, and they are critical in the electrical functioning of nerve cells. Membranes can segregate charged

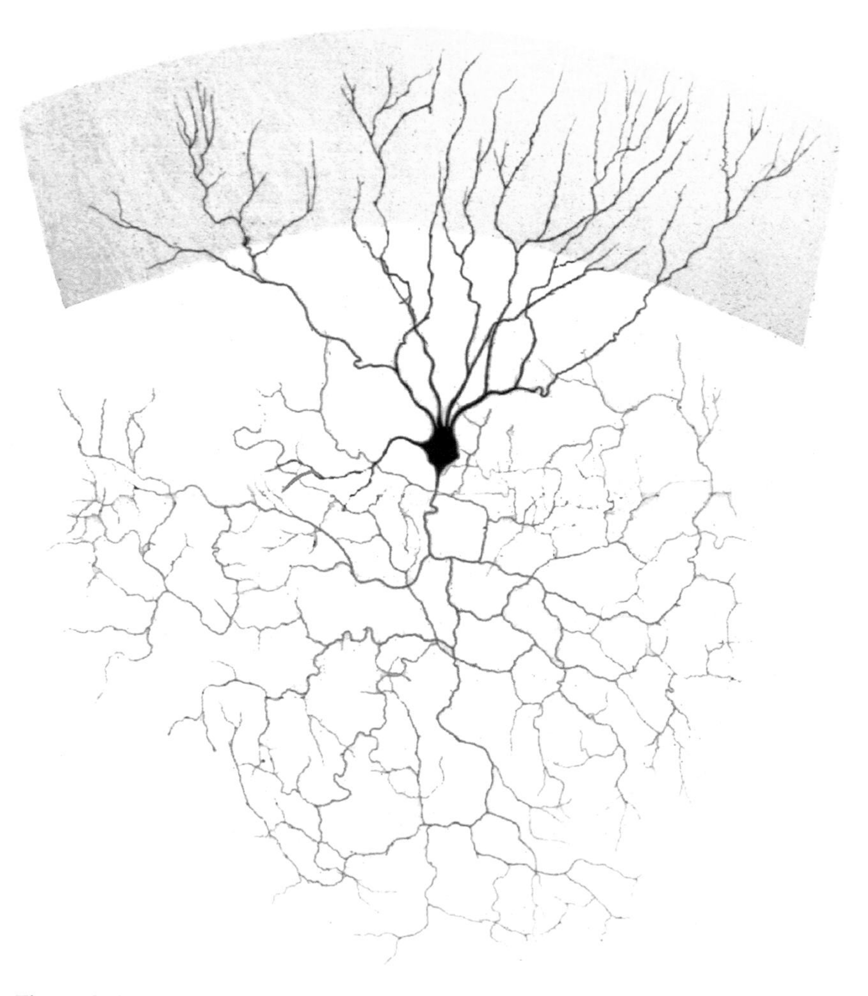

Figure 3.6
Golgi cell; Golgi stain. From Golgi's *Opera Omnia.* This is a beautiful illustration of the cell type in the cerebellar cortex first clearly described by Golgi that still bears his name. The dendrites of the Golgi cell extend toward the surface of the cerebellar cortex, the axons branch repeatedly, but, unlike Purkinje cell axons, they remain in the vicinity of the cell body

particles—ions of common elements—and thus, the cell can be electrically charged similar to a capacitor in an electronic circuit.

The nerve cell typically is functionally polarized. Inputs to the cell usually arrive at the dendrites and cell body. These inputs either activate or inhibit the cell. Dendrites are specialized principally for receiving incoming neural activity. In many types of neurons the dendrites are covered with tiny spines or thorn-like appendages. When these spines were first seen there was doubt as to whether they were truly part of the dendrite or merely represented some sort of artifact created in the staining process. The electron microscope confirmed that dendritic spines are, in fact, true components of normal nerve cells. They serve as specialized receptor surfaces for receiving connections from incoming axons. Such connections also are made at numerous terminals on the shafts, on the cell body, and even on axons.

Dendrites typically branch at gentle angles from parent dendritic shafts. The pattern of dendritic branching varies widely among cell types. In the cerebral cortex two principal types of nerve cells are recognized (figure 3.7). One is the stellate cell, so named because the radial pattern of its dendritic branching gives it a star-like appearance. The other cell types are named for the shape of their cell body. For example, the *pyramidal cell* takes its name from the pyramidal shape of its perikaryon, or cell body.

The form of the neuron's dendritic tree is of importance in its function. We can get an intuitive understanding of this relationship by considering the neurons in the retina, the network of cells in the eye that receives visual information, processes it, and transmits the processed signals to the next cell in the pathway and then to the brain (figure 3.8).

One class of retinal cells is bipolar. Each bipolar cell has just one principal dendrite—which may subdivide—and one axon. Bipolar cells in the retina receive their input from the light-sensitive cells—the rods and cones. In a well-focused eye each cone or rod is influenced by light from a tiny region in visual space. In vision, as in the other senses, neurons are the building blocks for sensation. In the eye and then in the brain, tiny, point-like stimuli are integrated into larger and larger units. In some bipolar cells the extent of the dendritic tree is very small, and hence they connect to only a very few rods or cones. As a consequence they are influenced from a very small region of the visual field. Some bipolar cells have a much more extensive dendritic tree; hence, they are influenced by a larger number of rods and cones, and therefore, they respond to a much larger portion of the visual field.

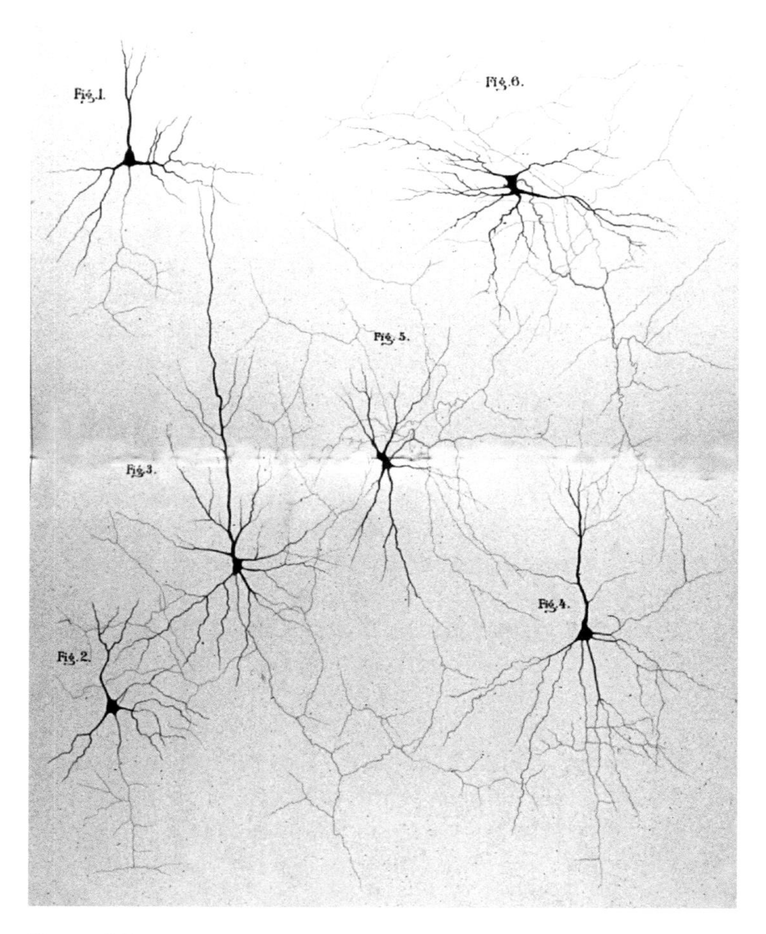

Figure 3.7
From Golgi's *Opera Omnia*, illustrating the two major nerve cell types in the mammalian cerebral cortex. Pyramidal cells (e.g., Fig. 1 and Fig. 4 above) have a characteristic triangular shape, with dendrites extending from the base and the apex of the triangle. Stellate cells (e.g., Fig. 5 above) have dendrites extending in all directions from a more rounded cell body.

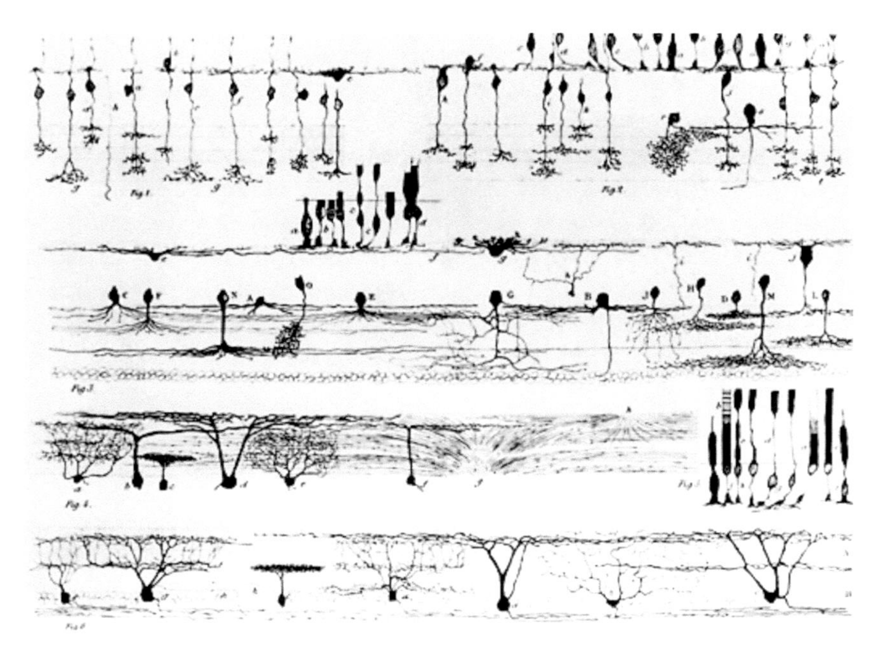

Figure 3.8
From Cajal's *Structure of the Retina*, published in 1892; Golgi-stained figure illustrating the great variety of retinal cell types. The retinal ganglion cells (with dendrites pointed upward) are illustrated in the lower two rows. One hundred years after Cajal's monograph, neurophysiological techniques began to allow understanding of the differences in function of each of these cell types and how they relate to vision. Courtesy Charles C. Thomas, Publisher.

Cytoplasm

The cell body of a neuron contains, between its membrane and nucleus, an outer rim of material called the cytoplasm. Within the cytoplasm are clumps of material that stain darkly with certain basic dyes. These dyes were first systematically applied to sections of nerve tissue by the German histologist Franz Nissl in the nineteenth century. This class of basic dyes is still called *Nissl stains*, and the large clumped structures in the cytoplasm that they stained are known as *Nissl bodies*. When they were first seen, it was not clear whether the Nissl bodies represented true functional elements in the nerve cell or were merely clumping of material within the cell. The electron microscope established that they are real. Nissl bodies represent the elements of a system of tube-like structures. Attached to these tubes, or near them, are small darkly staining particles called ribosomes.

What we now know as *rough endoplasmic reticulum* corresponds to the Nissl bodies of the light microscopist. The nucleus of most large-sized neurons is much paler than its cytoplasm when stained with basic dyes, although a prominent, very darkly stained sphere called the nucleolus is visible in sections that are cut through the center of the nucleus.

In Nissl-stained material, Nissl bodies are abundantly present in the cell bodies of neurons and in the base of the dendrites near the cell body. Correlated with this at the electron microscopic level, there is an abundance of rough endoplasmic reticulum that extends throughout the soma and into the base of dendrites. Dendrites have Nissl bodies at their base, but axons do not.

The Axon

Axons transmit information from one neuron to its target: they are the output element of neurons. The absence of Nissl bodies (rough endoplasmic reticulum) provides yet another means of differentiating between dendrites and axons under the microscope. In summary, axons differ from dendrites in number (a neuron has only one axon, although it may have many dendrites), shape (dendrites taper), length (axons sometimes are more than a meter in length) and the angle at which they branch (dendrites branch gradually; axons at right angles).

Because it lacks Nissl bodies, the axon cannot be seen in Nissl-stained preparations. In neurons stained by some of the Golgi methods, the axon often shows up well, thus revealing another difference between axons and dendrites. Whereas dendrites are usually covered with spines or swellings, axons are typically smooth.

The axon originates on the nerve cell body from a gentle swelling, the *axon hillock*. Sometimes an axon extends to its target—a gland, a muscle or another neuron—and ends there without extensive branching. More commonly, axons branch near their ending into a number of fine fibers before making contact with their target. Axons also may branch close to the cell body or along their course. A single axon may make contacts with several successive targets before it ends.

The Synapse

Typically, the axon makes contact with another neuron via that neuron's cell body or one of its dendrites. Such contacts are called *synapses*. The synapse is of extraordinary importance in all communication between nerve cells. It also has been the subject of two of the most fascinating—and acrimonious—controversies in the history of the neural sciences. To understand both the structure of the synapse and the debates that raged around it, let us consider the theories that prevailed until the late nineteenth century about the basic structure of the brain.

In those days most scientists believed that nerve cells were physically connected to one another. It looked that way under the light microscope. Clusters of neurons appeared to be a trellis-like network, or reticulum, in which axons were literally fused to dendrites, cell bodies, or one another in a continuous web. This point of view was known as the *reticular theory*. Its leading proponent was the Italian anatomist Camillo Golgi, who invented the staining method that bears his name. Ironically the Golgi stain was so effective that it enabled the Spanish anatomist Santiago Ramon y Cajal to support a conflicting theory about the connections between nerve cells.

Based on his studies using the Golgi stain, Cajal began to provide systematic support for what was then a radical new theory soon to be known as the *neuron doctrine*. The neuron doctrine states that the brain and spinal cord are made up of individual elements that touch one another but do not fuse. The English neurophysiologist Sir Charles Sherrington, Professor of Physiology at Oxford University, quickly recognized the importance of the concept and was the first to use the term synapse for the point of contact. The term was actually suggested to him by a professor of Greek at Oxford. The Greek roots of the word synapse mean fasten together.[1]

The debate between the proponents of the reticular theory and the neuron doctrine was a bitter one. Cajal could only infer the nature of the contacts between nerve cells because their structure is far too small to be resolved by the light microscope. Because Cajal could not prove the existence of synapses, his old foe Camillo Golgi clung to the reticular theory. In 1906 the two men shared the Nobel Prize in Medicine, but the bitterness between them did not abate. Cajal, in his autobiography, viewed Golgi's acceptance speech as an attempt to revive the reticular theory. It is said that although they had nearby rooms in a Stockholm hotel, the two scientists passed without speaking.

Nearly half a century later, the electron microscope revealed the true nature of the synapse—and vindicated Cajal's insight. Nerve cells do not fuse. Most synapses have a cleft of about 20 to 30 nm between nerve cells. The electron microscope also helped to resolve another decades-old controversy: What happens at the synapse?

This issue—sometimes referred to as the "spark versus soup" controversy—was hotly argued for many years even after most neuroscientists agreed with Cajal that nerve cells are individual and separate elements. The "spark" theorists believed that synapses worked by direct electrical coupling across the synaptic cleft. The "soup" theorists believed that a chemical transmitter was the key to synaptic function.

We now know that both were right, although the soup theorists were more right. The evidence in favor of the soup theory is discussed in the next chapter. It is now clear that the great majority of synapses in the vertebrate nervous system are chemical, but there are also demonstrable examples of electrical synapses. In those cases in which the presence of electrical synapses has been proven, the electron microscope shows that the synaptic cleft between nerve cells is much narrower than in chemical synapses. Such electrical synapses are seen, for example, in certain giant neurons in the brainstem of a fish.

Most synaptic contacts in the mammalian nervous system consist of axons synapsing on dendrites. In such instances the axon is the presynaptic element, and the dendrite is the postsynaptic element. The prefixes indicate the direction in which chemical activity flows across the synaptic cleft. Axons may also terminate on cell bodies, a pattern known as axosomatic contacts, and in some cases on other axons, axoaxonic contacts. Axons are not the only presynaptic elements. Cell bodies and dendrites can also be presynaptic elements. Some neurons in the cerebral cortex each make more than a thousand synapses with other nerve cells.

Structure of the Chemical Synapse

With the insight provided by the electron microscope, we can construct a simplified portrait of a typical chemical synapse (figure 3.9). We see an axonal ending, which is the presynaptic element, and a dendrite, which is the postsynaptic element. The membranes of these two elements are separated by a synaptic cleft of about 20 nm. On the postsynaptic side the dendrite's membrane is noticeably thickened. On the presynaptic side the axon is swollen into a knob. Inside this knob is an assemblage of tiny spheres called synaptic vesicles. These vesicles, each surrounded by a membrane, contain an all-important chemical, a neurotransmitter.

The *neurotransmitter*, in effect, carries the signal across the cleft between neurons. When the axon is active, the vesicles inside the axon become mobilized, fuse with the axon's membrane, and then release a tiny quantity of neurotransmitter into the synaptic cleft. Then, the

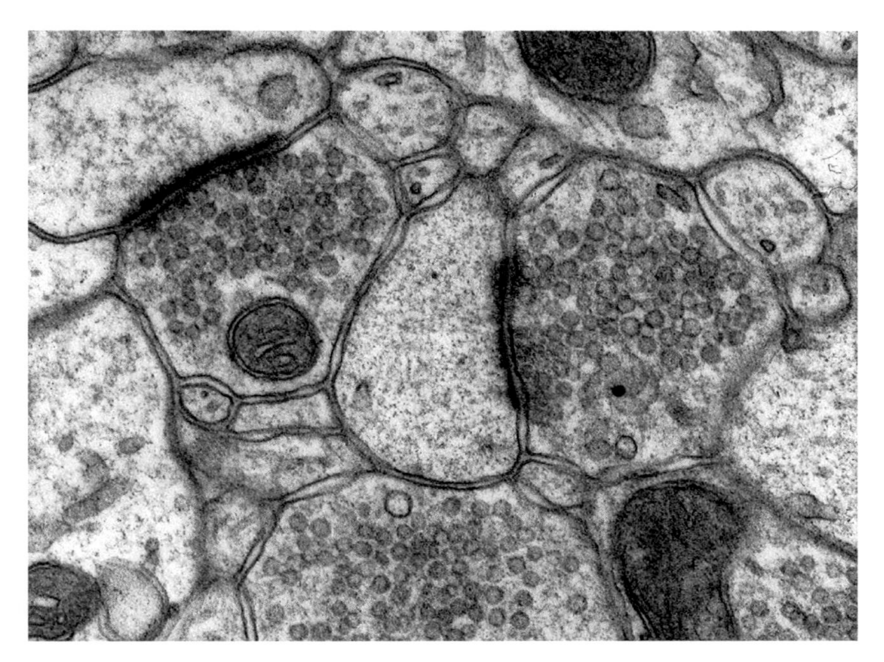

Figure 3.9
A synapse in the cerebral cortex as revealed by the electron microscope. The presynaptic endings have small, round vesicles within them; note also the characteristic thickening of the postsynaptic membrane at synaptic sites. In life the vesicles contain small quantities of transmitter substance. Courtesy Professor John Parnavelas, University College London.

neurotransmitter diffuses across the cleft and produces an effect on the postsynaptic nerve cell.

We study the properties of such synaptic mechanisms in more detail in chapter 5. It should be noted here, however, that there are as many as forty or more different known or suspected neurotransmitters. Although the small size of the synaptic vesicles makes it impractical to directly measure their chemical constituents, we can infer a good deal about the properties of the synapse from physiological experiments. There are correlations between the types of vesicles that can be seen in a synapse and their probable function. Whereas the majority of synaptic vesicles so far described are rounded with a diameter of 30–60 nm, the vesicles in some synapses are larger (70–150 nm) and have a prominent darkly stained region in their center when seen with appropriate fixatives. Synaptic vesicles of different shape, round versus flattened, are usually associated with different types of transmitters.

In addition to size differences among synaptic vesicles, there are variations in shape. Although the exact shape of the vesicles seen under the electron microscope is influenced by the nature of the procedures used in preparing the tissue for study, under standard conditions some synaptic terminals are seen to contain rounded vesicles, and others contain flattened ones. There is evidence that synapses with rounded vesicles may have different effects on the postsynaptic cell than those that contain flattened vesicles.

Supporting Tissues

What else is there in the nervous system beside neurons? Like most parts of the body, the brain and peripheral nerves have a blood supply. There is also an elaborate web of supporting cells. In the peripheral nervous system the supporting cells are called Schwann cells—after Theodor Schwann, who first described them. In the central nervous system these supporting cells are called neuroglia or glia—from the Greek word for glue. Glial cells appear to seal the spaces between neurons, literally sticking them together.

Glial cells outnumber neurons: there are ten times as many glial cells as neurons in the brain. Although all of the functions of glial cells are not known, one of their roles has been established elegantly and conclusively by electron microscopic study: Schwann cells in the peripheral nervous system and a particular class of glial cells in the brain and spinal cord produce myelin, the fatty white sheathing that surrounds axons. They create the myelin sheath by fastening onto the bare axon and wrapping around it again and again.

Schwann Cells

If we look at an appropriately stained cross section of a peripheral nerve under the light microscope, we can see a large array of roughly circular profiles in black (figure 3.10). In the example chosen here—the sciatic nerve—the myelinated fibers can be seen to vary greatly in size. Axons vary widely in diameter and may or may not be coated with myelin. Note that in this section the myelin appears as homogeneous black rings. But under the much more powerful electron microscope, the myelin is seen to be composed of many concentric rings of alternating dark and light bands (figure 3.11).

The first clear picture of how myelin is formed came in the 1950s from Betty Ben Geren's study of the developing myelin in nerves of

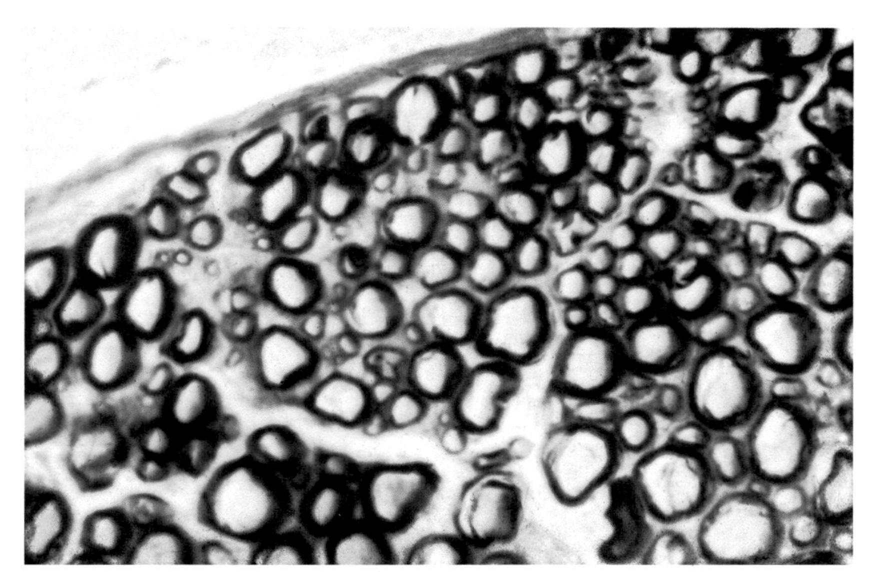

Figure 3.10
Cross section of a peripheral (sciatic) nerve stained with osmium. This is a mixed nerve containing both sensory and motor fibers. The variability in fiber size is apparent. It is generally true that fibers of different caliber have different functions.

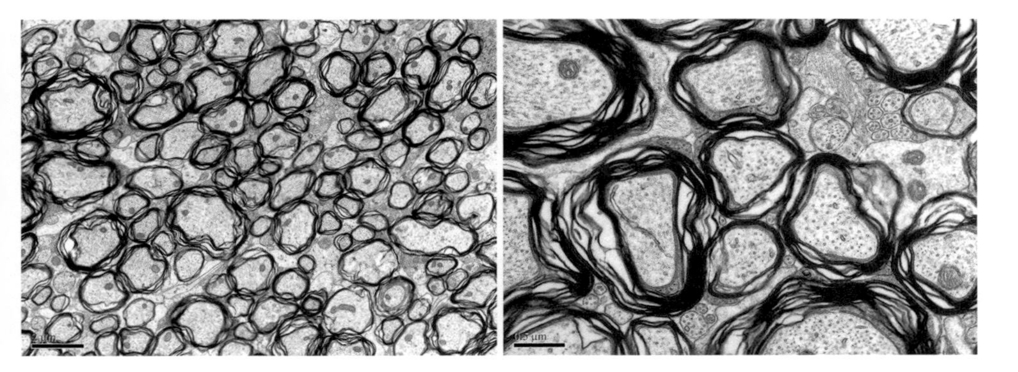

Figure 3.11
Electron microscopic images of the pyramidal tract magnified up to ×10,000. It is only with such great magnification that the smallest nerve fibers can be seen.

the chick. Myelin is actually successive paired membranes of the Schwann cell.

Imagine a long soda straw (representing an axon). Now take a small, closed-off bag (representing the Schwann cell). Collapse and flatten the bag so that its two sides are touching for most of its length. Now attach one end of the closed bag to the soda straw and wrap the bag around and around the straw, perhaps a dozen or more times. If you cut across, you would see the straw in the center, and successive wrapping of the two sides of the bag.

The Schwann cell winds itself around the axon in wrap after wrap to form a double-membrane covering. Light microscopy could not resolve the way myelin is formed or the fact that myelin is actually composed of the membrane of Schwann cells. Older textbooks often treated the myelin and Schwann cell layers as if they were separate and successive coatings of peripheral nerves.

Each Schwann cell, together with the myelin coating that it makes, covers the axon for only 1 mm or so. Thus, covering a long axon requires hundreds of Schwann cells. Because each Schwann cell coats only a millimeter or so of the axon under the microscope, the myelin sheath is seen to be interrupted at regular intervals. The gaps between successive Schwann cells are called the *nodes of Ranvier*, after the nineteenth-century French anatomist Louis Antoine Ranvier, who first described this periodic interruption of the myelin. Nodes of Ranvier are of great importance in regulating the conduction of nerve impulses along an axon. Even axons that do not have a myelin coating are also covered by a single layer of Schwann cell membrane.

Glial Cells

Glial cells provide the structural framework for neurons in the brain and spinal cord. There are three major types of glial cells: the *oligodendroglia*, so called because of its few dendrite-like expansions; *astroglia* (also called astrocytes), so named because of their resemblance to a star when stained by certain methods; and *microglia*, which are very small. Glial cells in the central nervous system have several known functions, although some functions may still be learned. The oligodendroglia are abundantly present in white matter of the central nervous system, and they act like Schwann cells in peripheral nerves, forming the myelin coating on axons. Astrocytes and oligodendroglia support the cells of the central nervous system. A special class of glial cell, the radial glia, prominent in the brain of embryos, directs the pattern of growth of the developing brain and spinal cord.

Another function of glia is evident when an injury occurs in the brain: glial cells—astrocytes and microglia—proliferate rapidly in the affected region. These cells serve as phagocytes (from the Greek: phago, eating; cytes, cells) that eat up the debris left by the damage. In appropriately stained sections taken just after a brain lesion, microglia can be seen much enlarged with bits of ingested material inside them. Such phagocytized material is eventually broken down.

Tracing the "Wiring Diagram"

So far, we have looked at a typical neuron, shown how individual neurons make contact with one another at the synapse, and suggested some functional relationships of the glia and Schwann cells. Next, we consider ways of determining precisely how neurons are "wired up" to form the links between the various structures of the brain. Anatomists trace connections from one part of the brain to another to determine its "wiring diagram." Over the years they have developed and applied increasingly precise techniques for following the intricate courses taken by nerve cells and their fibers.

The first anatomic method for tracing fiber tracts was simple dissection. In the mammalian brain, as noted in the previous chapter, it is easy to see a large bundle of white fibers emerging from the back of the eyes as a single optic nerve. We can follow this nerve, observing its course and the point at which it joins the brain. This method of simple dissection tells us something about the wiring diagram on a very crude level. It is rather like seeing an electric power cable composed of many different wires entering a huge factory. The human optic nerve consists of about a million individual nerve fibers, none of which can be seen by the naked eye. Moreover, nerve fibers are often intricately intertwined. Even when they are stained and viewed under the microscope, it is usually impossible to sort out individual axons and trace them to their connections with other nerve cells.

Degeneration Methods

Clearly, mapping nerve fibers required methods to allow the anatomist to distinguish one axon from another and to follow that axon through a maze of nerve cells. One way to identify the course of a group of nerve fibers was through degeneration methods, which are based on the fact that neurons in the brain and spinal cord and their myelin coating ordinarily degenerate and die if they are damaged. Different phases of degeneration take from one or two days to several weeks after injury to the neuron. During this period the axon and cell body typically change in their appearance and in their affinity for various stains. Cutting the axon, for example, causes degeneration, which can be then detected by proper staining methods.

If an axon is cut, degeneration occurs in the portion of the axon on the distal or far side of the cut; that is, degeneration occurs from the site of the lesion to the axon's terminal (figure 3.12). This is called *anterograde degeneration* (in effect, "going forward") or sometimes *Wallerian degeneration* (after the English scientist Augustus Waller, who first described the process).

The first technique to take advantage of the chemical changes that accompany anterograde degeneration was the *Marchi method*, which selectively stains the degenerated myelin sheath surrounding the severed axon. In an experiment using the Marchi method, a lesion would be made in the brain of an experimental animal and, about ten days later, the animal would be killed and its brain sectioned. The sections were then stained so that only the degenerating myelin of the severed axons will show up under the microscope as black and beaded droplets (figure 3.13), enabling the anatomist to identify the damaged fibers.

The Marchi method had major disadvantages and is now only of historical interest. The principal problem was that its effectiveness was limited to fibers coated with myelin. And even in the instance of myelinated fibers, it was only partially effective because the axons that had no myelin at the very tip could not be followed to their very end.

A major improvement in the use of degeneration to stain was achieved by Walle Nauta and his colleagues in the 1950s. Nauta perfected a method that is capable of selectively impregnating the degenerating axons themselves. In the Nauta method a lesion would be placed in the brain of an experimental animal. Five to ten days later the animal would be anesthetized and killed, and the brain would be removed and frozen. Sections of tissue would then be sliced from the frozen block, stained with the Nauta method, and viewed under the light microscope. The results showed degenerating axons rather clearly as broken and irregularly beaded. The Nauta method was applicable to both myelinated and unmyelinated fibers, and variants of the technique were capable of staining almost to the terminals of a degenerating nerve fiber.

Another degenerative process that enables anatomists to track neural connections involves changes in the nerve cell body that follow after an axon is severed. This process is known as *retrograde degeneration* because it occurs behind or backward from the severed axon. When sections are

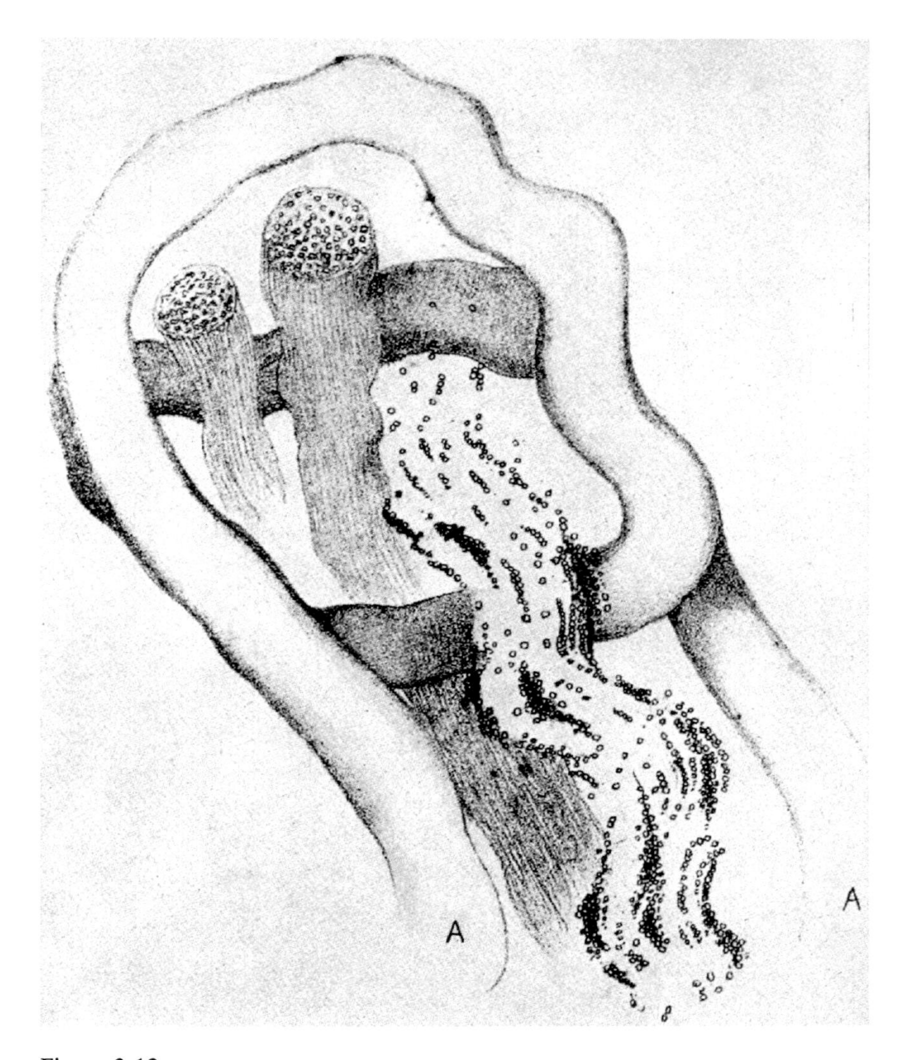

Figure 3.12
Augustus Waller (1816–1870) cut the nerve to a frog's tongue. He killed the animal a few days later and described the appearance of degenerating nerve fibers. The figure shows the characteristic fragmentation of the nerve fiber distal to the cut. Courtesy Wellcome Library, London.

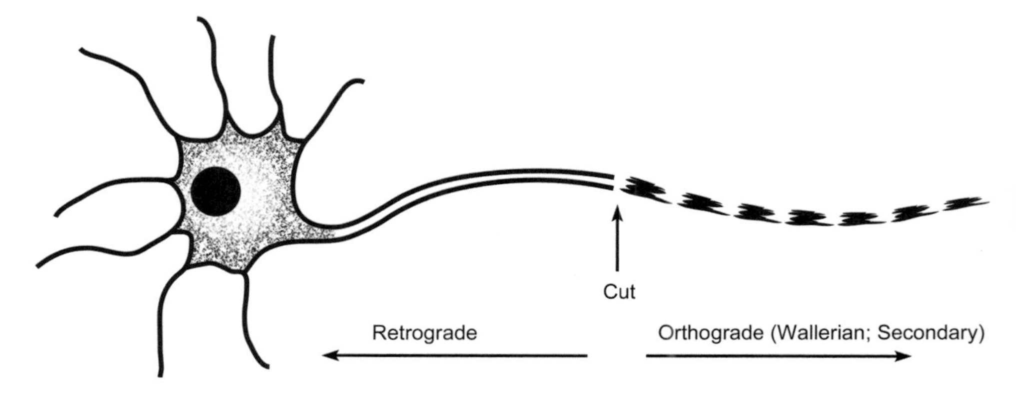

Figure 3.13
Schematic drawing of a neuron with a cut axon to illustrate degeneration that follows cutting of the axon. Degeneration of the axon and cell proximal to the cut is called *retrograde* degeneration. Distally the change is called *orthograde, secondary*, or *Wallerian* degeneration; all three are equivalent terms.

stained to show cell bodies, retrograde changes in the cell can be seen under the microscope. Because the stained Nissl bodies often appear to break up into much smaller particles, the changes are called *chromatolysis*—chroma for colored material (that is, the Nissl bodies) and lysis for liquefaction. After a period of chromatolysis and swelling (figure 3.14), the cell body often shrinks, dies, and is removed by the glial phagocytes.

Thus, in a typical experiment using retrograde degeneration, a lesion would be made in one brain region, and the degenerating changes in the cells in the location that sent their axon to that region would be studied. For example, study of retrograde degeneration allowed anatomists to begin clarifying which divisions of the thalamus project to a particular target in the cerebral cortex. Lesions were made in the cortex, and stained sections taken from the thalamus revealed the locations of degenerating cell bodies.

But there are major drawbacks in the use of these tracing techniques based on nerve cell degeneration. It is often difficult to limit a lesion to one specific group of nerve cells or tract of fibers in order to trace anterograde degeneration from that nucleus. Also, not all cells show clear retrograde degeneration when branches of their axons are cut.

Current Techniques

Over the past forty years, new techniques have been developed for tracing connections in the nervous system. Instead of relying on selective

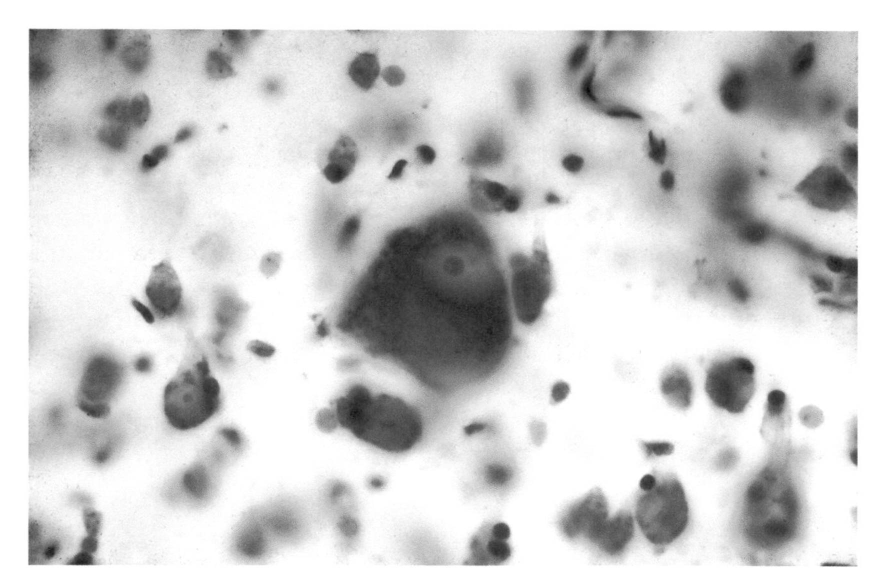

Figure 3.14
Retrograde degeneration in a neuron whose axon had been cut; Nissl stain. The Nissl bodies are not seen in the center of the neuron: the nucleus is shifted to an eccentric position. The cell is somewhat swollen and appears rounder than normal, all characteristics of retrograde degeneration.

staining of degenerating nerve cells or nerve fibers, these methods are based on so-called transport mechanisms that occur normally within the healthy neuron. Certain substances are injected into or near a neuron. These become incorporated into the nerve cell and travel to its other parts, which can then be identified and traced by histochemical staining or specialized photographic techniques.

One of these techniques, *autoradiography*, first developed by Max Cowan and his colleagues at Washington University in St. Louis, makes use of the cell's normal transport mechanisms in a particularly ingenious way. It is known, for example, that if an amino acid such as leucine or proline is injected near a nerve cell, it will be taken up by the cell and incorporated into the cell's normal protein. In the case of retinal ganglion cells, these proteins are then pumped down the axons that form the optic nerve. In order to identify the presence of the amino acids that were injected, they must first be labeled in some way. Labeling is accomplished by replacing some of the hydrogen in the amino acid molecule with tritium, a radioactive isotope of hydrogen. Thus, when the protein containing the modified amino acid is pumped down the axon, it bears a radioactive label.

In a typical autoradiographic experiment, a tritiated amino acid is injected into the appropriate region of the brain. A few days later the animal is killed and its brain cut into thin sections. These sections are then mounted on slides and coated with photographic emulsion. If these slides are kept in a darkroom, the radioactive proteins that were injected into the brain eventually expose the silver grains in the emulsion and reveal their locations, and hence those of the labeled axons. When the slides are developed like a photograph and fixed, the radioactive labels show up as concentrated regions of black dots overlying labeled areas. Autoradiography thus makes it possible to follow the course of an unknown pathway without killing the cells of origin or cutting the fiber tracts.

Another widely used method for following anterograde and retrograde connections uses the enzyme horseradish peroxidase (HRP), first described for identifying retrograde transport by Jennifer and Matthew LaVail in 1972. HRP is injected in a region containing fiber terminals of interest. The fibers pick up this material, which is pumped retrograde, or backward, to the cell body. With appropriate sectioning and histochemical staining methods, not only the locus but also the specific cells that give rise to the fiber tract can be identified. HRP may also be taken up by cell bodies and transported in both retrograde and anterograde directions.

A recently developed method, extensively used by Peter Strick, allows tracing circuits beyond a single synapse. A stainable substance is linked to a virus. The virus might be injected into a muscle, for example. The complex of virus and stainable material is transported back to the motor neuron controlling that muscle. The virus then kills the motor neuron and releases the complex of virus and tracer, which is then transported back to the cells that connect to that motor neuron.

These new anatomical methods, together with the physiological techniques that are discussed in the following chapter, have already revolutionized our understanding of brain connections. They give promise of immeasurably deepening our understanding of how the brain is wired.

4

Electrical Transmission in the Nervous System

The nervous system works electrically. Warmth, cold, touch, and pain stimuli are received by specialized structures in the skin and internal organs. The sensations are conveyed by sensory nerves and tracts from skin and viscera via the spinal cord to the brain. Movement is controlled by motor neurons in the brain and spinal cord, whose axons go out into the body to activate muscles. Nerve fibers within the brain link sensory to motor areas.

All of these fiber systems act electrically. The way in which electricity is generated and transmitted in the nervous system differs from the household current that is used to light a room or the battery that powers a flashlight. In the nervous system the electricity is generated and transmitted by ions. Storage and transmission are typically slower and less powerful than the more familiar household electricity. This chapter describes some of the experiments that led to our current understanding of electrical transmission in the nervous system.

By the eighteenth century, manifestations of electricity were well known. Amber and glass, when rubbed, attracted small bits of paper. Sailors saw shimmering lights—St. Elmo's fire—at the top of the masts of their ships. Lightning was observed and described. Certain fish could sting most painfully. In the eighteenth century the idea that these diverse phenomena are based on the same underlying mechanism was not recognized.

Luigi Galvani (figure 4.1) was an Italian scientist and physician who was born in 1737 and was educated and died in the city of Bologna some sixty years later. Galvani was among the first scientists to propose and attempt to demonstrate that nerves and muscles act electrically. It had been known for some years that frog legs continue to twitch even in animals that had been recently killed. Galvani interpreted such actions as based on a form of internally generated electricity within the animal.

Figure 4.1
Luigi Galvani (1737–1798) was among the first to explain nerve and muscle action as being based on "animal electricity." His experiments, although not conclusive, were consistent with the role of electricity in the functioning of nerve and muscle. Courtesy Wellcome Library, London.

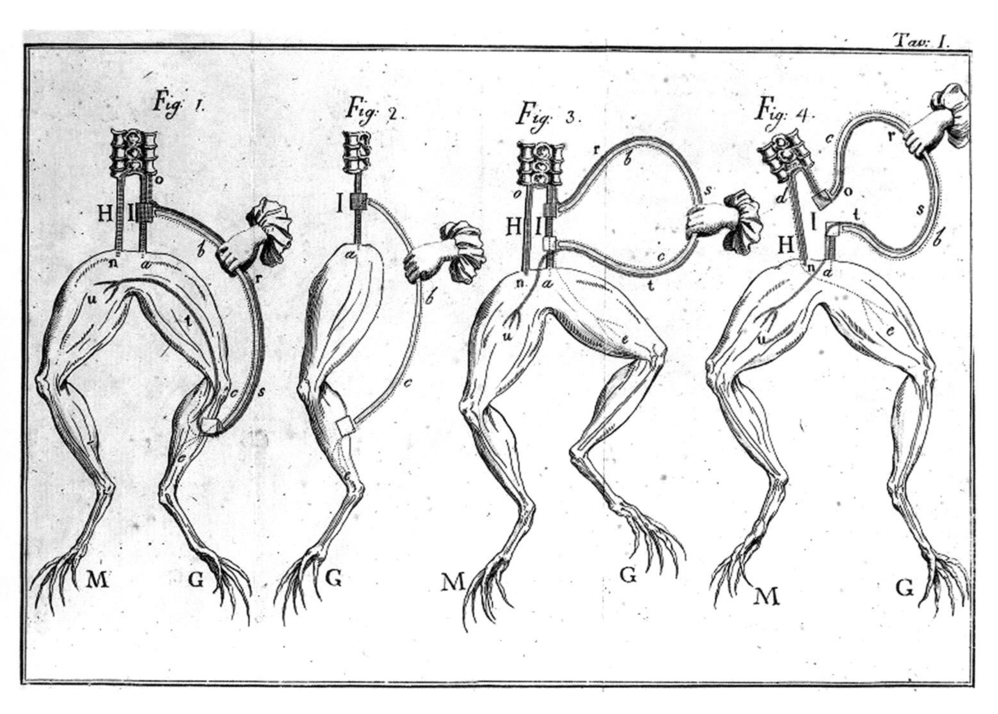

Figure 4.2
Galvani studied the activity of frog spinal cord. In one of his many experiments he impaled a frog spinal cord on a brass hook. When the animal's leg touched the iron fence, it responded with a kick. Galvani's contemporary, Alessandro Volta, argued that the junction of a brass hook with an iron fence would have been the source of a voltage, and it was probably the electricity so generated that served to stimulate the nerve. Courtesy Wellcome Library, London.

In one of his experiments Galvani placed the back end of a killed frog on a brass hook that had been mounted on an iron fence (figure 4.2). Whenever the frog's leg touched the iron fence, the muscle in the leg would contract. Although Galvani believed that his experiments demonstrated the existence of animal electricity, his contemporary and rival, Alessandro Giuseppe Volta, disagreed. Volta argued that an electrical current would be produced by the disparate metals. The junction of a brass hook against an iron fence would act like a primitive battery. When the frog's leg touched the fence it would close the circuit causing a current to flow. Volta suggested that it was the current produced in this way that activated the nerves and muscles, causing the leg to kick.

In the absence of units of electricity and instruments for its measurement, it would have been difficult to measure directly the nature of that electricity. Although Galvani's observations were not accepted as definitive proof for the existence of animal electricity, his work formed the basis of subsequent experiments that did.

The frog nerve-muscle preparation had been used for studying the physiology of the nervous system for many years. At first, experiments that relied on mechanical recordings of muscle contraction were used. In later years, with the increasing recognition of the distinction between voltage and current and the development of instruments for measuring them, it became possible to make direct measurements of electrical properties of nerve and muscle.

Measuring the Speed of Conduction in Nerves

A prominent nerve, the sciatic, exits from the lower spinal cord and connects to muscles in the leg. One leg muscle that can be easily dissected is the gastrocnemius, which is particularly prominent in frogs. In a killed frog the sciatic nerve and its attached muscle can be removed with both remaining functional. Electrical stimulation of the nerve leads to contraction of the muscle.

By the middle of the nineteenth century, it was accepted that nerve fibers conduct electrically, but there was no agreement about how fast the electrical impulse travels down the nerve. Some believed the speed of conduction to be very fast, similar to that of conduction in a copper wire, and thus impossible to measure with the available equipment. But they were wrong. Measuring conduction speed was not only possible, but the measurements were also a first step toward understanding in more detail how nerve fibers work. Hermann von Helmholtz, a great German physician and scientist (figure 4.3), addressed the question directly and solved it using the frog sciatic nerve-gastrocnemius muscle preparation.

Each time that Helmholtz stimulated the nerve, the muscle would contract, and the contraction was recorded on a fast-moving chart paper. When the muscle contracted it would pull on a pen that made a mark on the paper. The nerve was mounted so that it could be stimulated at two different points along its course. The muscle would respond identically to stimulation at either site, but there was a slightly greater delay following stimulation by the more distant electrode. Helmholtz measured the difference in the time of the muscle's response to stimulation of the closer and the more distant sites (figure 4.4). Because he knew

Figure 4.3
Hermann von Helmholtz (1821–1894) was one of the great physicists and physiologists of the nineteenth century. His ingenious experimental methods allowed him to measure the velocity of the nerve impulse. Courtesy Wellcome Library, London.

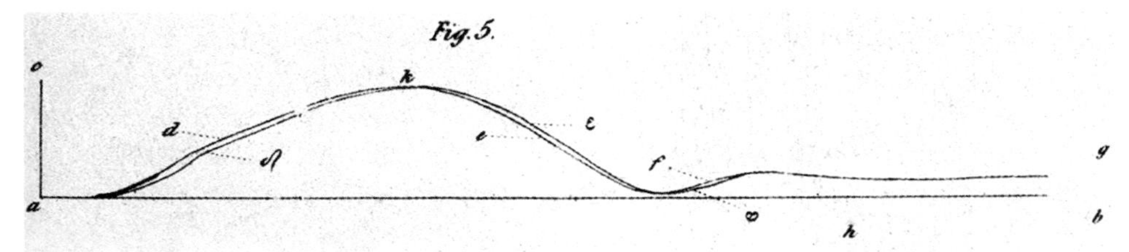

Figure 4.4
From Helmholtz's 1852 paper. Helmholtz activated an excised frog muscle by stimulating its attached sciatic nerve. Stimulating points were placed at different distances along the nerve. When the point of nerve stimulation was farther from the muscle, it contracted a bit later than the response to stimulating a nearer point on the nerve. Note the two slightly displaced curves. Dividing the distance between the electrodes by the time difference between the muscle responses allowed Helmholtz to calculate the velocity of nerve conduction.

the distance between the electrodes, he could calculate the speed of conduction as distance divided by time. His number of around 27 m/s for conduction in the frog sciatic nerve is close to the results you would obtain using a modern, sophisticated electronic apparatus.

With the limitations of the available apparatus, it was hard to measure the voltages and currents that are associated with nerve and muscle activity. In the nineteenth century various types of apparatus were developed that allowed detection of weak electric currents. If current is flowing in a circuit, a nearby magnet will be attracted by the wire that carries that current. For recording, a wire in the recording circuit can be stretched between the poles of a magnet. Movement of the magnet or the wire can be observed under a microscope. In some instruments a needle would detect the flowing current, and in others a tiny mirror was cemented to the wire, and a light beam directed toward it. The beam of light was reflected by the mirror toward a distant wall, hence amplifying the movement. Although these sorts of apparatus were a major advance, the accuracy of such measurements was limited by the fact that a weak current must move a physical object, so the record of electrical activity was distorted and slowed.

Despite the limitations of the available instruments, Carlo Mateucci in Florence and Emil du Bois-Reymond in Berlin—contemporaries of Helmholtz—managed to establish two facts about the electrical nature of nerve and muscle action. First, there is a resting voltage that can be measured from nerves and muscles; second, there is an electrical change associated with activity in the nerve or muscle.

Ionic Conduction

The relatively slow speed of nerve conduction showed that it is not at all like conduction down a copper wire. The current in nerve must be carried by ions, not by electrons as in a wire. Toward the end of the nineteenth century, experiments showed that ions may be distributed unequally across the nerve and muscle membrane, leading to an electrical potential difference between the inside and the outside of cells.

The membrane behaves like a capacitor in an electronic circuit, capable of maintaining an electrical charge across it. The mechanism of conduction now could be understood. If a segment of the axon becomes depolarized it would then activate the next segment. The mechanism of axon conduction is similar to the action of a firecracker fuse. Burning the tip of the wick ignites the next patch of fuse. An *action potential*—a sudden change of voltage across the axon membrane—activates the next patch on the axon.

Initially it was assumed that nerve conduction might be associated simply with a brief breakdown of the membrane, like the short-circuiting of a capacitor. The change in voltage associated with the action potential might be enough to serve as a stimulus to activate the next patch of axon. But the action potential is not just the short-circuiting of the axon membrane. As we shall see, the story is more complex—and more interesting.

Size and Speed

The word "nerve" can be confusing. It can be used to refer either to an individual axon or to a group of axons connected into a bundle. Peripheral nerves like the sciatic are usually made up of a large number of axons of varying diameter. If the word "nerve" is used, the reference to either a single axon or a mixed nerve carrying several axons is usually clear from the context. Figure 3.10 in the previous chapter shows a cross section of a mammalian sciatic nerve as seen at high power under a light microscope. The individual axons within the nerve can be seen as circular profiles of various sizes.

As should be obvious from the figure, axons vary in diameter. Does that matter? Helmholtz's measurement of 27 m/s would have reflected the conduction speed of the largest fibers in the mixed nerve, that is, only the fastest-conducting axons. Smaller axons conduct more slowly. It would have been difficult to measure the conduction speed of the smaller fibers using the methods that were available to Helmholtz and his contemporaries.

In the early twentieth century a new and far superior method for electrical measurements, the cathode ray oscilloscope (figure 4.5), became available. Currents in a nerve or muscle could now be greatly amplified. The amplified signal is applied across two electrodes in the oscilloscope, causing a stream of electrons to be deflected with a record of the deflection appearing on a screen.

In the 1920s the American physiologists Joseph Erlanger and Herbert Gasser (figure 4.6), working at Washington University in St. Louis, used the cathode ray oscilloscope to record nerve activity and measure conduction speed. The oscilloscope gave far more accurate measurements of conduction speed than had been possible with earlier devices. Helmholtz's technique was too insensitive to enable him to detect variability

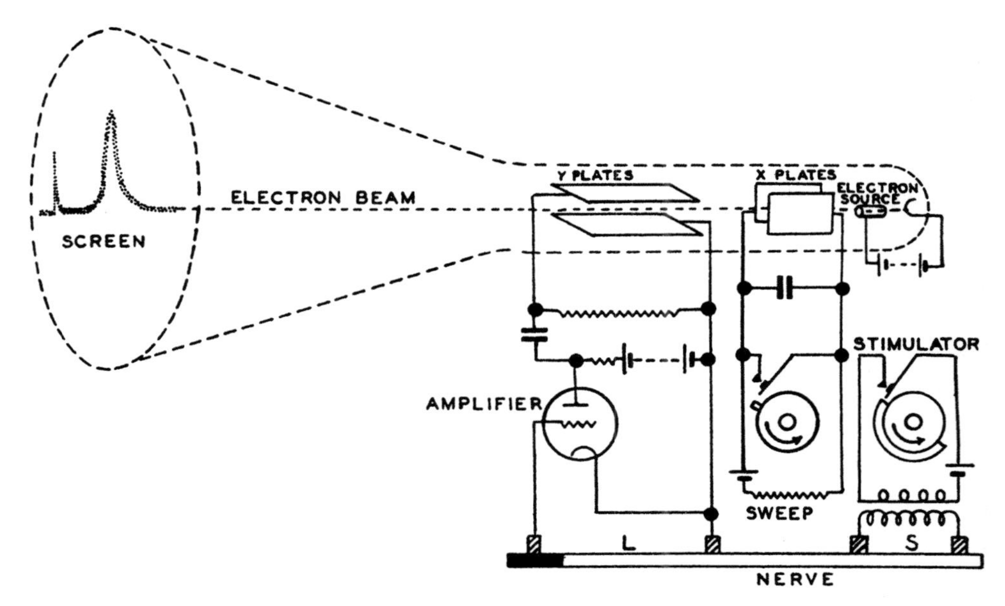

Figure 4.5
Gasser and Erlanger used the cathode ray oscilloscope to measure conduction velocity in a compound nerve. Previous methods like the one using the string galvanometer were relatively insensitive and did not allow for precise timing. Courtesy University of Pennsylvania Press.

in the conduction speed of different axons within the nerve. The increased precision permitted by the oscilloscope, however, allowed Gasser and Erlanger to observe and measure such differences.

The sciatic nerve is composed of many axons of different diameters. With the great amplification and precise timing that the oscilloscope makes available, the response of all of the fibers in a compound nerve like the sciatic could now be recorded. The logic of the experiment is similar to the situation that might be involved in judging a foot race.

Suppose a group of runners sets off from the same point along a straight track. The judges stand some distance away from the start and see the runners go by. If one runner is faster than the others, the judges would see that person pass first followed by the others in order of their speed. The farther away from the starting line that the judges stood, the greater the time that would elapse between the appearance of the first and subsequent runners. In a compound nerve like the sciatic there are axons of many different diameters, and these conduct at different speeds. How was this established?

Figure 4.6
Joseph Erlanger (1874–1965) on the left; Herbert Gasser (1888–1963) on the right. Early pioneers in electrical recording in the nervous system. Courtesy Wellcome Library, London.

Gasser and Erlanger delivered a strong electrical stimulus repetitively to the sciatic nerve at a single point. The stimulation was strong enough so that it initiated an action potential in every axon within the nerve. The action potentials so elicited were conducted down the nerve to the recording electrode. The oscilloscope showed a series of deflections—the so-called compound action potential—with each successive wave in the compound action potential, like runners arriving at the checkpoint at successive times (figure 4.7).

It soon became apparent that in peripheral nerves there is a relatively precise relationship between conduction speed and fiber diameter. Thick axons conduct fastest; thin axons conduct more slowly. It is also clear that fibers of different caliber have different functions. Each of those different classes of nerve fibers carries a different message, be it sensory or motor in function.

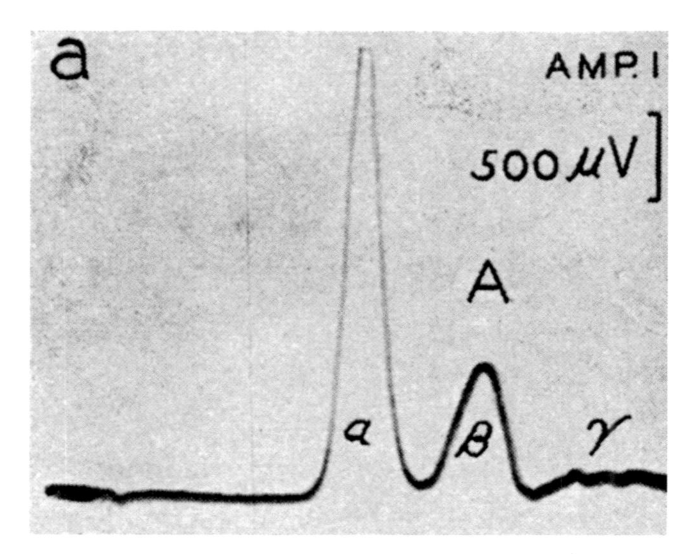

Figure 4.7
Gasser and Erlanger's record of the response of a compound nerve to electrical stimulation. Three labeled alpha, beta, and gamma peaks are seen corresponding to their relative speeds of conduction. Alpha is associated with the thickest axons in the compound nerve. Courtesy University of Pennsylvania Press.

In general the speed of conduction is related to the diameter of the nerve fiber. For peripheral nerves the conduction velocity of component axons can be approximated by multiplying the fiber diameter in micrometers times six. Thus, a 20-μm axon conducts at 120 m/s. Conduction across a synapse is much slower. Studies of synaptic delay were essential in the analysis of how information is transmitted across the synapse and in working out the mechanism of the simplest reflexes. Chapter 12 deals with this in more detail.

Stimulus Strength and Firing Rate

In the early 1930s the American physiologist H. Keffer Hartline studied the mechanism of action of light on the compound eye of the horseshoe crab, *Limulus. Limulus* is a remarkable animal from several points of view. It has existed in its present form for almost half a billion years. Like many invertebrates, it has a compound eye made up of many individual small elements. That eye has served as a useful structure for measuring the effect of varying light intensity on sensory response.

Hartline recorded the response of a single nerve fiber from one such compound eye to light of varying intensity. Figure 4.8 shows the orderly relationship between the intensity of the light and the rate of nerve firing. It is also apparent that the stronger light initiates firing earlier than weaker lights. This experiment establishes a basic principle of sensory function: intensity of a stimulus is coded by the frequency of firing.

A similar principle is true for the motor system. A stronger command to a muscle is coded by a higher frequency of firing by its motor fibers.

Hodgkin, Huxley, and the Giant Squid Axon

Physiologists initially assumed that the action potential reflects a simple breakdown of the voltage across the membrane. If this were the case, at the height of the action potential the voltage across the membrane would read zero. Because mammalian axons are so small, it would have been difficult to observe the mechanism of the action potential in a single axon by direct measurement. The anatomist and zoologist J. Z. Young in 1936 discovered and described the fact that squids and some related marine animals have a giant axon. The relatively enormous axon, a millimeter or more in diameter, allowed Alan Hodgkin and Andrew Huxley to record electrical activity from an electrode placed inside it (figure 4.9). They showed that the mechanism of the action potential is more complex than a simple breakdown of the voltage across the membrane.

Hodgkin and Huxley began collaborating as young students at Cambridge University and at a marine biology station in Plymouth, England. Their experiments were interrupted by the Second World War. Both men were technically gifted. Hodgkin, for example, was one of the scientists who cooperated in the design and improvement of radar, which proved invaluable to British air defense during the war. Huxley was the grandson of Thomas Huxley, one of the early and great defenders of Charles Darwin and evolution, and the half brother of Aldous Huxley, the novelist.

As we described earlier, neurons are surrounded by a very thin membrane that is continuous around the entire cell, including its dendrites and its axons. The structure of this membrane is crucial to the normal functioning of nerve cells. Membranes regulate the ionic composition of the cell that they enclose, and thus the amount of water contained within the cell, and hence its size. There is a difference in electrical potential between the inside and outside of the membrane, which is called the *resting potential.*

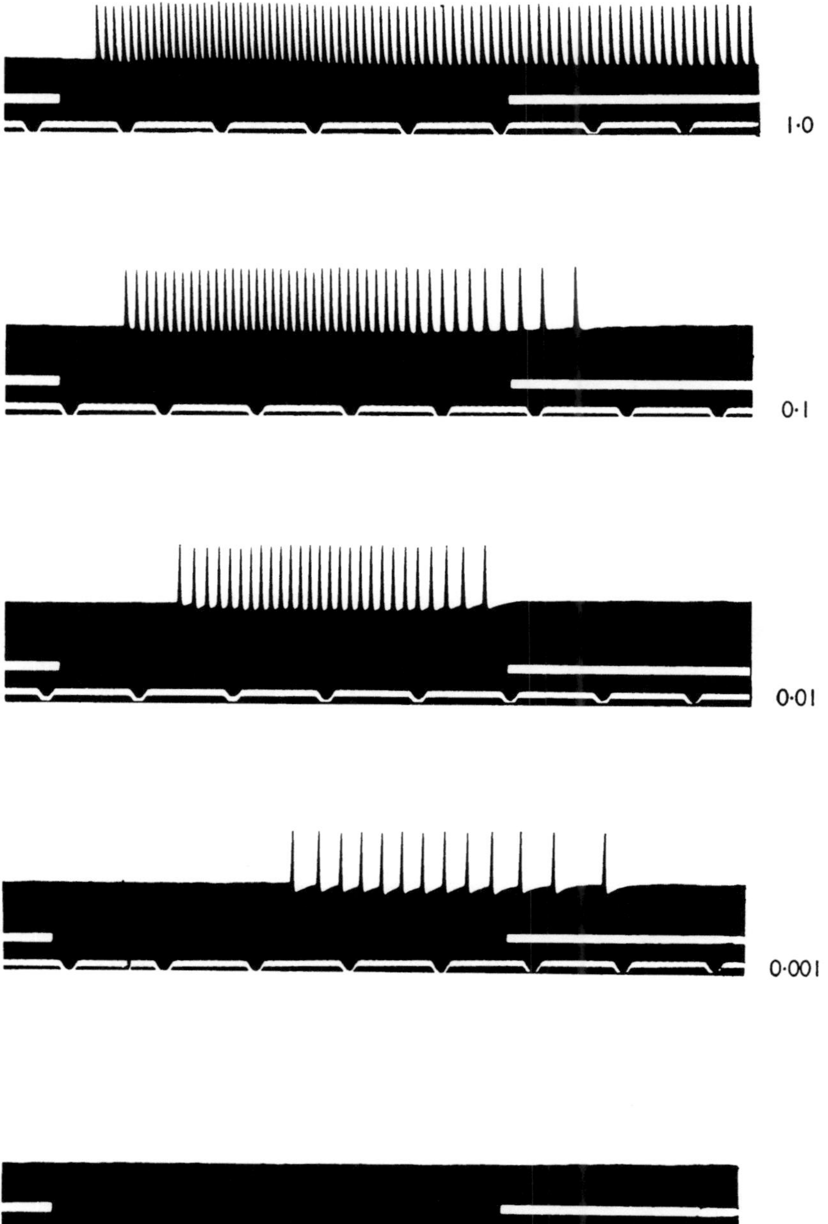

Figure 4.8
Neuronal response recorded by Hartline from a single element in the compound eye of the horseshoe crab *Limulus*. The intensity of light striking the eye was varied by successively denser filters, labeled from the bottom .0001 to 1.0. The intensity of the light in the top trace is 10,000 times brighter than that shone at the eye in the bottom trace. The eye responded earlier and at a higher frequency to the brighter light. © 1934 The Wistar Institute of Anatomy and Biology.

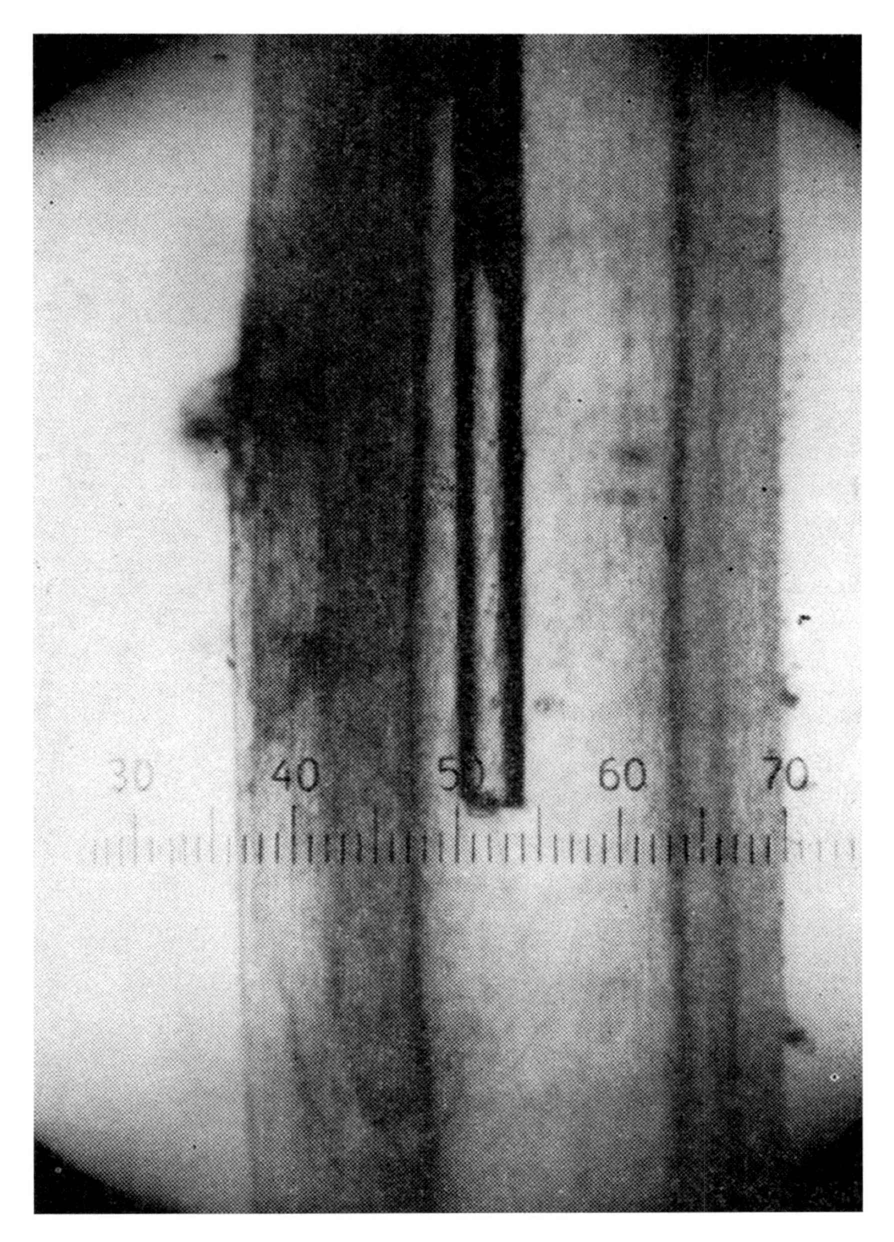

Figure 4.9
The giant axon of the squid is much larger than that of mammals, so Hodgkin and Huxley were able to place a recording electrode inside the axon. Courtesy Nature Publishing Group.

The resting potential across a nerve cell's membrane is an essential prerequisite to impulse conduction—the basic mechanism by which axons interconnect cells in the nervous system. A preparation containing healthy nerve cells or a nerve fiber is mounted in the apparatus. The giant axon of a squid is an especially useful preparation for this type of study in that it is much larger than a vertebrate axon and hence easier to impale with an electrode for electrical recording. One electrode is advanced toward the axon while a reference electrode is placed in the salt water that surrounds it. The electrode must be constructed so that it is electrically insulated except for its tip. In the example shown in figure 4.9, the recording electrode is made out of a tapered glass tube filled with a salt solution. Ions in the solution can carry currents that are monitored by the recording apparatus. In the example shown, the axon is bathed in a salt solution whose resistance is relatively low. With both electrodes outside the cell there is no current between the recording electrodes, so the recording device reads zero. As long as the tip of the recording electrode remains in the medium surrounding the axon, we record a zero voltage. If we now lower the recording electrode toward and into the axon, when it penetrates the membrane there is a sudden and dramatic voltage shift. The meter now records a steady potential difference between the inside and the outside of the cell.

In the example shown here the inside of the cell is –70 mV or 0.07 of a volt negative relative to the surrounding medium. The exact voltage may vary, but in healthy neurons there is always a resting potential difference across the cell's membrane. The inside is always negative relative to the outside in the resting state. The difference is maintained by the cell's segregating potassium and sodium ions.

The Action Potential

There is a difference in potential from the inside to the outside of nerve cells due to a differential distribution of charged molecules across the nerve membrane. Let us consider the sequence of events that occurs when an action potential is generated. At rest, there is a resting potential of 70 mV across the nerve cell membrane with the inside negative. The axon is said to be *polarized.*

Suppose that we connect a battery across the nerve. If we close a switch we change the polarization of the membrane. For example, we could increase the resting membrane potential by putting the anode, the positive terminal of the battery, outside the axon, with the negative terminal inside. The membrane would now be *hyperpolarized.* If we placed the cathode, the negative terminal of the battery, outside the nerve membrane with the anode inside, and closed the switch we would now *depolarize* the membrane.

Suppose that we could easily change the voltage of our battery and hence regulate the voltage across the membrane. We now begin to depolarize the membrane by applying a slowly increasing voltage across it. At a certain level of depolarization the electrical record from the axon would show a sudden and dramatic change. The voltage across the membrane would no longer simply reflect the voltage applied to it from the battery. It is as if a switch within the axon were suddenly thrown, and the voltage across the membrane began to change rapidly of its own accord. The membrane now depolarizes further.

If it were simply reflecting a breakdown in the membrane's voltage, the measurement would stop at zero. But it does not stop. The inside of the axon becomes *positive* relative to the outside for a very brief period. Within a millisecond, the voltage returns rapidly to its resting level, with the inside of the axon again at about 70 mV negative relative to the outside. This sudden and rapid sequence of voltage changes across the membrane is the action potential.

The action potential is initiated by depolarization across the nerve cell membrane. There is an electrical threshold of depolarization necessary to produce the action potential. Once that threshold is reached the parameters of the action potential are independent of the depolarization that initiated it, and it takes about one-thousandth of a second to go through its complete cycle. The sequence of voltage changes across the membrane is nearly identical in every action potential. The switch-like nature of the action potential once threshold has been reached is one of the most important and characteristic aspects of the functioning of nerve cells. Action potentials are not graded in intensity—they either occur or they do not. This basic property of nerve cells is called the *all-or-none law.*

The all-or-none law is one of the fundamental concepts to keep in mind when you are thinking about how the nervous system works. We will see some of the implications of the all-or-none law especially when we consider coding of sensory messages within the nervous system. Because the voltage change associated with the action potential is always the same in a given axon, the strength or intensity of a sensory input cannot be coded by differences in the amplitude of the action potential. In most cases the *intensity* is signaled by the *frequency* of the action potentials. The *quality* of the sensation, whether it is signaling light or sound, pain or pressure, is coded by *which* axon carries the message. Optic nerve axons signal visual stimuli; auditory nerve axons signal sounds.

In the late 1940s and early 1950s, Hodgkin and Huxley were able to resume the experiments that had been interrupted by the war. They knew that sodium ions (Na^+) are in much higher concentration outside the cell, and potassium ions (K^+) are in much higher concentration inside the cell. They showed that the initial rising phase of the action potential is due to a sudden and dramatic shift in the permeability of the nerve membrane to Na^+ ions (figure 4.10). When the threshold for the action potential is reached there is a sudden increase in Na^+ permeability. Because Na^+ ions now enter rapidly, the voltage across the membrane decreases very rapidly. The change in the permeability of the membrane to Na^+ sharply reduces the membrane voltage.

But the response does not stop at zero. The change in voltage continues so that the inside of the axon becomes *positive* relative to the outside

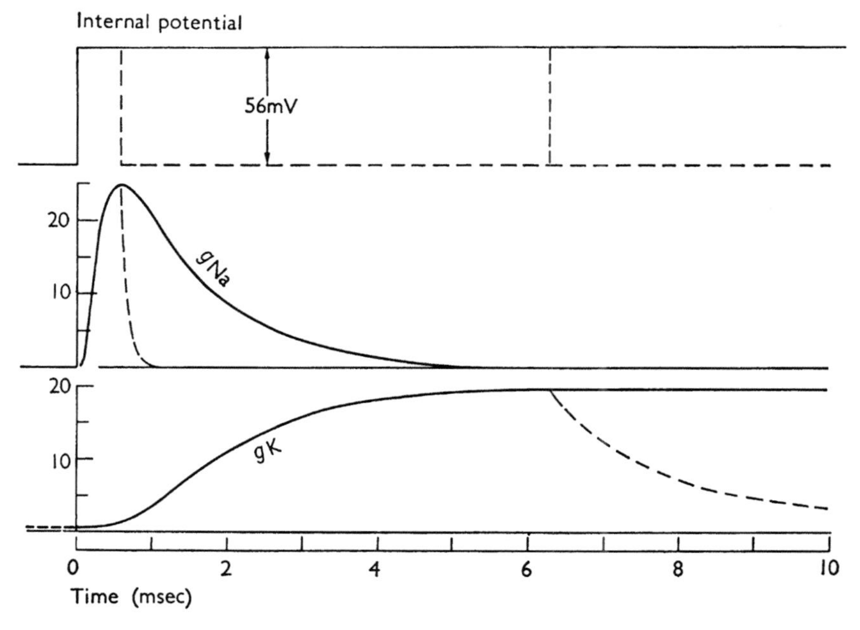

Figure 4.10
Hodgkin and Huxley measured the change in permeability over time to sodium (gNa) and potassium (gK) in axons in association with the action potential. Courtesy Royal Society Publishing.

for a very brief period of time. Within a fraction of a millisecond this permeability to sodium is turned off and is followed by an increase in the permeability of the membrane to potassium ions (K^+). Because K^+ is more highly concentrated inside the cell than outside, and because the inside of the cell is now positive relative to the outside, K^+ rushes out, and the membrane is quickly repolarized to its original state. The axon is ready to fire another action potential.

Although the movement of Na^+ from outside to inside the axon produces the rising phase of the action potential, and the movement from inside to outside of K^+ the falling phase, the actual amount of Na^+ and K^+ required to cross the membrane and produce the action potential is small. It is only a very tiny fraction of the concentration difference of these two ions. Hence, axons can fire thousands of action potentials and continue to work normally. Ultimately the differences in concentration of Na^+ and K^+ must be restored. Na^+ and K^+ must be actively pumped across the membrane to restore the normal concentration differences. The pump is a slow process that can work when the fiber is at rest. It is unrelated to the permeability changes associated with the action potential.

Saltatory Conduction

Nerve conduction demands energy and can be relatively slow. Another process—known as *saltatory* conduction from the Latin *saltare* (to hop or leap)—increases velocity irrespective of axon thickness. In axons covered with a myelin sheath, depolarization skips between the periodic gaps in the myelin—the nodes of Ranvier. This sequential generation of the action potential from node to node economizes on energy use and speeds the process of transmission.

The importance of myelin to normal function of axons becomes dramatically evident in the disease multiple sclerosis. The disease attacks the soft myelin sheath and replaces it with a hard fibrous tissue. As a consequence, timing of the nerve impulse is thrown off. There are often temporary remissions in the symptoms of the disease, but over time the axons whose myelin is affected may die.

In this chapter we saw how the nature of nerve conduction in a single axon came to be understood. In the chapter that follows we consider how nerves connect to other nerves and to muscles.

5

Chemical Transmission and the Mechanism of Drug Action

How does the firing of an action potential activate or inhibit the nerve cell or muscle to which it connects? Until about 1950 physiologists disagreed about how nerves activate the next neuron in a pathway. Some, the "soup" people, believed that nerves activate muscles and communicate from neuron to neuron by releasing a minute amount of a chemical neurotransmitter. Others, the "spark" people, believed that the transmission from neuron to neuron must involve direct electrical coupling. They pointed out that electrical coupling is immediate; diffusion is much slower. The spark people could not believe that a method that is based on the release of a chemical spread by diffusion could be fast enough to account for the short time that it takes for one neuron to activate a muscle or influence the next neuron in a pathway.

As noted earlier we now know that both were right, although the "soup" theorists were more right. The question of how the link from nerve to muscle or neuron to neuron works was only solved by degrees.

The Discovery of Neural Transmitters

The first evidence came from studies of the autonomic nervous system. The autonomic system is involved in functions that are continuous and automatically regulated, like the control of heart rate, blood pressure, and the contraction of gut and bladder.

In the early twentieth century people in several laboratories were studying the effects of certain chemicals on muscles innervated by the autonomic nervous system. Epinephrine (called adrenaline in the United Kingdom), a substance extracted from the adrenal glands, seemed to produce the same effects on the heart and smooth muscles as did activating a sympathetic nerve to those organs. T. R. Elliott was a young physiologist working at Cambridge in the early twentieth century. Citing the

PROCEEDINGS

OF THE

PHYSIOLOGICAL SOCIETY,

May 21, 1904.

On the action of adrenalin. By T. R. Elliott.
(*Preliminary communication.*)

In further illustration of Langley's generalisation that the effect of adrenalin upon plain muscle is the same as the effect of exciting the sympathetic nerves supplying that particular tissue, it is found that the urethra of the cat is constricted alike by excitation of the hypogastric nerves and by the injection of adrenalin. The sacral visceral nerves, on the other hand, relax the urethra of the cat. But while the hypogastric nerves relax the tension of the bladder wall in the cat, they do not cause any similar change in the dog, monkey, or rabbit: and though, as is well known[1], adrenalin inhibits the cat's bladder, this reaction is the exception in the mammalian bladder, for adrenalin does not produce any change in those of the three animals named above.

I have repeated the experiment of clean excision of the suprarenal glands and find that the animal, when moribund, exhibits symptoms that are referable to a hindrance of the activities of those tissues especially that are innervated by the sympathetic. They lose their tone; and may even fail to respond to electrical stimulation of the sympathetic nerves. The blood-pressure falls progressively, and the heart-beat is greatly weakened. And at the latest stage previous to death, though the nerves of external sensation and those controlling the skeletal muscles are perfectly efficient, the sympathetic nerves exhibit a partial paralysis of such a nature that nicotine, when injected, is unable to effect through them a rise of blood-pressure or to cause dilatation of the pupil.

[1] Lewandowsky. *Centralblatt f. Physiol.* p. 433. 1900.

PROCEEDINGS OF THE PHYS. SOC., MAY 21, 1904. xxi

This marked functional relationship of the suprarenals to the sympathetic nervous system harmonises with the morphological evidence that their medulla and the sympathetic ganglia have a common parentage[1]. And the facts suggest that the sympathetic axons cannot excite the peripheral tissue except in the presence, and perhaps through the agency, of the adrenalin or its immediate precursor secreted by the sympathetic paraganglia.

Adrenalin does not excite sympathetic ganglia when applied to them directly, as does nicotine. Its effective action is localised at the periphery. The existence upon plain muscle of a peripheral nervous network, that degenerates only after section of both the constrictor and inhibitory nerves entering it, and not after section of either alone, has been described[2]. I find that even after such complete denervation, whether of three days' or ten months' duration, the plain muscle of the dilatator pupillæ will respond to adrenalin, and that with greater rapidity and longer persistence than does the iris whose nervous relations are uninjured[3].

Therefore it cannot be that adrenalin excites any structure derived from, and dependent for its persistence on, the peripheral neurone. But since adrenalin does not evoke any reaction from muscle that has at no time of its life been innervated by the sympathetic[4], the point at which the stimulus of the chemical excitant is received, and transformed into what may cause the change of tension of the muscle fibre, is perhaps a mechanism developed out of the muscle cell in response to its union with the synapsing sympathetic fibre, the function of which is to receive and transform the nervous impulse. Adrenalin might then be the chemical stimulant liberated on each occasion when the impulse arrives at the periphery.

Figure 5.1
Thomas Elliott (1877–1961) suggested that autonomic nerves might act by releasing a small amount of a transmitter substance.

similarity of the response to electrical stimulation and directly applied epinephrine, Elliott made the bold suggestion that sympathetic nerves might act by releasing a chemical stimulant when the impulse arrives at the target organ (figure 5.1).

Elliott went on to a career as Professor of Medicine at University College London. Although his evidence had made it seem likely that autonomic nerves work by liberating a small amount of a chemical transmitter, definite proof was lacking. The proof came from studies of the connection between the vagus nerve and the heart, one of the many organs that this cranial nerve innervates.

A frog heart will continue to beat for hours in a dish of saline after removal from the body. In the isolated heart, action potentials are initiated within a specialized pacemaker region, and these impulses are carried through the rest of the heart's cells by an intrinsic conduction system. Note that this capacity of a heart to beat on its own makes heart transplants possible, for a transplanted heart would certainly have no functioning nerve connections. The vagus nerve can be dissected along with the heart. Electrical stimulation of the attached vagus nerve slows the rate of heart beats and reduces the force of contraction. In normal life this effect of vagus stimulation is part of the normal mechanism for reflex control of the rate at which the heart beats and the force it develops. It was suspected that the vagus might act by liberating a chemical substance that acted on the heart to slow it and weaken its force of contraction. But conclusive evidence was needed.

The evidence that this chemical action was the correct explanation emerged in the early 1920s from the basement laboratory of an Austrian pharmacologist, Otto Löwi. The idea for the crucial experiment came to Löwi one night in a dream. He woke up, scribbled down his inspiration, and went back to sleep. The next morning he could not decipher his own handwriting. That night he had the same dream, woke up while the insight was still clear in his mind, and rushed down to his basement to do the experiment.

Löwi worked with a frog heart that was beating in a dish of saline. When he stimulated the attached vagus nerve, the heart slowed as expected, and the force of its contractions became weaker. After prolonged and repetitive stimulation of the nerve, he collected the fluid bathing the heart. When he injected that same fluid back into the same or another frog's heart, it had the identical effect of slowing the heart rate and weakening the force of each beat. Löwi gave the released material the arbitrary name *Vagus Stoff* or vagus material. *Vagus Stoff* was soon identified as acetylcholine.

A few years after Löwi's experiment William Alexander Bain, a Scottish pharmacologist, did an elegant experiment demonstrating the principle of transmitter release that had been discovered by Löwi.

Bain connected the output from one frog heart to a second heart (figure 5.2). When he stimulated the vagus nerve that was attached to the first heart, it slowed, and the force of its beat became weaker. A short time later the same effect was seen in the second heart. Löwi's experiment and Bain's demonstration confirmed that there is chemical transmission from nerves to muscles.

At the time of Löwi's work, physiologists and pharmacologists recognized that many of the organs of the body had a dual autonomic supply, receiving connections from both sympathetic and parasympathetic nerves. The two sorts of input typically produce opposite effects when stimulated. In addition to its parasympathetic input, the heart, like

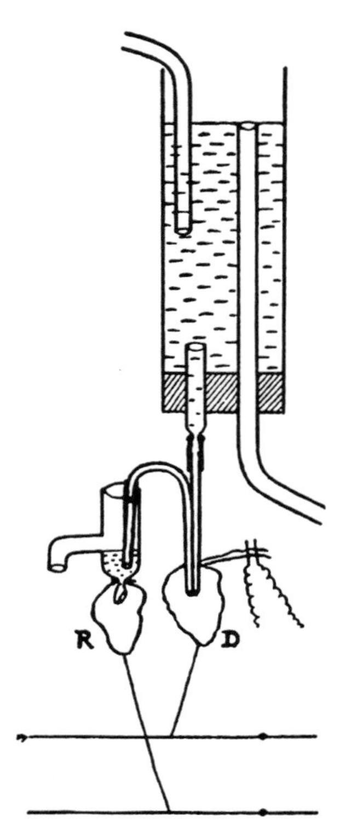

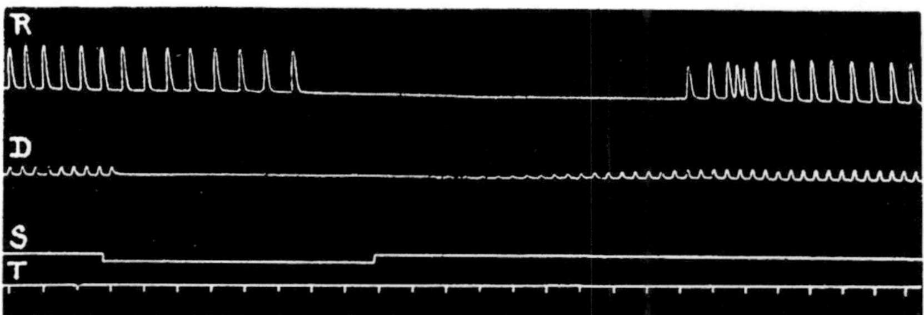

Figure 5.2
William Alexander Bain's 1931 replication of Otto Löwi's discovery of the release of a chemical when the vagus nerve is stimulated. Bain excised frog hearts with the vagus nerve attached to one of them. Stimulation (S) of the vagus nerve that is connected to the first heart D, leads to slowing of and reduction in the force of its beat. Fluid is pumped from the first to the second heart (R). After a delay, the second heart slows and the force of its beat is reduced. Acetylcholine, released when the first heart was stimulated, acts on the second heart. Courtesy John Wiley & Sons.

other organs, has an input from the sympathetic nerves. Löwi himself showed that, in contrast to the effect of parasympathetic vagus stimulation, which slows the heart rate, when the sympathetic nerve to the frog heart is stimulated, a substance is released that produces an increase in the rate and force of the heart's contraction.

We now know that the substance released by stimulation of the sympathetic nerves is either epinephrine or the closely related *norepinephrine* (British, *noradrenaline*). The American and British terms are often used interchangeably; both mean "near the kidney," one in Latin, the other in Greek. Thus, there were at least *two* proven transmitter substances in the autonomic nervous system: acetylcholine and norepinephrine.

In 1938 Löwi was driven out of Austria by the Nazis and moved to the United States. Although he had pioneered in the discovery of chemical transmission in the autonomic system, he was skeptical about its possible role for activation of voluntary muscles. He once said, "For myself, I don't believe a word of it." In 1936 Henry Dale (figure 5.3), Wilhelm Feldberg, and Marthe Vogt, working in London, demonstrated that transmitter release also activates voluntary, striated muscle. Henry Dale was among the most eminent of British pharmacologists. Like Löwi in Austria, Wilhelm Feldberg and Marthe Vogt had left Germany because of the Nazi regime.

The design of their experiments was similar to those of Otto Löwi. They reasoned that if rapid activation of the nerve caused the release of a transmitter substance, and if the transmitter could be protected from circulating enzymes that might destroy it, they should be able to collect fluids that contain the transmitter, and the collected fluid should have predictable physiological effects. They isolated a single muscle along with its motor nerve, keeping the blood vessels to the muscle intact. They perfused the muscle continuously with fluids. Rather than using blood for this perfusion, they used a saline solution so that enzymes that break down acetylcholine would not be present in the fluid. Using a very sensitive assay of strips of leech muscle, they demonstrated that acetylcholine is released when the motor nerve is stimulated.

Chemical Transmission between Neurons

The idea of chemical transmission between nerve cells was not easy for many physiologists to accept. Although they granted that such a mechanism must operate for the connections from nerve to muscle, they were

Figure 5.3
Henry Dale (1874–1968), on the left, and Otto Löwi (1873–1961), pioneers in the study of neurotransmitters, at a meeting in Oxford. Courtesy Wellcome Library, London.

slow to accept the idea of a chemical substance that mediated the transmission from one neuron to another. Two sorts of objections were raised to the idea of chemical transmission between neurons: (1) synaptic action seems too fast to be mediated chemically; (2) only a tiny amount of transmitter substance seems to be released when a motor nerve is stimulated. That was not enough, it was felt, to activate the next neuron in a pathway.

With the advent of newer experimental methods, these reservations were swept away. When recordings are made from pre- and postsynaptic nerve cells in the spinal cord of a mammal, or in certain invertebrate neurons, there is almost always a period of irreducible delay from the time that the impulse arrives at the presynaptic terminal to the time that the postsynaptic neuron is activated. This delay is consistent with the relatively slow diffusion of a chemical transmitter substance across a narrow synaptic gap. Furthermore, detailed study has shown that neurons are indeed sensitive to the very tiny amounts of transmitter released by a single nerve impulse.

The great majority of synapses in the vertebrate nervous system are chemical, but there are also demonstrable examples of electrical synapses. In those cases in which the presence of electrical synapses has been proven, the distance between the successive nerve cells is much smaller than it is in chemical synapses. Such electrical synapses are seen, for example, in giant neurons in the brainstem of a fish.

It is now clear that most synaptic transmission between neurons in the vertebrate nervous system and the transmission from motor neurons to muscles are mediated chemically. When an action potential reaches the end of an axon, it usually acts on the next nerve cell in the pathway or on a muscle cell by releasing a small amount of a transmitter substance into the synapse. The synapse is the narrow space that separates the end of the axon from the membrane of the next cell. Transmitter substances may either activate or inhibit the next cell.

Neural Transmitters in the Central Nervous System

So far we have identified two substances, acetylcholine and norepinephrine, as chemical transmitters released by autonomic neurons. It is now known that in addition to their actions in the autonomic nervous system, norepinephrine and acetylcholine are also transmitters at certain synapses within the brain and spinal cord of mammals. There are many more transmitters.

The search for transmitters and analysis of their action together constitute one of the central problems of neurobiology. How could we determine whether a suspected substance, a "putative" transmitter, is indeed a transmitter substance in the brain or spinal cord? Several criteria have been established by which the possible role of a substance as a chemical transmitter can be evaluated.

1. The substance must be physiologically active; that is, it must affect nerves or muscles when applied directly.
2. It must be present at specific sites in the nervous system along with enzymes that can synthesize it and enzymes that can break it down.
3. The substance must be recoverable from the appropriate region of the nervous system during the period when a nerve is stimulated, but not in other periods. (Recall Löwi's procedure with the frog heart and Dale, Feldberg, and Vogt's experiment with voluntary muscle.)
4. The substance must produce physiological responses identical to those of nerve stimulation when it is applied to the appropriate nerve, muscle, or gland.
5. The response of the presumed transmitter substance must be modifiable by drugs in the same way that such drugs modify the effects of nerve stimulation.

These criteria for identifying a transmitter were developed in the course of study of transmitters in the autonomic nervous system. The same criteria are applied to transmitter candidates in the central nervous system. But the great complexity and dense packing of neurons in the brain makes the definite identification of neural transmitters more difficult. An organ like the heart or a piece of smooth muscle from the gut can be perfused through its blood vessels, and the fluid that is collected during periods of stimulation can be compared with fluid recovered during periods in which there is no stimulation. But in the brain even small blood vessels often supply several different brain areas. Nearby cells in the brain may contain totally different transmitters, and it would be difficult to stimulate selectively only a single group of nerve fibers within a specific region of the brain.

Although study of neural transmitters in the brain is difficult, it is not impossible. There are a number of experimental maneuvers that can be used to help.

1. A small known region of the brain of an experimental animal can be removed and analyzed for the presence of a suspected transmitter substance.

2. Minute amounts of drugs that mimic the effects of a suspected transmitter can be applied through a tiny pipette, and the effect on individual cells can be tested. Especially useful are methods that allow the experimenter to release a tiny but measured amount of substance just outside a target cell and then to record the response of that cell to the injection using an electrode whose tip is just adjacent.
3. Specialized chemical techniques can be used to reveal the presence of certain transmitter substances and the enzymes that synthesize them.

Research on neurotransmitters has been helped by development of techniques whereby the cells and fibers that contain certain transmitters can be seen directly in microscopic sections of the brain. Sections of the brain are treated with formalin vapors that cause certain transmitters to form compounds that are highly fluorescent. When illuminated by ultraviolet light those regions containing a given transmitter will glow with a characteristic color. Ultraviolet light is normally invisible to the human eye, but the fluorescent glow that it produces in the tissue is at a wavelength that can be seen. If the tissue is properly prepared and illuminated with ultraviolet light, cells containing two known transmitters, dopamine and norepinephrine, appear bright green when viewed under the microscope. Tissue containing the neurotransmitter substance serotonin fluoresces a bright yellow.

A group of nerve cells in the brain may share a common transmitter substance, and those cells and their axons can be followed to their terminations. If the cell bodies are destroyed, a pathway that is normally found in the brain will not be seen.

These techniques have revealed connections that had been previously unknown. For example, the *locus coeruleus* (Latin for the blue place), a small nucleus in the brainstem, has very widespread connections to the cerebral cortex, brainstem, spinal cord, and cerebellum—none of which had been suspected before the use of histofluorescence to trace its pathways.

Another method for identification of neurons containing a given transmitter is borrowed from immunology. In almost all cases in which a neuron manufactures a given transmitter, there is usually a unique enzyme contained within the neuron that is essential for making that transmitter. By use of immunological techniques it is possible to develop antibodies that will bind specifically to that enzyme and no other. If the antibodies are attached to a compound that can make them visible in a section by fluorescence or a histochemical reaction, we can see directly all of the neurons that contain within them a given transmitter substance.

Acetylcholine and norepinephrine are now firmly established as transmitters in the brain as well as in the autonomic nervous system. There are also many more substances that are possible transmitters but for which the evidence may not yet be fully complete.

There are two basic effects that one nerve can exert on another. Liberation of transmitter substance can *excite* the postsynaptic neuron, or it can *inhibit* it. Some substances, such as glutamate, seem to function almost exclusively as an excitatory transmitter substance. One transmitter, gamma-amino butyric acid (GABA), is excitatory in the developing brain but is inhibitory in the adult brain. In all adult brains in which a cell is known to release GABA from its terminals, that action inhibits the postsynaptic cell. Although some transmitters have only a single function, many have more. The same transmitter can be either excitatory or inhibitory in its effects depending on the nature of the postsynaptic site—the region of the cell onto which it is released.

The Idea of a Receptor

In 1906 J. N. Langley, Professor of Physiology at the University of Cambridge, was studying the effects of certain drugs on whole animals and on isolated nerve–muscle preparations. Langley noted that certain poisons had profound physiological effects on an isolated nerve–muscle preparation, although they did not act directly on the nerve or its endings. Nicotine, for example—the active agent in tobacco—seemed to be acting on the muscle itself. He confirmed this view by showing that nicotine produced powerful contractions even in preparations in which the muscle had been entirely deprived of its motor nerve supply prior to testing. Langley concluded that there must be on muscles ". . . one or more substances (receptive substances) which are capable of receiving and transmitting stimuli, and capable of isolated paralysis . . ."

Langley himself refused to speculate on the exact nature of the "receptive substance" or receptor, but it is now clear that receptors are regions of the postsynaptic cell that are specialized for mediating the responses of the cells to transmitters. The concept of a receptor has been a fundamental one in interpreting normal physiology and drug effects since it was first put forward by Langley.

The idea of a receptor can help to explain some facts that would otherwise be puzzling. For example, it was well known that certain drugs could simulate the effects on target organs of acetylcholine. Henry Dale, a student of Langley's and a friend and supporter of Löwi, noted that two such drugs—*muscarine* and *nicotine*—only partially mimicked the effects of acetylcholine. Muscarine, a drug that was first extracted from the mushroom *Amanita muscaria*, produces vigorous contraction in smooth muscles of the gastrointestinal tract. Nicotine has its principal effect on autonomic ganglia and voluntary muscles, where it is a stimulant at low doses and blocks activity at higher doses. Muscarine and nicotine each give some but not all of the effects of acetylcholine. It seemed likely that there are at least two kinds of receptors in target organs, both responsive to applied acetylcholine; one is also responsive to nicotine, and the other to muscarine. It is conventional to differentiate between the *nicotinic* and the *muscarinic effects* of acetylcholine.

Receptors and Drug Actions

The idea of a receptor is a powerful, central, and productive one for understanding the mechanism of neural transmission and the actions of drugs. Raymond P. Ahlquist (figure 5.4), an American working at the University of Georgia, showed a similar distinction in the action of epinephrine. In a paper published in 1948, he showed that there are two strikingly different kinds of response to epinephrine and similar compounds, depending on the receptor type in the target organ. Ahlquist called these *alpha* and *beta receptors*. Within a few years of Ahlquist's discovery, drugs had been devised to block one or the other class of receptor for control of blood pressure and other conditions.

How can we locate these receptors and study their distribution in the body? The problem of receptor localization might be understood by an analogy. Suppose we had a long, thin-walled tube with a fluid flowing through it. Suppose that within the walls of that tube there were occasional magnets, but that these small magnets, although powerful, could not be recognized from the surface appearance of the tube's walls.

One possible way to locate the magnets would be to take a handful of iron filings and suspend them in the fluid circulating through the tube. The filings would be attracted to the magnets and stick there. If we now looked into the tube or cut it open, we should see a number of small regions along the wall that marked the position of the underlying magnets by an accumulation of iron filings.

By analogy, we might inject a labeled drug or transmitter substance into an experimental animal and then see where it "sticks" in the brain. In one method a substance labeled with a radioactive tracer is injected,

Figure 5.4
Raymond Ahlquist (1914–1983) first proposed that epinephrine has different effects on two distinct types of receptors which he called *alpha* and *beta* receptors. Courtesy Lasker Foundation.

and the brain is then studied to see whether the radioactivity is bound selectively to some definite regions in the brain.

These techniques for finding the locus of receptors have been applied in the study of a number of known transmitter substances. In 1973 two American scientists, Candace Pert and Solomon Snyder, used the techniques to study the site of action of the drug morphine. (Pert was a graduate student at Johns Hopkins with Snyder.) Radioactively labeled morphine was injected into experimental animals. By appropriate processing it was possible to determine that morphine is selectively bound to certain nerve cell membranes at definite regions of the brain and the spinal cord. Morphine appeared to be specifically active in regions of the nervous system that are known to be involved in transmission of painful stimuli (in the dorsal horn of the spinal cord) and the regulation of mood (in the limbic system of the brain).

These experiments revealed the curious fact that there appear to be specialized morphine receptors in the brain. This finding led to fascinating questions: Why are the receptors there? Does evolution provide us with sites in the brain that could elevate our mood or alleviate pain just in case we should happen to encounter a drug from poppy juice? Or are the receptors there because they are built to respond to a naturally occurring substance? Could there be morphine-like compounds manufactured and secreted by nerve cells that serve as hormones or neural transmitters and play some role in regulation of mood and response to pain?

Two years after the opiate receptor was discovered, investigators made another extraordinary discovery. They found that the brain made its own substances that bind to opiate receptors. The first two of these substances manufactured by nerve cells were called *enkephalins* from the Greek for "in the head." The enkephalins are present at a number of sites in the brain. The effects of the enkephalins when administered bear a powerful resemblance to the action of morphine. Chemically, the enkephalins are a series of five amino acids, the basic building blocks for proteins.

A second type of compound, called *endorphin*, was soon discovered in extracts of the mammalian pituitary gland. Endorphin—a word formed by combining "endogenous" and "morphine"—also is highly similar to morphine in its effects. There are now naturally occurring morphine-like compounds—collectively called endorphins—that have been identified in the mammalian brain.

Endorphins behave like opium and its derivatives in many ways. All are addictive, and all seem to share with opiate drugs the abilities to modify response to pain and to have a pronounced effect on mood. The endorphins show a regional variability in distribution that is similar to the distribution of opiate receptors, and there is also evidence that endorphins are especially highly concentrated in certain nerve endings, which would be the expected location for a true neural transmitter.

The discovery of the endorphins stimulated a number of suggestions about the possible mechanism of response to drugs. One immediate suggestion was a possible mechanism for addiction and tolerance to morphine-like drugs. Tolerance refers to the fact that increasingly large doses are required to produce an equivalent effect by the drug. Addiction refers to the fact that habitual use of such drugs leads to a state of dependency, such that if the drug is withdrawn, acute and distressing withdrawal symptoms are produced.

It was suggested that, in the presence of exogenous opiates, those neurons that normally release endorphins would begin to produce less and release less. In order to maintain a constant state, to "stay even" more and more of the drug would then have to be taken to make up for the shutting off of the body's own supplies. Such a mechanism could well account for the tolerance for the drug experienced by habitual users.

A similar mechanism would account for distressing symptoms that an addict experiences when the drug is withdrawn. Without the drug there would be sharply lower amounts of endorphins in the brain. Withdrawal symptoms would reflect that state of depletion. Recent experimental evidence is not entirely consistent with these suggestions, and it may be that there are intermediary steps in the production of tolerance and addiction.

Transmitter Families

There are now a large number of substances that are proven or suspected to be neural transmitters. Many of these transmitters are chemically similar to one another; there are "families" of transmitters. The first transmitter to be identified was acetylcholine. Norepinephrine and dopamine, mentioned earlier, belong to another class of transmitters, the catecholamines.

A series of relatively well-understood metabolic steps relates each of the catecholamines to one another. Thus, dopamine, a known transmitter at certain central nervous system sites, is converted by the action of specific enzymes in the brain to norepinephrine. Norepinephrine is in turn converted by other enzymes to epinephrine. Each of these steps is typically catalyzed by a specific enzyme, and the presence of each enzyme is crucial for normal production of each transmitter.

Amino acids are the basic building blocks of all proteins, and there is now strong evidence that four of the amino acids—gamma-amino butyric acid (GABA, discussed above), glycine, glutamate, and aspartate—are themselves transmitter substances.

Peptides are chains of amino acids linked in a definite sequence. We have already discussed one class of peptides, the endorphins. Of the many other peptides that serve as transmitters in the brain, substance P is prominently found in neurons responsive to painful stimuli and seems to play an important role in transmission of pain signals to the brain.

Another peptide, angiotensin, causes drinking when injected into the appropriate site within the brain and probably serves as a transmitter in a circuit signaling thirst and initiating drinking (see chapter 13 of this volume).

Inactivation

When a transmitter substance is released at a nerve ending, where does it go? Why does it not continue to produce its effect? There must be some way of stopping the action of transmitters. The first identified neurotransmitter—acetylcholine—has a unique enzyme called cholinesterase that is associated with its target receptors. Cholinesterase is found in high concentration at those places in the brain that have nerve terminals containing acetylcholine. Cholinesterase splits the acetylcholine molecule into two parts, thus inactivating it. Cholinesterase is necessary if the synapse is to be used again.

At first it was anticipated that all transmitters would have a similar method of inactivation. But inactivation by splitting a transmitter molecule into two constituent chemicals seems to be unique to acetylcholine; most transmitter substances are not inactivated in this way. Rather, there is a mechanism in which the transmitters are rapidly and selectively taken back directly into the presynaptic cell and quickly "packaged" again without first being broken down into constituent molecules. There are specific proteins in the membrane that serve this reuptake function. A class of drugs often used to treat depression—the selective serotonin reuptake inhibitors (SSRIs)—act to increase the amount of transmitter remaining in the synapse by blocking transfer back into the neuron (see figure 5.5).

Neurotransmitters and Brain Disease

Understanding chemical transmission in the nervous system is crucial for interpreting the effects of drugs that act on the brain and certain brain diseases. The action of some drugs can be understood in terms of transmitter chemistry. Drugs may mimic a transmitter by binding to postsynaptic receptors. Some drugs produce their effects by altering the ability of the presynaptic neuron to synthesize, store, release, or retake up the normally present transmitter. As is discussed in chapter 19 of this volume, certain diseases of the brain are caused by interference with the normal production, activity, release, or storage of a transmitter substance.

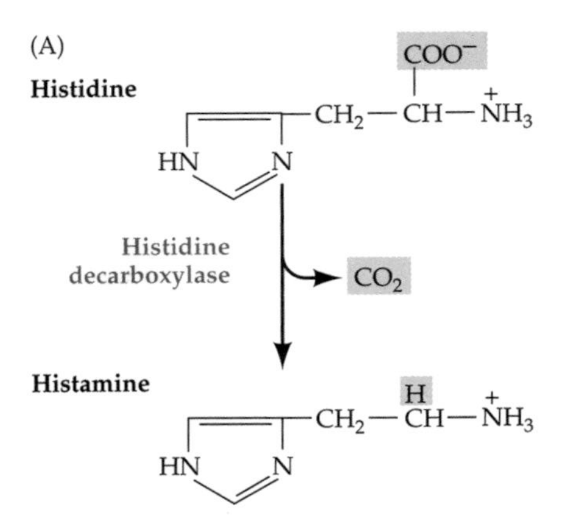

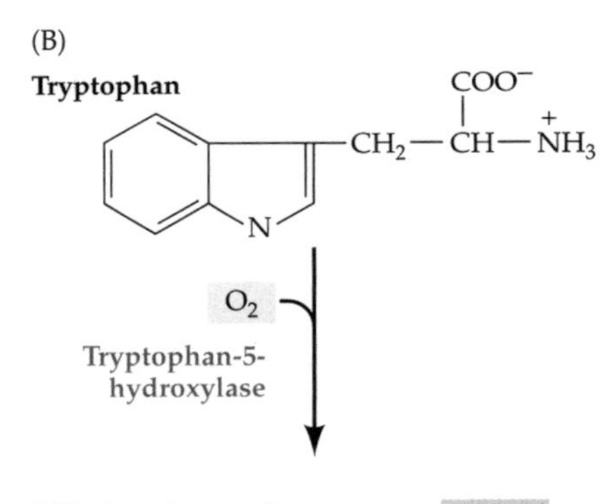

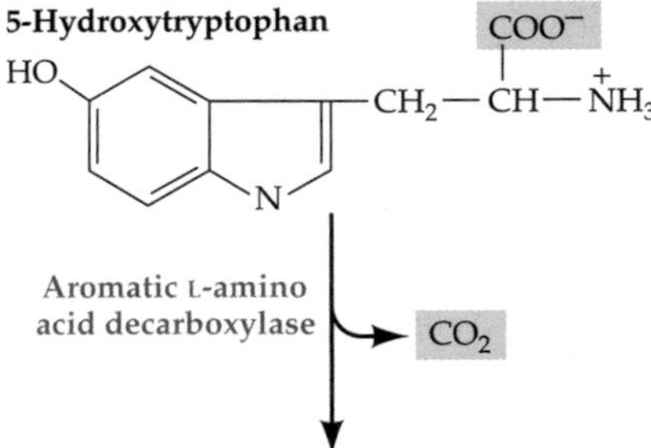

Figure 5.5
Biochemical pathway for synthesis of serotonin. Courtesy Dale Purves and Sinauer Associates.

6

Sensation

Everything that we hear, touch, see, taste, or smell—even the pain we feel from a stubbed toe—must first be converted into an electrical signal before it can be transmitted by nerve fibers to the brain. Sensory transduction is the process by which physical stimuli are transformed into electrical signals. Sensory perception refers to the more complex process by which we identify a familiar face as that of an old friend or interpret a certain odor as arising from coffee roasting. In this chapter we consider some aspects of sensory transduction that are common to all sensory systems. In later chapters we deal with some of the more complex problems of perception.

The Specificity of Sensation

We are immediately aware of what *sort* of sensation it is when we hear a noise in the bushes, see a car going by on the street outside our window, or feel the warmth of a coffee cup in our hand. We also evaluate the intensity of a sensory signal. We distinguish between the faint sound of the wind stirring the branches of a tree and the scream of a jet engine just before a plane takes off. We are also aware of the location of a stimulus—the golf ball lying to our left on the grass, for example, or the ache from a sprained right ankle.

How do we distinguish *between* senses? Why do we *hear* and not *see* a sound? One of the basic principles of sensory coding was formulated by Johannes Müller in the eighteenth century (figure 6.1). Müller, the son of a shoemaker, rose to become one of the most eminent physiologists of his day.

Müller formulated a principle that has been labeled "the law of specific nerve energy." The law simply says that only one kind of sensation is carried over a given nerve fiber. An axon in the optic nerve signals

Figure 6.1
Johannes Müller (1801–1858) formulated the law of specific nerve energies, the idea that a nerve fiber signals only a single type of message. Stimulation of the optic nerve by light or by any other route, for example, would always lead to the sensation of vision. Courtesy Wellcome Library, London.

only visual information. We hear sounds because they activate fibers in our auditory nerve.

Müller wrote:

Concerning the Characteristics of Individual Nerves
If the nerves were merely a passive link for a light stimulus, a tone stimulus, or odors, why is it that a specific nerve which responds to odors is able to carry only that type of sensation and no other, and likewise, a different nerve can not carry the sense of smell. The same is true of nerves which carry light or sound. . . . each sensory nerve has a unique responsiveness to a specific stimulus. They serve to convey a specific quality and no other.[1]

If you bang your elbow against a hard surface, you may feel a tingling extending to your fingertips because the impact has activated a sensory nerve from the hand on its way to the spinal cord. The principle that each sensory nerve carries a particular sensation can be demonstrated in another way. If you poke gently at the edge of your eye with the back end of a pencil, you will not only feel a sensation of pressure from the skin that surrounds the eye, but the pressure may also activate cells in the retina. Because you have activated a sense organ that normally responds to light, you would perceive a hazy blob of light even in total darkness. The visual sensation is called a *pressure phosphene.*

In normal life, sense organs are sensitive to only one type of stimulus, which is called the *adequate* stimulus. The adequate stimulus for the eye is electromagnetic radiation in the range of frequencies we call light, although, as we have seen, the eye can also be activated by pressure. The adequate stimulus for the ear is air pressure waves in the range of frequencies and intensity that we call sound.

Sensory Receptors

The marvelous devices that convert physical stimuli into nerve signals are called receptors. There are many kinds of receptors. Most receptors are specialized structures adapted to respond maximally to one kind of physical stimulus, transforming that stimulus into neural signals. In the skin, receptors may be nerve endings themselves or nerve endings with specialized capsules that surround them.

When a stimulus is applied to a receptor it produces a change in the resting potential of that receptor. This so called receptor or generator potential is a localized change in the membrane potential caused by the stimulus. The process by which the physical stimulus is converted to a receptor potential is called *transduction.* The receptor potential in turn initiates a train of impulses down the sensory fiber; in the case of very small cells it simply changes the rate of transmitter release from the cell. The conversion of the receptor potential to a pattern of nerve impulses or a change in transmitter release is called *encoding.* In most but not all cases the sensory input depolarizes the sensory nerve fiber.

The rods and the cones of the vertebrate eye are very small specialized cells, each of which contains light-absorbing pigment molecules. When the pigment molecules absorb light, the rod or cone *hyperpolarizes,* and transmitter release decreases. In other words the photoreceptors' response is the inverse of that of most other nerve cells.

The fibers that reach the skin and deep tissues usually branch repeatedly just before their endings. In the nineteenth century, histological study began to reveal nerve endings in human and animal skin that differ from one another in their structure. Some receptors, such as those that signal pain, may consist only of a sensory nerve fiber ending freely in a layer of skin without any obvious specialized structures attached to it. In contrast, some sensory nerve fibers in the skin do not end freely but are surrounded by elaborate connective tissue structures.

There are a variety of specialized sensory endings. The most striking of these is a wonderfully elaborate onion-like structure, the Pacinian Corpuscle, named for the Italian anatomist Filippo Pacini, who described it fully in 1840 (figures 6.2 and 6.3). Actually, it had been briefly recognized by Vater a hundred years earlier. Pacini was a medical student at the University of Pisa in Tuscany when he first recognized this structure. He wrote:

> From the time that I first began to do anatomical dissections, I saw certain small structures, either ellipsoid or globular, white-opaque or opal-like, about two thirds of a line in length situated along the digital branches of the median and ulnar nerves. At first I paid no great attention to them, thinking that they consisted of small lumps of cellular tissue. However I found more of these in the hand than in any other type of tissue. When I saw them again, I had the same interpretation. But on a later occasion when I wished to learn the distribution of nerves relating to the sense of active touch in more detail, I paid more attention to them. When it was possible to prepare the terminations of the median and ulnar nerves in the fingers, I again encountered the same globules even more clearly, which at the time, re-awakened an earlier memory.[2]

A Pacinian corpuscle may be over a millimeter in diameter, about one hundred times the diameter of the nerve fiber that it encloses (figure 6.4). Pacinian corpuscles are exquisitely sensitive to slight mechanical stimuli, making the attached nerve fiber responsive to high-frequency mechanical stimulation. On the basis of its position in joints and its mechanical properties, the Pacinian corpuscle plays a role in the sense of vibration that we feel when a vibrating object such as a tuning fork is placed against a joint.

Pacini's contribution was the first specialized skin ending to be described, and his contribution was quickly recognized. The German anatomist Wilhelm Krause (1833–1910), who later discovered another and different characteristic type of encapsulated nerve ending, wrote:

> I have now been teaching histology at the University of Göttingen for 25 years. When I have to start dealing with nerve terminals, I begin with Pacinian corpuscles, And every semester I read that Pacini, still as a student, with a small non-achromatic microscope and a wooden tube, like those sold at fairs, discovered the corpuscles that are now named after him, which represent the starting

Figure 6.2
Filippo Pacini (1812–1883), an early pioneer in the study of sensory endings in the skin and joints. Courtesy Wellcome Library, London.

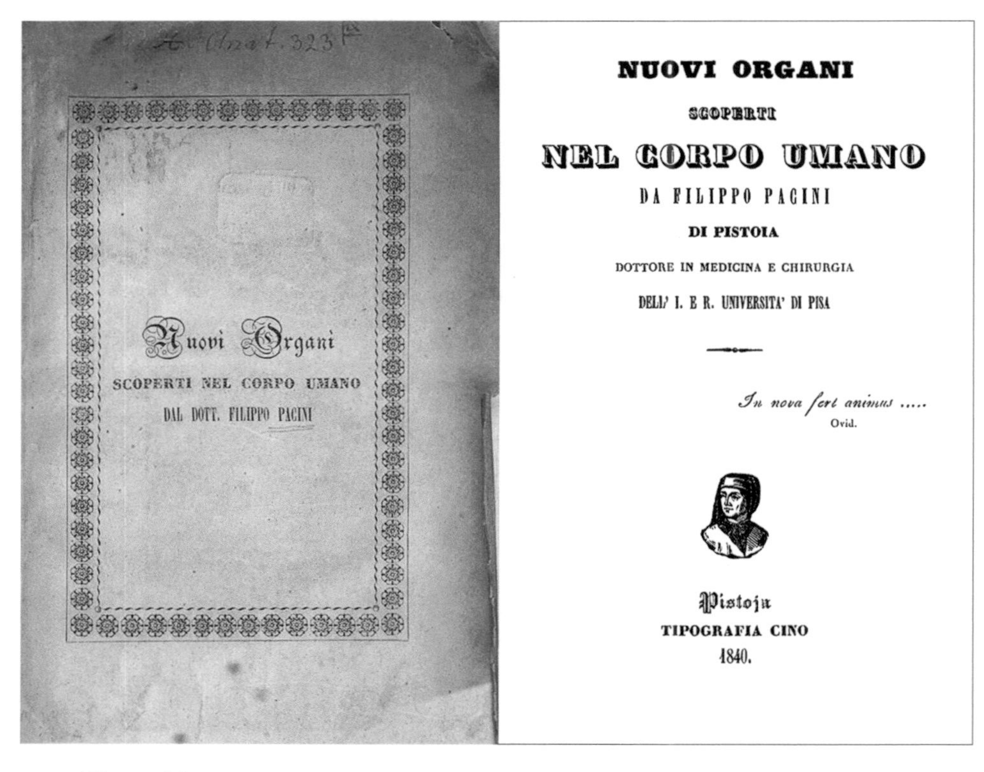

Nuovi Organi
SCOPERTI NEL CORPO UMANO
DAL DOTT. FILIPPO PACINI

NUOVI ORGANI
SCOPERTI
NEL CORPO UMANO
DA FILIPPO PACINI
DI PISTOIA
DOTTORE IN MEDICINA E CHIRURGIA
DELL' I. E R. UNIVERSITA' DI PISA

In nova fert animus .....
Ovid.

Pistoja
TIPOGRAFIA CINO
1840.

Figure 6.3
Cover and title page of the 1840 monograph in which Pacini described the location and appearance of the corpuscle that bears his name.

point for all that we know nowadays on sensory nerve terminals: you see, gentlemen, it is not the instrument but the observer who makes the discoveries.[3]

Sensory Coding

How is a physical stimulus converted into a generator potential? How does the generator potential cause firing of the sensory fiber? The Pacinian corpuscle was not only the first to be identified as a specialized structure, but it also provided a model system for studying the way in which sensory end-organs function. In many ways, sensory transduction is similar to synaptic mechanisms. At synapses, the liberation of a small amount of transmitter substance causes a graded potential change that excites or inhibits the next neuron in the pathway. Receptors are similar

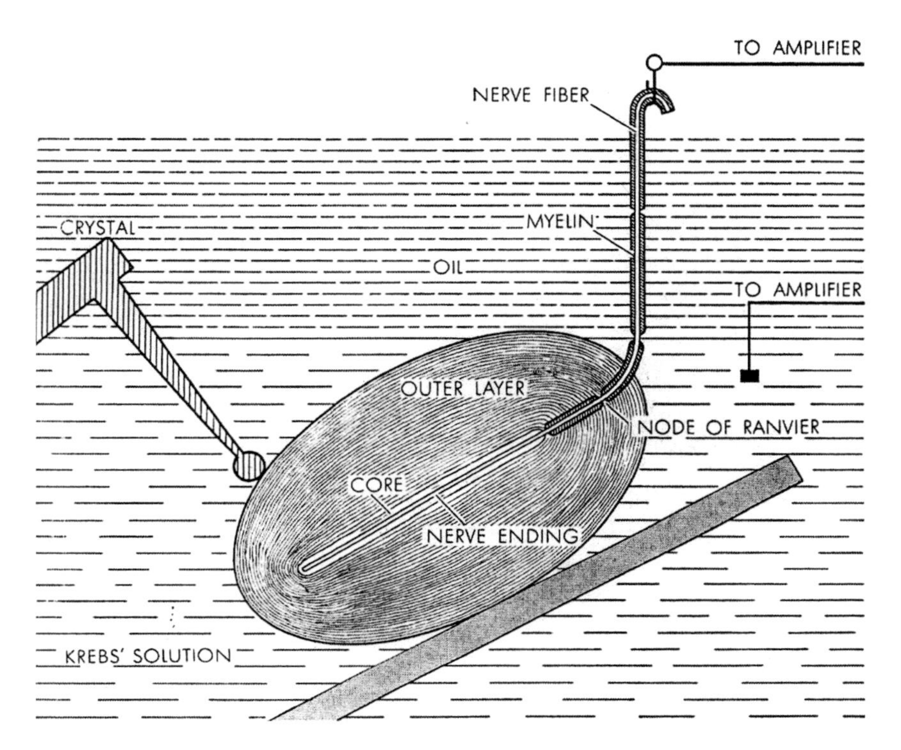

Figure 6.4
Diagram of the experimental arrangement to study the properties of the Pacinian corpuscle. The probe can deliver measured pressure on the Pacinian corpuscle. The response of the attached nerve fiber can be recorded. Illustration by Bunji Tagawa for *Scientific American*, 1960.

except that it is a physical stimulus that produces a graded potential. If that potential reaches threshold, it leads to a train of action potentials along the sensory nerve fiber. In the case of Pacinian corpuscles action potentials are produced in the sensory fiber by mechanical deformation of the tip of that fiber. The mechanical deformation changes the permeability of the axon's tip and thereby depolarizes it. This depolarization then leads to a succession of action potentials in the nerve fiber.

Figure 6.5 shows an experiment in which an isolated Pacinian corpuscle is studied along with its attached nerve fiber. A mechanical probe is placed on the surface of the corpuscle, with an electrode inside the fiber.

When the electrode penetrates into the inside of the sensory fiber, it records a voltage—the resting potential—with the inside of the axon

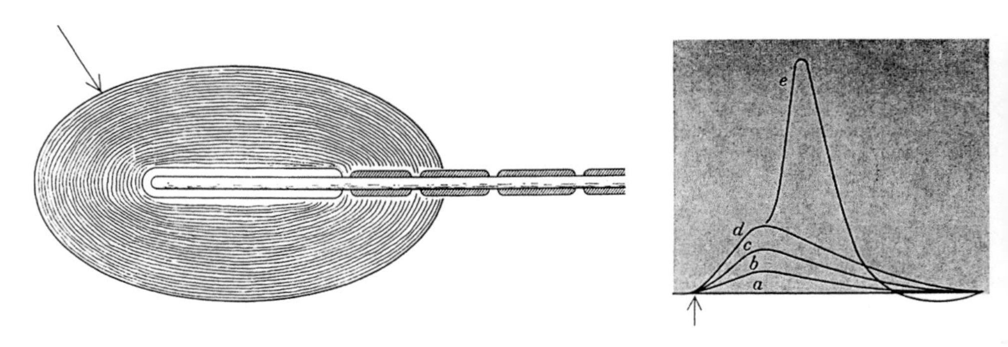

Figure 6.5
A Pacinian corpuscle is excised along with its attached nerve. A probe capable of rapid mechanical stimulation is placed on the corpuscle, and the response of the axon is measured. Successively stronger stimuli (*a*, *b*, *c*, *d*) produce successively larger generator potentials in the attached axon. The strongest stimulation (*e*) produces an action potential. Illustration by Bunji Tagawa for *Scientific American*, 1960.

negative with respect to the outside. If we now push on the corpuscle with the probe, the stimulus causes a change in the potential across the axon membrane.

The mechanical stimulus produces a change in resting voltage that is called the *generator potential.* The idea of a generator potential applies to all sensory mechanisms, in most cases in the direction of depolarization.

In addition to the Pacinian corpuscle there are other varieties of encapsulated nerve endings in skin. Each is associated with a characteristic microscopic appearance of its connective tissue capsule. We discuss the experiments that led to our understanding of the specialized functions of these end organs in chapter 10 of this volume. Although receptors may differ in structure, they have a common problem: all must convert physical energy into electrical signals that are the common currency of the nervous system.

Receptors vary a great deal in the extent to which they mirror faithfully the exact time course of a sensory stimulus. *Sustained receptors* will continue to signal for as long as a stimulus is applied. *Transient receptors* respond with a generator potential when the stimulus is first applied, but they quickly return to the resting potential even if the stimulus is maintained.

Receptors often have a mixed response. When the stimulus is first applied the receptor might show a transient generator potential, which then declines to a relatively steady level. When the probe is released, there is another brief generator potential followed by a return to the resting potential of the receptor.

How does the physical stimulus produce the generator potential? In the example given, the generator potential is produced by a change in permeability to ions in the membrane of the sensory fiber. The mechanical stimulus applied to the fiber's tip increases the permeability to all ions, depolarizing it. We record a voltage change, the generator potential. Although the mechanical stimulus changes permeability to all ions nonspecifically, in this case it is sodium that is principally responsible for the generator potentials. The generator potential is a link between the physical stimuli that impinge on sense organs and the sensory response.

Generator Potential and Spike Frequency

On its own even a very large voltage generator potential would never make it beyond a few millimeters of axon. Consider our hypothetical receptor with recordings made at three successive distances from the point of stimulation. When the probe is depressed, we would find that the generator potential grows weaker when it is recorded at successively distant points along the axon. The information would not reach the spinal cord, much less the brain. There must be a mechanism that allows sensory messages to be carried over great distances, such as the 5 or 6 feet required to relay to your brain the message that you have stubbed your toe.

Sensory receptors code the generator potential into spike discharges that can be conducted over the entire length of the axon to the spinal cord or to the brain without decrement. The generator potential acts like the excitatory postsynaptic potential at the synapse. There is a spike-generating zone on the receptor axon that produces an action potential when the generator potential reaches a threshold level. If the generator potential is maintained above this threshold level, then a series of action potentials is produced along the sensory fiber. Thus, if we were to record upstream of our hypothetical nerve ending at point B, far away from the sensory ending and at point A quite near it, we should record the same train of action potentials, differing only in the slight delay in the time of arrival following the stimulus.

What are the relationships among the intensity of a sensory stimulus, the generator potential, and the train of spikes that is produced by the generator potential? Typically the generator potential reflects the logarithm of stimulus intensity. The frequency of spikes in the sensory axon reflects the amplitude of the generator potential. So the logarithm of the stimulus intensity is roughly proportional to the frequency of spike discharge for most sensory cells.

In most cases the intensity of a stimulus is thus coded by the nervous system as the frequency of firing of a sensory axon. *What* is being signaled is typically determined by *which* nerve fiber is active. The intensity of the stimulus is coded by how fast the axon fires. *Where* the stimulus is located is usually coded by where the fiber ends in the central nervous system.

On the Range of Sensory Responses

Human senses can often work over an enormous range. We can see on a country road at night by the dim light of the stars. The same pair of eyes can see and track a boat in the distance on a bright day at the beach. The physical light intensity in these two cases differs by a ratio more than ten billion to one. Our senses are like scales that could accurately weigh a few grains of salt at one time and then a 10-ton truck. The scaling of sensitivity is done at several levels from the receptor to the brain. One way to allow for such an enormous range would be for the sensitivity of the receptor to vary with stimulus intensity. Mathematics allows us an easy description of one such process. For example, suppose sensory responses were proportional not to the absolute value of a stimulus but to its logarithm. Such a receptor would give an equal additional response when we went from 1 g to 2 g as when we went from 1 pound to 2 pounds. At low weights our scale would be a very sensitive one, but less so as the weight on it was increased. In most cases generator potentials reflect the logarithm of stimulus intensity rather than the absolute magnitude of a stimulus.

Psychophysics

Psychophysics is concerned with the description and analysis of sensory mechanisms in humans and animals. How are responses to a sensory stimulus determined by its physical or chemical properties? How weak can a stimulus be and still be perceived? Threshold is a general concept that applies to all sense modalities. How much sugar must we place in a glass of water before we can barely taste its sweetness? How strong must an odor be before we can barely smell it? In addition to absolute threshold, psychophysics is concerned with difference thresholds. How much more intense must we make the second of two tones before it sounds louder?

The problems of sensory thresholds have interested scientists for hundreds of years. In the nineteenth century useful techniques were evolved to measure them. The person who did most to codify and use these methods was the German physiologist and physicist Gustav Theodor Fechner (1801–1887).

One of these psychophysical methods is called "the method of limits." Suppose I set the intensity of a tone well below the level at which you hear it. I then raise the intensity in gradual steps until you say, "I hear it." I might raise it a few more steps to make sure and then stop. Now let's do the same thing in reverse. I set the tone well above the threshold for your hearing, and I decrease it in small steps until you say, "I don't hear it." In both cases we would cross a point at which your response changes from not hearing to hearing or vice versa. That point at which the stimulus becomes audible or just fails to be audible is the absolute threshold for sound intensity. If the stimuli were physically identical on every trial and if the observer were a perfect detector, then the threshold should not vary at all. There would always be a precise value of stimulus intensity, a sharp point below which you would never report hearing the tone and above which you would always report hearing the tone.

But the results of this sort of procedure seldom yield such a sharp threshold point. More typically the threshold seems to vary over a narrow range of physical intensities; hence, any single value for the threshold is to some extent an arbitrary one. There is a range of stimulus intensities over which the probability that you will hear the tone changes in a systematic way. Why does the absolute threshold vary? There are two sorts of reasons. One is the obvious variability that people have in fatigue, awareness, attention, and the like. But interestingly, especially in those modalities to which we are extremely sensitive—such as our ability to detect the presence of a very dim light—small variations in the stimulus itself contribute to the variability in threshold. Light can be thought of as made up of tiny particles of energy called quanta. A given light may contain two quanta or three but not two and a half. There is an inherent variability in the number of quanta given off by a light source at any time. Our eyes are so sensitive that this physical variance will contribute to variance in the measured threshold. The important and fascinating point is that even under optimal conditions some of the variability in human absolute threshold measurements is due to actual physical variance in the intensity of the stimulus.

Absolute thresholds are useful for thinking about how sense organs work. For example, the remarkable sensitivity of the human eye to minute amounts of light suggests that each individual light receptor probably responds to only a quantum or two of light.

In the example above we tested for the threshold of hearing at a single auditory frequency. In order to have a more complete picture, we would have to repeat the same sort of procedure at several frequencies throughout our range of hearing. Auditory testing of this sort is important not just for evaluating the range of normal hearing but also for assessing hearing loss. There are some clever methods that speed up such testing.

Suppose you sit in a quiet room with a pair of earphones on. There are two buttons on the table in front of you. If at a signal you hear a tone of any sort through the earphones, you press one of the two buttons. If you hear no tone through the earphones, you press the other button. The button presses are simply an automatic way of saying "I hear it" or "I don't hear it." Suppose a moderately intense tone were to come on in the earphone. You hold down the "I hear it" button. The apparatus is so arranged that as you hold down the "I hear it" button, the intensity of the tone is lowered in small steps. The intensity goes down until the sound is below your threshold. You release the "I hear it" button and press the other. Now the sound intensity is increased in gradual steps until you hear it again. If we were to keep this up for a little while, we could measure rather precisely your intensity threshold for sound.

Now suppose that we modify this apparatus further so that, in addition to the intensity changes, the frequency of the sounds presented is slowly raised from the very low, near 20 Hz, to very high, nearly 20,000 Hz. If you continued to play the "I hear it"/"I don't hear it" game as the frequency goes up, the apparatus could now record your hearing threshold for all frequencies. A device of this sort was invented by Georg Von Bekesy in the 1940s. The Bekesy audiometer is used for rapid clinical assessment of hearing and hearing loss.

Most psychophysical experiments are carried out with cooperative human subjects. But we can apply the same sorts of procedures to studying sensory threshold in animals. Psychologist Donald Blough was still a graduate student in psychology at Harvard when he set out to determine the absolute threshold of pigeons to lights. He trained pigeons in a box containing two keys much like the two keys of the Von Bekesy audiometer. The animal pecked at key A when a light was on and key B when the light was off. The circuit was arranged so that pecking at key A not only recorded the fact of pecking but also lowered the illumination of the light. Pecking at key B recorded the fact of pecking and increased the illumination. With this apparatus Blough accurately measured the visual sensitivity of the pigeon as well as the change of sensitivity in the dark. By a similar procedure others have measured the auditory threshold of animals and then have evaluated the effects of intense noise or drugs that impair hearing.

The Difference Threshold

Suppose that I give you two weights and ask you to say whether the second is heavier than the first. Suppose that you could tell a 51-ounce weight from a 50-ounce weight. Your difference threshold in this case would be 1 ounce. Could you tell apart two weights where one was 50 pounds and the second 50 pounds, 1 ounce? Probably not.

The nineteenth-century German scientist Ernst Weber studied systematically the ability of people to tell the difference between two stimuli over a wide range of stimulus intensity. Over a surprisingly wide range the difference threshold was precisely related to the magnitude of the stimulus. You can just tell the difference by some constant proportion of the stimulus. If it took an extra ounce for you to tell 50 ounces from 51 ounces, it would take an extra pound for you to tell the difference between 50 and 51 pounds. The difference threshold would always be some definite proportion of the comparison stimulus. This so-called Weber's law can be expressed as $\Delta I = kI$, in which ΔI means the change in intensity, I is the intensity of the reference stimulus, and k is a constant. Weber's law works for many sense modalities and over a wide range of intensities, although it is inaccurate for very small and very large stimuli.

Psychological Scaling

What does it mean to say one sound is twice as loud or that one light is twice as bright as another? If some physical quantity of sound energy appears to be of a certain loudness, would twice that amount of sound energy be twice as loud? The short answer is "No." Psychological intensity is not a simple multiple of physical energy. But people can decide that a light appears twice as bright or a sound half as loud. Is there any system to such judgments? If twice as much sound energy does not appear twice as loud, how much more intense would a sound have to be so that it would appear twice as loud? Is there some general principle whereby we can relate the physical intensity of a stimulus to its psychological magnitude? Weber's law gives some clues as to how we might construct such a scale.

Suppose that the difference thresholds or just noticeable differences (*jnd*s) were of roughly equally psychological magnitude. In the case of judgment of weight, suppose that 1 ounce added an equal amount of perceived "heaviness" to a 50-ounce weight as 1 pound did to a 50-pound weight. One added ounce would be equal in psychological magnitude to one added pound at different points in the scale. If psychological magnitude grew as the jnd of the stimulus, we could construct a scale in which equal increments of psychological magnitude for the entire range of lifted weights were based on the jnd at each stimulus value.

The German physicist and philosopher Gustav Fechner derived just such a scale in 1860. His solution climaxed a decade of Herculean effort in which he, by his own account, made no fewer than 24,576 separate judgments of difference thresholds using various small weights. Fechner also conducted numerous experiments investigating the jnds between varying stimulus intensities of both light and sound.

The scale that Fechner suggested was one in which the *logarithm* of the physical intensity of the stimulus rather than its absolute value determines its psychological magnitude. Note that such a logarithmic scale allows us to respond to a wide range of physical intensities. At the upper ends of physical intensity, such a logarithmic scale becomes more and more compressed.

Alternate formulations have been put forward since Fechner's time to describe the relationship between psychological magnitude and the physical intensity of a stimulus. But Fechner's relationship in which psychological magnitude is equal to the logarithm of stimulus intensity multiplied by a constant holds reasonably well for most sensory stimuli over a reasonably wide range of intensities. Fechner's law and all such psychophysical scales involve a compression of the physical scale. We see and hear over a vast range of stimulus intensities, and such scales reflect the fact that the nervous system compresses these intensities into a much smaller range that it can code and interpret.

7

Vision and the Eye

Our eyes form a picture of the world around us and relay important aspects of that picture to the brain. In daylight the world is richly colored with constantly changing shapes and movement. We can resolve the small print of a telephone book or recognize the neighbor's cat across the street. At night the colors seem to fade, and our vision for small details becomes poorer, and yet we still see well enough to guide our steps on an unlit country road by moonlight. What sort of message does the eye send to the brain, and how does the brain interpret that message?

This chapter discusses the history of our understanding of visual processing by the eye. It begins by considering how images are formed by the eye and recounts some early contributions to understanding the structure and function of the retina. The next chapter then discusses the connections from the retina to the cerebral cortex by way of a relay in the thalamus and describes studies that began to reveal the multiplicity of visual areas outside of the primary visual cortex and some of their functions.

Image Formation

We can see objects if they emit light themselves, like a flame or the sun, or if they reflect light. In order for us to see there must be cells in the retina that detect the presence or absence of light. Vision requires more than just a light-receiving cell; in order to see an object, the eye must form an image of it. There are at least three different solutions for the problem of image formation. In one form of eye the image formed is upright. In pinhole eyes the image is inverted, as it is in our own refracting eye.

Consider how light-detecting cells might work if they did not have a mechanism for forming an image. Imagine two small spots of light and a single photoreceptor—a cell that registers the presence and intensity of a light shining on it. In the example illustrated in figure 7.1, light would radiate from sources A and B in all directions.

Some of the light from A would strike the photoreceptor, and some of the light radiating from B would also strike it. The photoreceptor would report if either light were turned on or off, but there could be no *directional* signal. It would not be possible to tell if a light was above or below, to the left or right of the photoreceptor. We might also be able to tell which of two equally intense lights was closer—but not the direction of the lights. An image would be needed to allow us to tell the relative position of an object or if that object were moving.

It might help to understand the nature of image formation by thinking about an extremely crude device: an imaginary eye. Suppose that we collect a large number of drinking straws, paint all of them black inside and out and then glue them together in a square array (figure 7.2, top).

The device would select a ray from each point on an object and form an image. Although such a device can form an upright image, for all practical purposes it would be useless. The "eye" would have to be the same size as the object it was meant to image. Although the soda-straw eye would be of no use, many invertebrate animals have an eye that forms an image by ray selection. Instead of parallel tubes the compound eye is made up of a large number of tiny cones, called *ommatidia* (figure 7.2, bottom). Bees and house flies solved the problem of seeing the wider world by evolving a small compound eye. The image can now be smaller than the object that gives rise to it, and it is upright.

The compound eye selects light rays to form an image. There are other ways to do the same thing. Imagine a sphere with a pinhole cut into it. The inside of the sphere is painted black, and the pinhole is pointed toward a lighted object. Each point on the object would fall on a single point at the back of the sphere. Together all of the points would form an image (figure 7.3). Like the compound eye, the image formed by a pinhole can be smaller than the object. Unlike the compound eye, the image in the pinhole camera will be *inverted*. The left side of the object will be imaged to the right side of the retina; the top of the object will be imaged toward the bottom of the retina. Some invertebrate animals do have such eyes, but a small pinhole severely limits the amount of light that can be used to form an image. In order to be useful in dim environments, much more light must be admitted.

Vertebrate eyes have a much larger opening, the pupil, and a lens system that allows for focusing the image at the back of the eye (figure 7.4).

Figure 7.1
A single photoreceptor and two light sources. The photoreceptor could not distinguish between differing forms, and it would confuse intensity, distance, or direction of a light source.

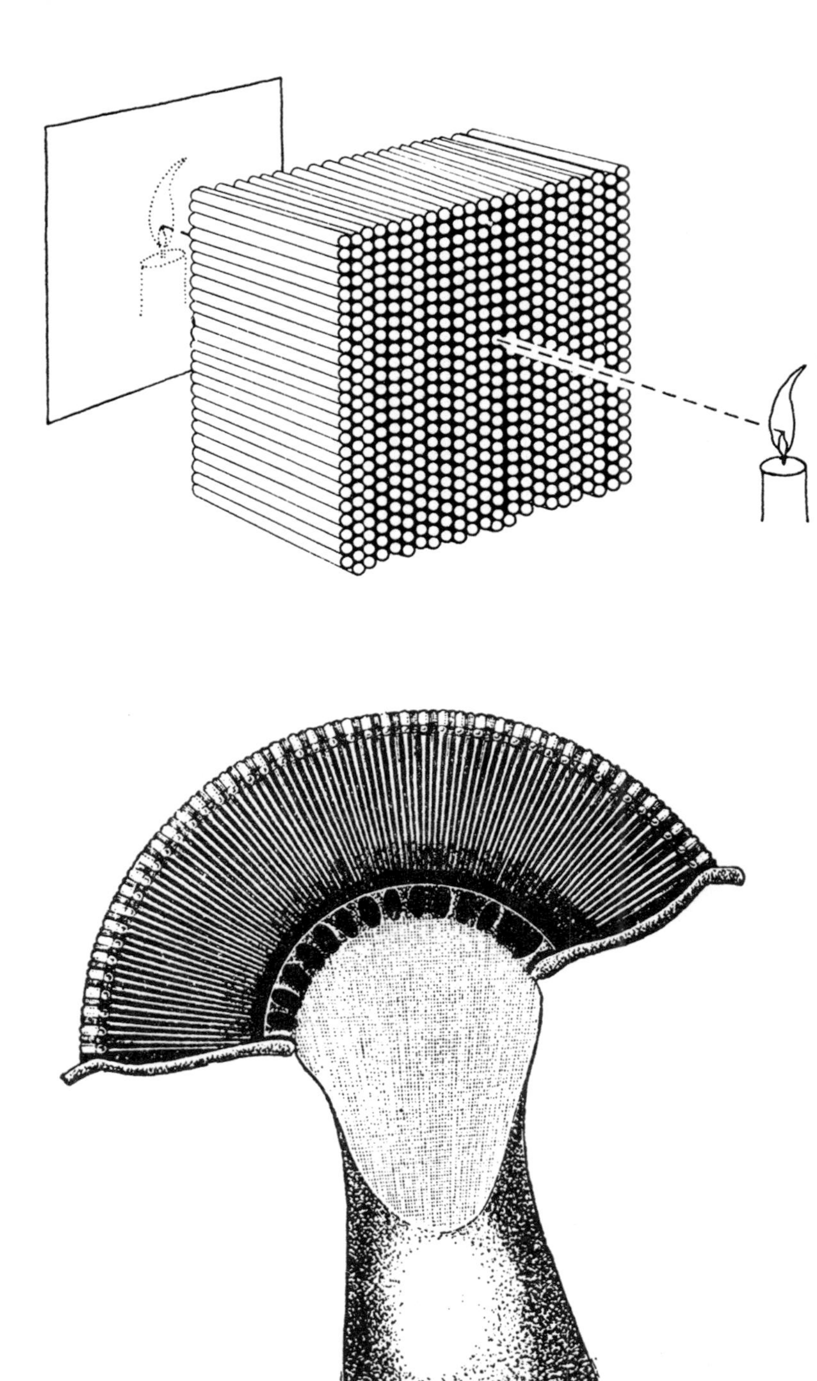

Figure 7.2
(Top) Imaginary "soda-straw eye" is capable of forming an image, but the eye would have to be as large as the object it wishes to see. (Bottom) Changing the shape of each element of a compound eye into a cone allows an upright image to be formed, with the eye now able to see objects much larger than itself. (After H. J. A. Dartnall.) Courtesy Elsevier.

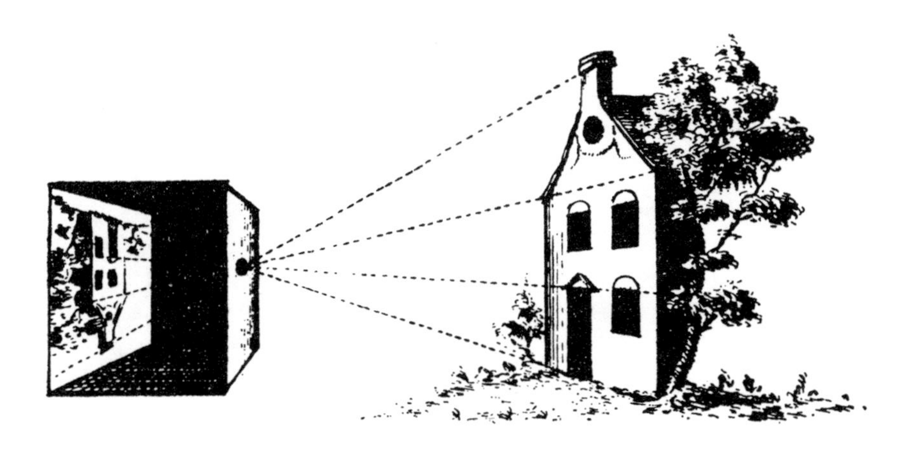

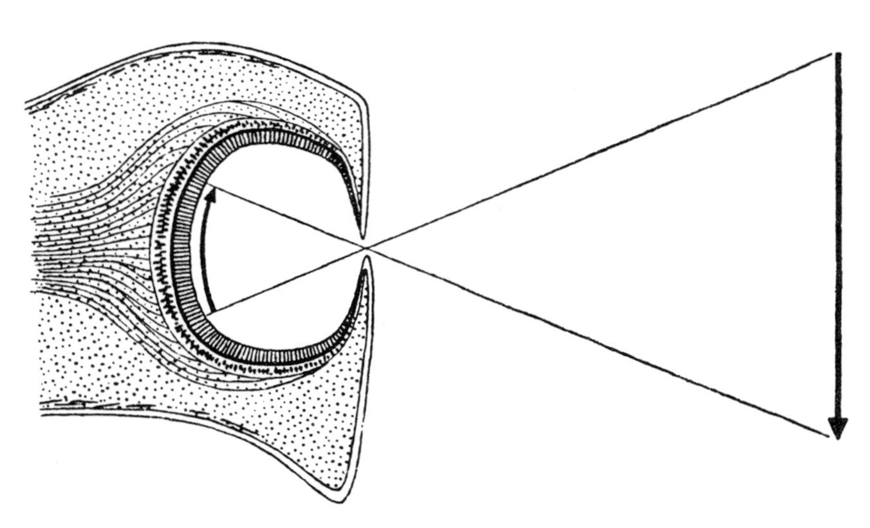

Figure 7.3
(Top) "Camera oscura" forming an inverted image. (Bottom) "Pinhole" eye, which works on a similar principle. (After H. J. A. Dartnall.) Courtesy Elsevier.

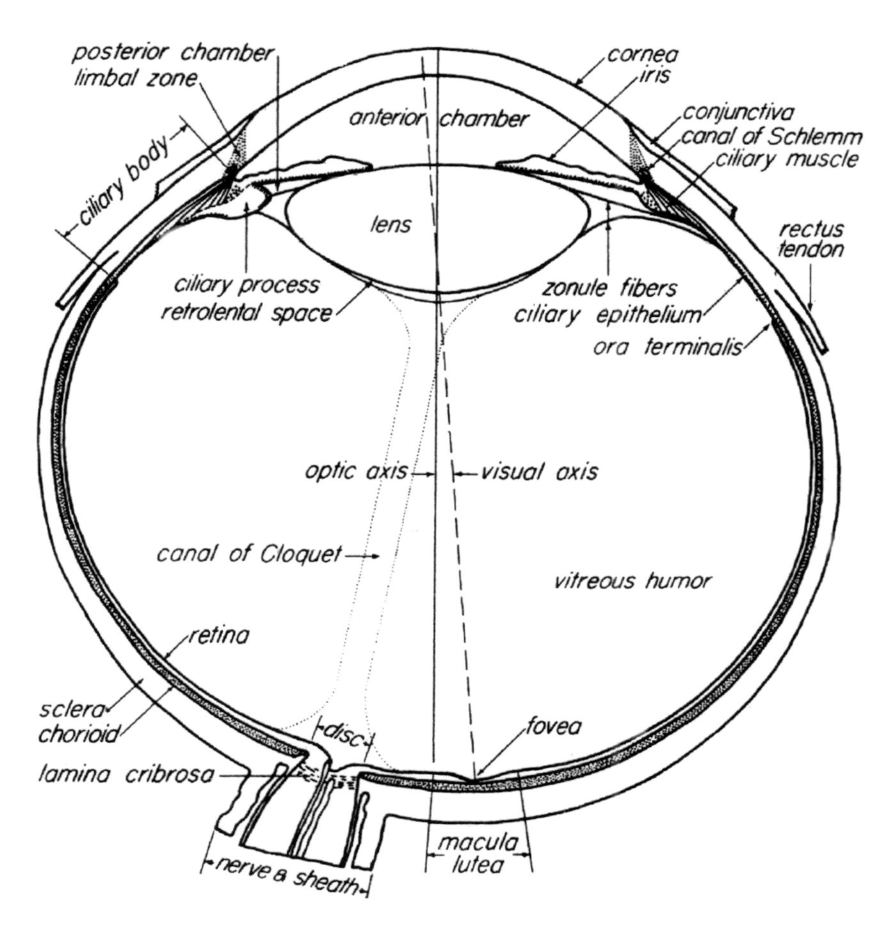

Figure 7.4
Diagram showing the major structures in a human eye. Adapted from Walls (1942).

A lens-based system with inverted images is used by the eyes of all vertebrates and a few invertebrates. Like the pinhole, eyes of this kind also form an inverted image.

The cornea and lens bend the light to form an inverted image of the visual scene at the back of the eye. Early scientists could not accept the idea of an upside-down image at the back of the eye. In the eleventh century, Ibn Al-Haithem, an Arab scholar, described the principles of image formation by the *camera oscura*, a dark chamber with a pinhole aperture that admits light and forms an inverted image of an illuminated scene.

Even though the optics of the *camera oscura* (figure 7.3, top) appeared to be similar to that of the eye, the concept of an inverted image was universally rejected. Leonardo da Vinci tried to construct a scheme whereby an inverted image would be received first by the lens, and then reinverted to form an upright image at the back of the eye. He wrote: "No image, of however small a body, penetrates into the eye without being turned upside-down and, in penetrating the crystalline sphere, it will be turned the right way again."[1]

The true nature of image formation by the eye was first put forward on theoretical grounds by Johannes Kepler (figure 7.5) in 1604 in his *Ad Vitelionem.*

Kepler's suggestion was confirmed experimentally by Christopher Scheiner (figure 7.6), who removed some of the opaque tissue at the back of an excised eye and directly demonstrated the inverted image.

Figure 7.5
Johannes Kepler (1571–1630). By tracing the pathway of light rays, Kepler showed that the image formed in the human eye must be inverted. Courtesy Wellcome Library, London.

Figure 7.6
Christopher Scheiner (1575–1650) was a Jesuit priest. He removed the pigment layer from the back of an excised animal eye, and could observe the image that it forms. He confirmed the fact that an inverted image is formed at the back of the eye by the cornea and lens.

Following Kepler's analysis and Scheiner's demonstration, the fact that the image at the back of the eye is inverted became widely accepted. The nature of image formation was beautifully illustrated by Descartes (figures 7.7 and 7.8).

The Structure of the Retina

Anatomical techniques were limited at the time; hence there was only fragmentary knowledge of retinal structure or how the pattern of light and darkness in the image is converted into a signal by the eye. One of the obvious problems was that of color vision. People can distinguish a great number of colors, and it seemed unlikely that there are specialized receptors for each one. In a remarkably prescient article in 1802 Thomas

Figure 7.7
Rene Descartes (1596–1650) was a mathematician and scientist. He beautifully illustrated the inverted nature of the retinal image. Courtesy Wellcome Library, London.

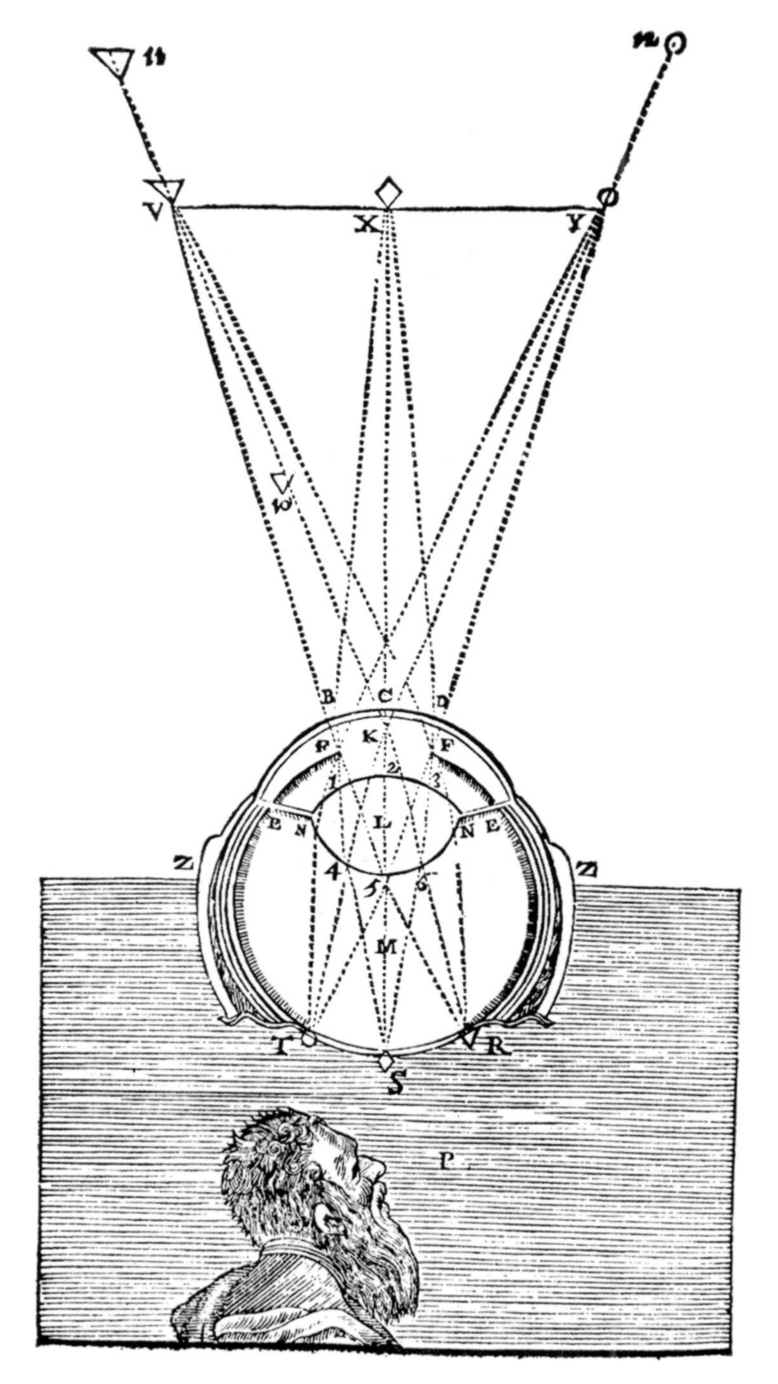

Figure 7.8
Descartes's diagram illustrating the way in which the image is formed on the human retina. Courtesy Wellcome Library, London.

Figure 7.9
Thomas Young (1773–1829). One of the great scientists of his day, Thomas Young contributed to several fields of knowledge. Courtesy Wellcome Library, London.

Young (figure 7.9) speculated that there must be a finite number of receptor types; he suggested three, which would be sufficient to account for human color vision (figure 7.10).

Young's suggestion of three types of color receptors was remarkably prescient. It is now clear that in the normal human eye there are three sorts of cones, responding differentially to light of different wavelength—one maximally sensitive to the long, one to the middle, and one to short wavelength light. Input from these three cone types is combined to produce the large number of hues that humans can distinguish.

as it is almost impossible to conceive each sensitive point of the retina to contain an infinite number of particles, each capable of vibrating in perfect unison with every possible undulation, it becomes necessary to suppose the number limited, for instance, to the three principal colours, red, yellow, and blue, of which the undulations are related in magnitude nearly as the numbers 8, 7, and 6; and that each of the particles is capable of being put in motion less or more forcibly, by undulations differing less or more from a perfect unison; for instance, the undulations of green light being nearly in the ratio of $6\frac{1}{2}$, will affect equally the particles in unison with yellow and blue, and produce the same effect as a light composed of those two species: and each sensitive filament of the nerve may consist of three portions, one for each principal colour.

Figure 7.10
Thomas Young suggested that there must be a finite number of receptors for color. He made the remarkably prescient proposal that these might be as few as three.

Young was a universal genius. In addition to his prescient speculations on the nature of color vision, he did a fundamental experiment demonstrating the wave properties of light and was the first to use the Rosetta Stone to interpret Egyptian hieroglyphics.

In the years following Young's remarkable speculation, two sorts of studies revealed some puzzling properties of human color vision. In one type of experiment colored stimuli were combined, often by rotating a disk with two or more different colors at a speed that was high enough so that the colors fused into a single percept. If different colored lights are blended, what is the resultant color? It was found that a mixture of green and red light looks yellow, never greenish-red. When blue and yellow lights are combined, the mixture looks white, never bluish-yellow.

In the other type of study, people explored the subjective possibility of thinking of different color blends. We find that it is impossible to think of a greenish-red but not of a greenish-blue. Although such studies revealed many of the subjective properties of color vision, the structure of the retinal elements remained poorly understood and could not account for these findings. This required, among other things, the invention of the compound microscope in the early nineteenth century, which led to an explosion of new knowledge about biological structures in general and of the retina in particular. Among the most important of the early contributors was the German histologist Max Schultze (figure 7.11).

Schultze described the three major cell layers of the retina, with special attention to the distribution and appearance of the receptors. He noted that there is a predominance of thin, rod-like receptors in the retina of

Figure 7.11
Max Schultze (1825–1874) described the difference in structure between visual receptors and posited that the two receptor types, rods and cones, work at different levels of illumination.

strongly nocturnal animals and thicker cone-like receptors in diurnal animals (figure 7.12). On the basis of comparative evidence and the distribution of receptors in the human eye, Schultze suggested that the two distinct classes of receptors might be associated with vision under two different conditions of illumination.

After the rods and cones there are four successive cell types that further process the retinal image. Schultze and his contemporaries were vague about the internal connections within the retina. The final stage, the retinal ganglion cells, transmits the output from the retina to the brain. The prevailing view of brain connections in general, and the retina in particular, was that they comprised a net-like structure, a reticulum, in which successive elements are continuous and fused. The prominent cell layers of the retina were thought of as swellings on optic nerve fibers.

The staining method of Camillo Golgi and the anatomic research of Santiago Ramon y Cajal changed that view. In the 1870s Golgi devised a method for staining brain and retina so that only a small subset of the cells would be stained (see chapter 3 of this volume). The Spanish anatomist Santiago Ramon y Cajal (figure 7.13) used the Golgi stain for a detailed study of the retina and its elements.

Cajal first saw an example of the then-new Golgi staining technique when he visited a colleague in Madrid in 1878. Struck by the beauty and promise of the method, he began to apply the Golgi staining technique systematically to the study of the vertebrate retina and brain (see figure 3.8). Cajal's classic monograph on the retina was published in French in 1892. Cajal was convinced that the reticular theory of organization of the nervous system was wrong. The retina as well as the brain and spinal cord, he argued, are made up of the individual elements now known as *neurons*. Neurons may touch one another, but they do not fuse.

In his monograph, Cajal described in detail the major cell types in all of the retinal layers. He emphasized that the direction of conduction is from the receptors, through the horizontal, bipolar, and amacrine cells of the inner nuclear layer, ultimately to the ganglion cells, whose axons constitute the optic nerve. Cajal's descriptions have remained the basis for all subsequent anatomical studies.

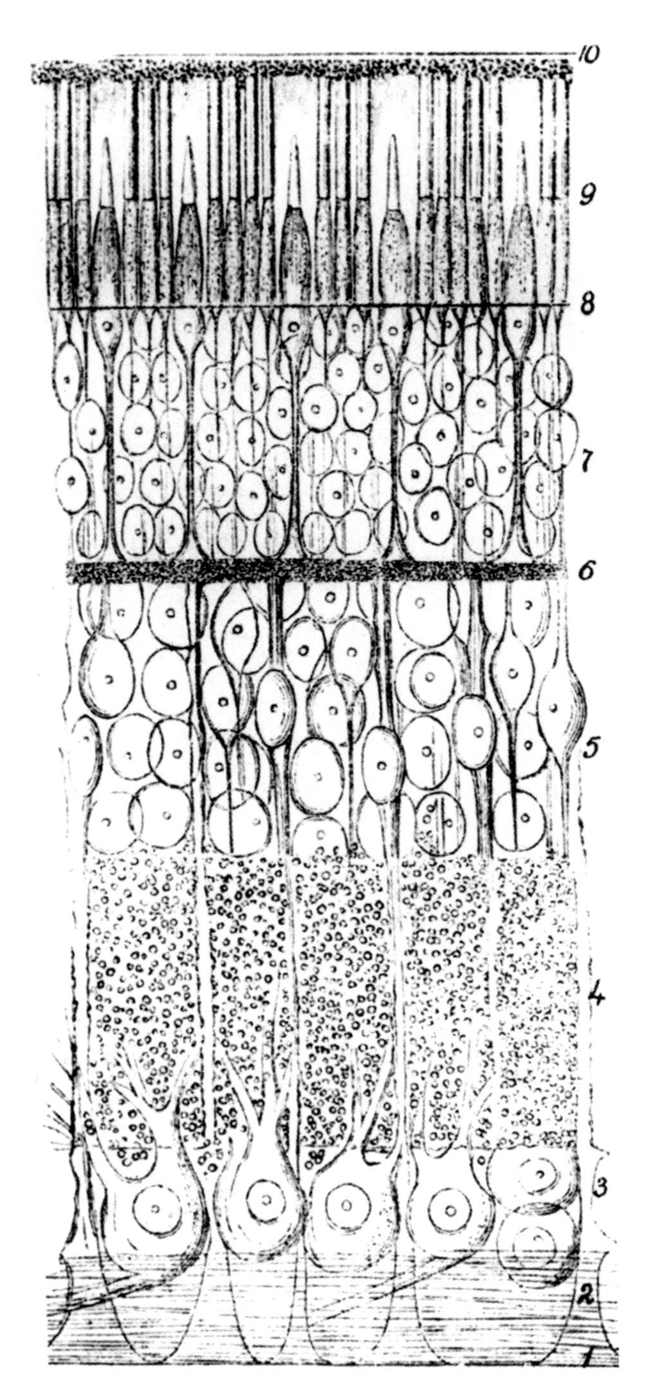

Figure 7.12
Schultze's figure of the retina illustrating the two distinct classes of receptor.

Figure 7.13
Santiago Ramon y Cajal (1832–1934). Cajal was among the greatest contributors to our understanding of the structure of the eye and brain. His monograph on the retina formed the basis for all subsequent study of its structure and function. Courtesy Cajal Legacy, Instituto Cajal (CSIC), Madrid, Spain.

8

Vision: Central Mechanisms

In the previous chapter we described the rods and cones, the structures that convert light into neural activity. Rods and cones are linked by bipolar cells to ganglion cells. The axons of the eyes' ganglion cells form the optic nerve. The optic nerves connect to the cortex by way of a relay in the thalamus. Figure 8.1 shows the anatomical pathway. This chapter describes the experimental and clinical evidence that led us to our current level of understanding.

The Optic Chiasm

Early anatomists saw a prominent nerve exiting from the back of each eye directed toward the brain. The fibers from each eye appeared first to unite and then continue across the midline in the X-shaped optic chiasma, from the Greek X-shaped letter "chi." With earlier techniques of crude dissection it was not clear whether some or all of the fibers from each eye crossed over to the other side. The true picture was not accepted until the late nineteenth century when better techniques made it possible to see that in the human brain axons from ganglion cells in the lateral retina remain uncrossed, whereas axons from ganglion cells in the nasal retina cross. Because of the inversion of the image on the retina, the left half of the visual field is sent to the right side of the brain. The right visual field is represented on the left side of the brain.

Curiously, the correct arrangement of the fibers at the chiasm had been suggested quite early, but it was neither recognized nor accepted. Isaac Newton (figure 8.2) in his second book on optics (1704) wrote:

> Are not the Species of Objects seen with both Eyes united where the optick Nerves meet before they come into the Brain, the fibres on the right side of both Nerves uniting there, and after union going thence into the Brain in the Nerve which is on the right side of the Head, and the fibres on the left side of

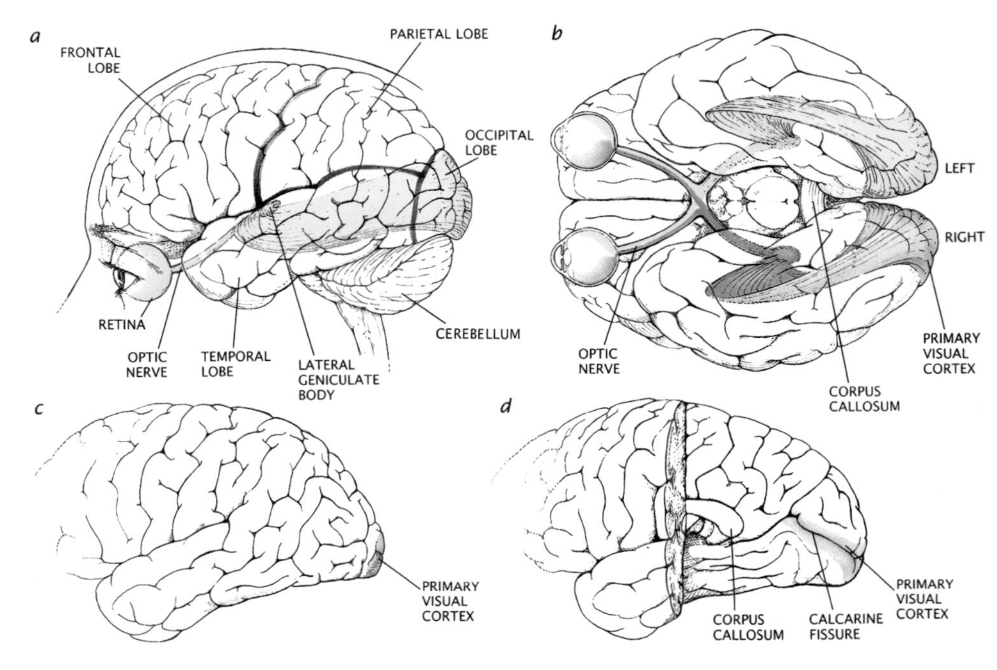

Figure 8.1
The diagram traces the origin and course of the visual fibers from the eye through the LGN to the primary visual cortex. Courtesy *Scientific American.*

both nerves uniting in the same place, and after union going into the Brain in the nerve which is on the left side of the Head, and these two Nerves meeting in the Brain in such manner that their fibres make but one entire Species or Picture, half of which on the right side of the Sensorium comes from the right side of both Eyes through the right side of both optick Nerves to the place where the Nerves meet, and from thence on the right side of the Head into the Brain, and the other half on the left side of the Sensorium comes in like manner from the left side of both Eyes. For the optic Nerves of such Animals as look the same way with both Eyes (as of Men, Dogs, Sheep, Oxen &cet.) meet before they come into the brain, but the optic Nerves of such Animals as do not look the same way with both Eyes (as of Fishes and of the Chameleon) do not meet, if I am rightly informed.[1]

Although he incorrectly assumed that the origin of the optic nerves was within the brain rather than within the eye, Newton described correctly the course of the optic nerves in the optic tracts. He was also aware of differences in the pattern of crossing in animals with laterally placed eyes. Despite Newton's scientific authority, the truth failed to penetrate the medical or biological literature. Over one hundred years later, William

Figure 8.2
Isaac Newton (1642–1727) was a great mathematician and physicist. In his *Optics* Newton suggested that the optic nerve fibers might partially cross, so that each half the visual field is represented in opposite half of the brain. The suggestion seems not to have been noticed by successive authors. Courtesy Wellcome Library, London.

Wollaston (1766–1828) (figure 8.3), an English chemist and physicist describing his own temporary partial blindness, wrote:

> It is now more than twenty years since I was first affected with the peculiar state of vision, to which I allude, in consequence of violent excercise I had taken for two or three hours before. I suddenly found that I could see but half the face of a man whom I met; and it was the same with respect to every object I looked at. In attempting to read the name JOHNSON over a door, I saw only SON; the commencement of the name being wholly obliterated to my view.

Unaware of Newton's suggested scheme for the course of the optic nerves, Wollaston wrote:

Figure 8.3
William Henry Wollaston (1766–1828) suffered from time to time from a reversible hemianopia, a transient loss of vision in half of the visual field in each eye. On the basis of his self-observation, he correctly suggested the way that the optic nerve fibers partially cross at the chiasm. Courtesy Wellcome Library, London.

It is plain that the cord, which comes finally to either eye under the name of the optic nerve, must be regarded as consisting of two portions, one half from the right thalamus, and the other from the left thalamus nervorum opticorum. According to this supposition, decussation will take place only between the adjacent halves of the two nerves. That portion of the nerve which proceeds from the right thalamus to the right side of the right eye, passes to its destination without interference: and in a similar manner the left thalamus will supply the left side of the left eye with one part of its fibres, while the remaining half of both nerves in passing over to the eyes of the opposite sides must intersect each other, either with or without intermixture of their fibres.[2]

Wollaston rediscovered the pattern of decussation—the crossing of visual fibers across the midline—by observing his own transient partial loss of vision. Despite Newton's scientific authority and Wollaston's evidence, the pattern of crossing was still largely disputed. Wollaston's report was cited 1 year later by a news item in the Boston Medical and Surgical Intelligencer as an isolated curiosity in the course of describing a boy who allegedly saw a candle flame upside down. As late as 1880, H. Charlton Bastian, Professor of Pathological Anatomy and Medicine at University College London, could still write: "Although the subject is by no means free from doubt and uncertainty, the weight of the evidence seems now most in favour of the view that 'decussation' at the Optic Commissure is as complete in Man as it is known to be in lower Vertebrates."[3]

In spite of Dr. Bastian's opinion, within a few years the picture was clarified. Figure 8.4 is a fiber-stained section through the optic chiasma, showing the true pattern of crossing. By the time that the neurologist William Richmond Gowers wrote his *Manual of Diseases of the Nervous*

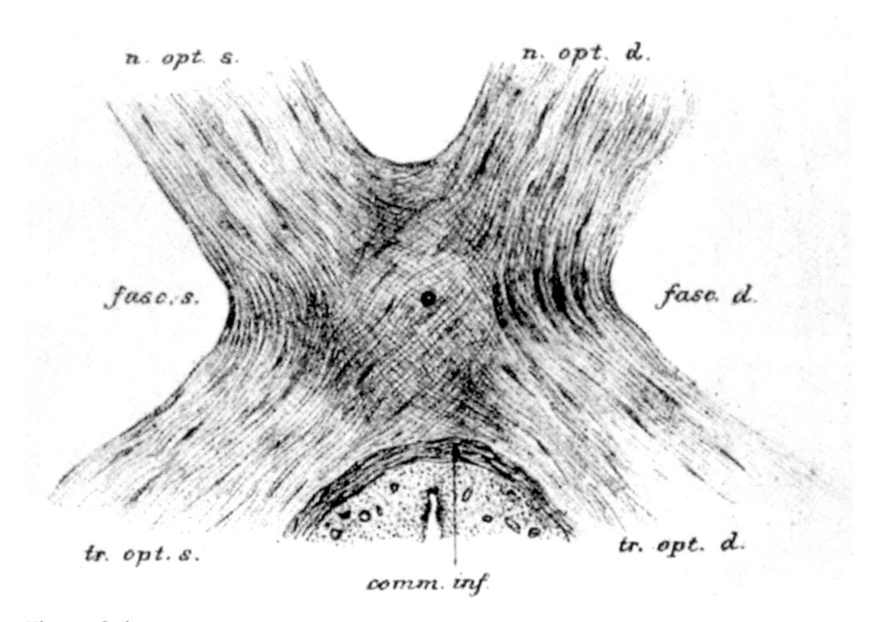

Figure 8.4
Bernhard Von Gudden (1824–1886) was a psychiatrist and anatomist who did careful dissections of the human brain. He studied in detail the pattern of fiber crossing at the optic chiasm.

System in the late 1880s, the pattern of crossing at the optic chiasma explained the visual loss that is experienced by people who suffer damage to the visual pathways after the crossing. Gowers wrote:

> Each optic tract contains the fibres from the same named half of each retina, i.e. from the temporal half of the retina on the same side, and the nasal half of the retina on the other side, and hence disease of the tract causes loss of vision in the opposite half of each field, the temporal half of the one, the nasal half of the other. This is termed homonymous hemianopia, or lateral hemianopia.[4]

The Lateral Geniculate Nucleus

Vision, like all sensory inputs except for olfaction, is relayed to the cerebral cortex by way of the thalamus. The thalamic relay for vision is the lateral geniculate nucleus (LGN). The LGN in humans and Old-World monkeys has an obvious striped appearance in anatomical sections, with six layers of neurons separated by interleaved fiber layers (figure 8.5).

Although it was clear that the eye projects to the LGN, the pattern of termination of optic tract fibers was not well understood. A better

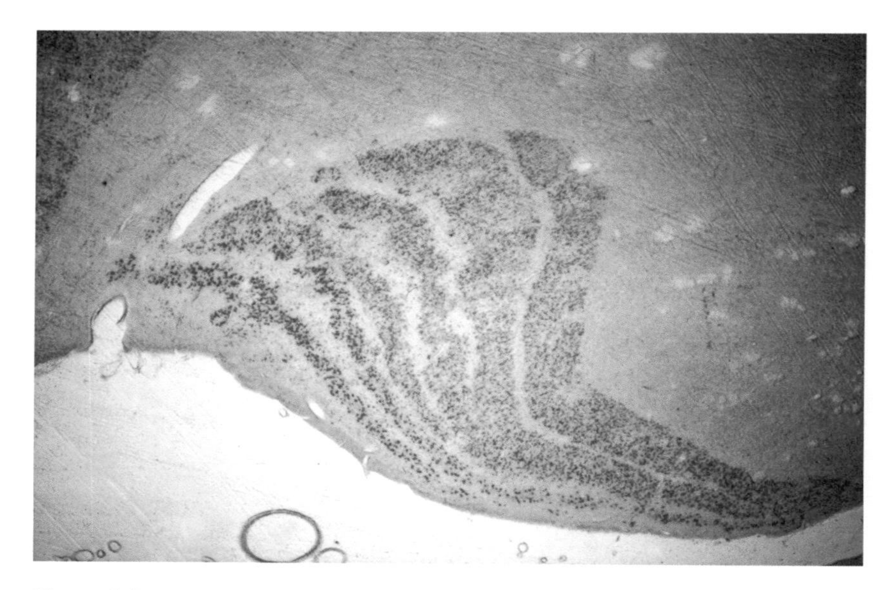

Figure 8.5
The human lateral geniculate nucleus (LGN) from geniculate (Latin, *little knee)* is a prominently striped visual relay nucleus in the thalamus. Six cell layers, numbered from 1 through 6, can be identified. Each eye connects to three independent layers.

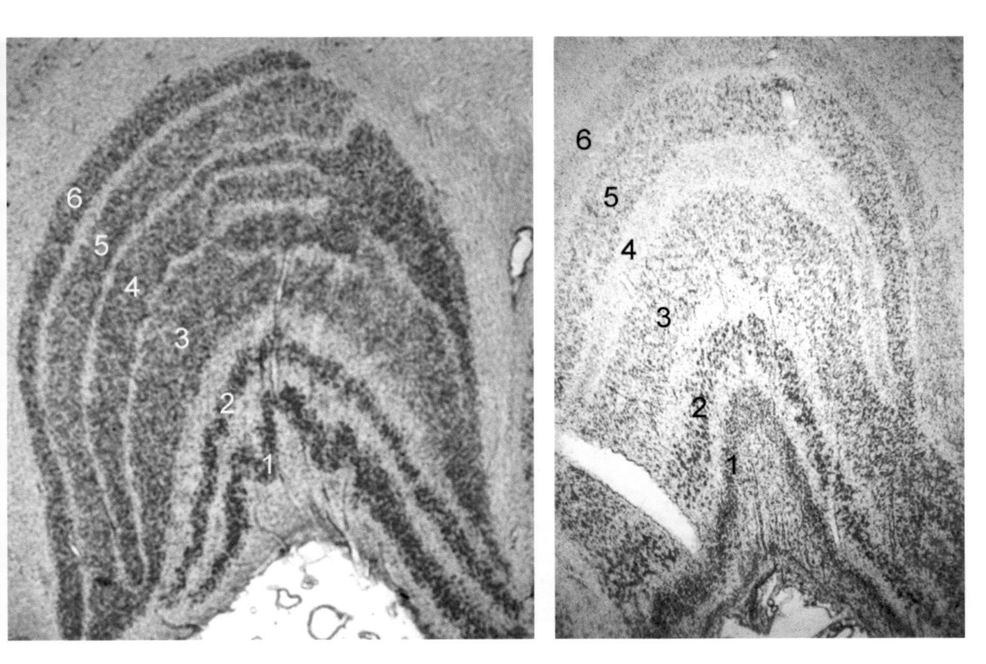

Figure 8.6
The monkey LGN is six-layered like the human geniculate. If a monkey or human loses the input from one eye, layers 1, 4, and 6 on the LGN opposite degenerate; layers 2, 3, and 5 on the same side degenerate.

understanding was provided by study of transneuronal atrophy and degeneration. Cells in the LGN that are deprived of their input from the eye shrink or die. Mieczyslaw Minkowski, working in Zürich in 1912, studied the LGN of a monkey that had one eye removed 8 months earlier and that of a 75-year-old woman who had suffered near blindness in one eye for 38 years before she died (figure 8.6). Minkowski saw that in the LGN layers opposite to the blind eye, three layers—1, 4, and 6—were atrophied. In the LGN on the same side side, layers 2, 3, and 5 were affected. Thus, the alternating layers of the lateral geniculate each receives an input from only one eye.

The technique of studying transneuronal atrophy has revealed the organization of the LGN in a large number of mammals. The six-layered pattern is virtually identical in the apes and the Old-World primates. In some cases the existence of a hidden laminar pattern can be revealed by transneuronal atrophy. For example, in the squirrel monkey, *Saimiri*, the dorsal, parvocellular region of the LGN is not obviously laminated. One year after removal of one eye, a clear six-layer pattern emerges that is similar to that of the Old-World primates.[5]

In the 1920s and 1930s anatomists such as Bernard Brouwer and W. P. C. Zeeman studied projections from the retina to the LGN. They made restricted retinal lesions and identified the distribution of fiber terminals in the LGN using a stain for degenerating fibers. Geniculocortical projections were studied by making small lesions of the primary visual cortex and mapping retrograde degeneration of cells in the LGN. A small set of adjacent retinal ganglion cells project to a restricted region in the LGN. A small lesion in the visual cortex leads to retrograde degeneration in a small area of the LGN.

These anatomical studies confirmed that there is an orderly projection from the retina to the LGN and from LGN to the visual cortex. Neighboring points in the visual fields are represented at neighboring points in the LGN and on the cerebral cortex. Later studies, such as those of J. M. Van Buren in 1963, discovered that, in addition to retrograde degeneration in the LGN caused by cortical lesions, *transneuronal* degeneration occurs in the retinal ganglion cell layer after a long time.

Mapping the Primary Visual Cortex

By the end of the eighteenth century the gross structure of the cerebral cortex was beautifully illustrated in anatomical texts, but the cortex was portrayed as structurally homogeneous. One part of the cortex was depicted as looking like any other. The first recognition that the cerebral cortex is not uniform in structure was made by an Italian medical student, Francesco Gennari, working in the newly refounded University of Parma.

Gennari packed brains in ice, enabling him to make clean, flat cuts through them. He noted a thin white line—sometimes two lines—within the cortex, running parallel to, and about half-way between the surface and the white matter. The lines coalesce into a prominent single stripe in the caudal part of brain, "in that region near the tentorium." Gennari first saw the stripe in 1776 and described it in his monograph *De Peculiari* (1782) some six years later (figures 8.7 and 8.8).

Gennari was poor but known to be a most promising young anatomist. Accordingly, a request was made to the Duke of Parma for a stipend for him to allow him to continue his researches (figure 8.9).

Gennari's monograph was published in limited number. It came from what was then an obscure university and was largely ignored. The same

FRANCISCI CENNARI

PARMENSIS

MEDICINAE DOCTORIS COLLEGIATI

DE PECULIARI

STRUCTURA CEREBRI

NONNULLISQUE EJUS MORBIS.

PAUCAE ALIAE ANATOM. OBSERVAT.

ACCEDUNT.

PARMAE

EX REGIO TYPOGRAPHEO

M. DCC. LXXXII

CUM APPROBATIONE.

Figure 8.7
Title page of the 1782 monograph in which Gennari described the characteristic striped appearance of a region of cerebral cortex in the occipital lobe.

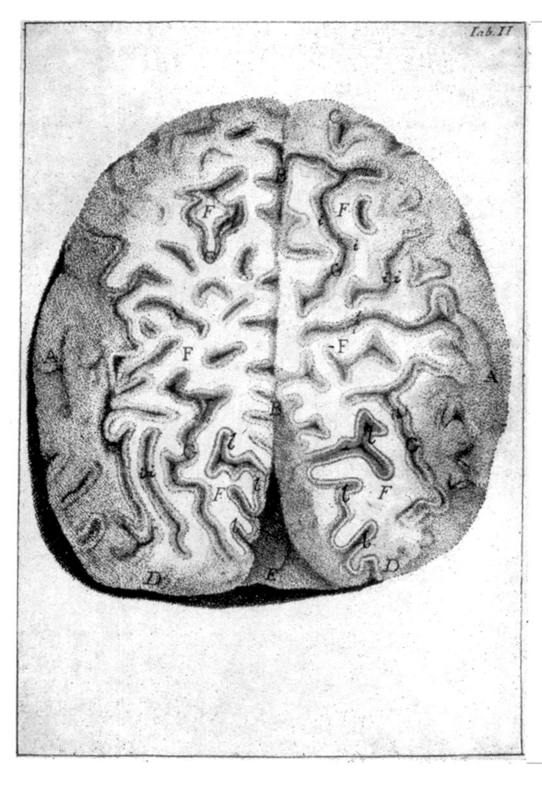

TABULA II.

Cerebrum humanum repraesentat sua in sede constitutum a quo prima superiora strata ablata sunt, ut Tertia, *sive* Nova *ejusdem substantia, ostendatur.*

A A	Hemisphaeria bina
B B	Eorum divisio.
C C	Lobi anteriores bini.
D D	Posteriores totidem.
E	Tentorium.
F F	Substantia medullaris.
G G	Corticalis substantia.
h h h h	Ejusdem divisio.
i i i i	*Tertia*, sive *Nova* cerebri substantia, quae describitur §. XLVI., quae in anteriori sinistro lobo
* *i i* *	*Duplex* apparet.
l l l	Eadem substantia, quae prope tentorium semper in *Albidiorem Lineolam* coacta intra corticem ipsum elegantior occurrit.

Figure 8.8
Human brain cut horizontally. Gennari identified a white stripe within the cortex, parallel to its surface. The stripe appeared most marked in the occipital lobe, a region labeled *(l,l,l,l)* in the figure. Gennari's striped area was later recognized to be coextensive with the primary visual cortex.

cortical stripe was discovered independently a few years later by the more eminent anatomist Vicq D'Azyr. The stripe was described in his Traité D'Anatomie (1786) three years later. It was the Austrian anatomist Heinrich Obersteiner who in 1888 found Gennari's earlier description of the white line and christened it: "the stripe of Gennari."

By the early nineteenth century regional differences in cortical structure were accepted, but there was no agreement about possible differences in the functions of different cortical areas. Two of the major authorities at the time, Franz Gall and Pierre Flourens, held opposing views. Gall died in 1828, whereas Flourens lived on until 1867, but their lives overlapped enough for them to attack one another.

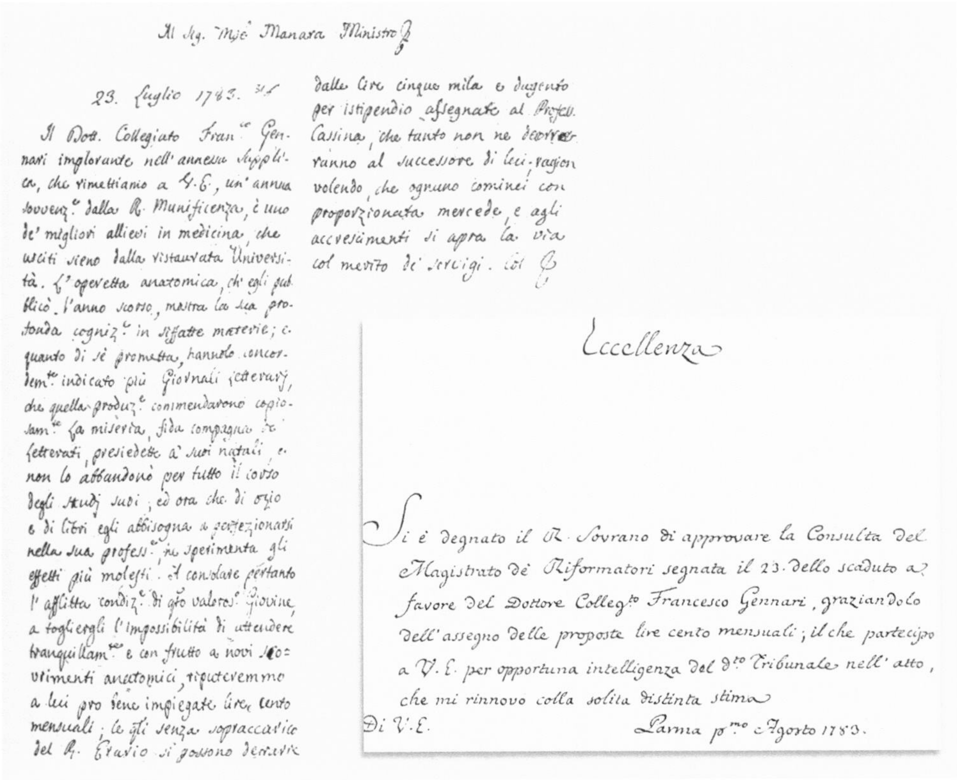
Al Sig. Mse. Manara Ministro

23. Luglio 1783.

Il Dott. Collegiato Franc.co Gennari implorante nell'annessa supplica, che rimettiamo a V. E., un'annua sovvenz.e dalla R. Munificenza, è uno de' migliori allievi in medicina, che usciti sieno dalla ristaurata Università. L'operetta anatomica, ch'egli pubblicò l'anno scorso, mostra la sua profonda cogniz.e in siffatte materie; e quanto di sè prometta, hannolo concordem.te indicato più Giornali letterarj, che quella produz.e commendarono copiosam.te La miseria, fida compagna de' letterati, presiedette a' suoi natali, e non lo abbandonò per tutto il corso degli studj suoi; ed ora che di ozio e di libri egli abbisogna a perfezionarsi nella sua profess.e, ne sperimenta gli effetti più molesti. A consolare pertanto l'afflitta condiz.e di q.to valoros.o giovine, a toglierglì l'impossibilità di attendere tranquillam.te e con frutto a novi ritrovimenti anatomici, riputeremmo a lui pro bene impiegate lire cento mensuali; le quali senza sopraccarico del R. Erario si possono detrarre dalle lire cinque mila e dugento per istipendio assegnate al Profess. Cassina, che tanto non ne decorreranno al successore di lui, ragion volendo, che ognuno cominci con proporzionata mercede, e agli accrescimenti si apra la via col merito de' servigi.

Eccellenza

Si è degnato il R. Sovrano di approvare la Consulta del Magistrato de' Riformatori segnata il 23. dello scaduto a favore del Dottore Colleg.to Francesco Gennari, graziandolo dell'assegno delle proposte lire cento mensuali; il che partecipo a V. E. per opportuna intelligenza del d.to Tribunale nell'atto, che mi rinnovo colla solita distinta stima

Di V. E.

Parma p.mo Agosto 1783.

Figure 8.9
A petition to the Duke of Parma to provide Francesco Gennari with a modest stipend of 100 lire/month to allow him to continue his research. The document on the right says that he was awarded the stipend.

As discussed in chapter 2, Gall and his followers, the cranioscopists/phrenologists, asserted that the cerebral cortex is made up of a number of individual areas, each associated with a specific personality characteristic. Flourens granted that vision, movement, and thought are each functions of the cerebral cortex, but he denied any localization of those functions.

By the late nineteenth century, evidence had begun to accumulate that different areas of the cerebral cortex are specialized for different functions. There was clear clinical evidence for brain localization of speech (see chapter 17 of this volume) and experimental evidence for the localization of motor functions (see chapter 13 of this volume).

There had been indications that lesions in the caudal part of the brain are associated with visual deficits. The clearest and most direct evidence for the visual function of the occipital lobe was provided by Hermann Munk (figure 8.10), Professor of Physiology in the Veterinary Institute in Berlin. In a series of experiments in 1881 Munk made lesions in the occipital lobe of monkeys. When he destroyed one occipital lobe the monkeys became hemianopic (the loss of vision in half of both eyes; figure 8.11). Bilateral lesions caused complete blindness.

Munk's discovery focused the attention of clinicians and scientists on the role of the occipital lobe in vision. In 1890, the Swedish neurologist Salomon Henschen (figure 8.12) summarized the findings in a group of patients who had suffered hemianopia as a result of a stroke and whose brains were studied postmortem (figure 8.13).

In humans most of the striate cortex is contained on the banks and depths of the calcarine fissure, a prominent fissure on the medial face of the hemispheres. Only a small region extends onto the cortical surface at the occipital pole. The cortical lesions causing hemianopia in humans all involve the striate cortex. Patients with a comparable loss of brain tissue that did not involve the striate cortex were not hemianopic. Henschen's results confirmed the location of the primary visual area, and he suggested a scheme for the way in which the visual fields are mapped on the primary visual cortex. He recognized that the left hemisphere receives its input from the right visual field and the upper bank of the calcarine fissure from the upper retina, hence the lower visual field. But Henschen also suggested that the periphery of the visual field is projected onto the caudal end of the striate cortex with the fovea represented anteriorly. In this, he was in error.

Henschen's error is understandable. Most of the lesions in the brains that he studied were the result of strokes, and hence the damage tended to be diffuse. A more accurate spatial mapping was made by studying partial field defects caused by smaller, subtotal lesions of the striate cortex. Such lesions, regrettably, arise in wartime. One of the earliest clear pictures of the representation of the peripheral-central visual field representation was made by a young Japanese ophthalmologist, Tatsuji Inouye (figure 8.14).

Inouye was in the medical service of the Japanese army during the Russo-Japanese War of 1904–1905. His responsibility was to evaluate the extent of visual loss in casualties of the war. Inouye used the opportunity to study the visual field defects caused by penetrating brain injuries. In that war the Russians had used a newly developed rifle that fired

Figure 8.10
Hermann Munk (1839–1912) was Professor of Physiology in the veterinary school in Berlin. He studied the effects of brain lesion on vision in monkeys.

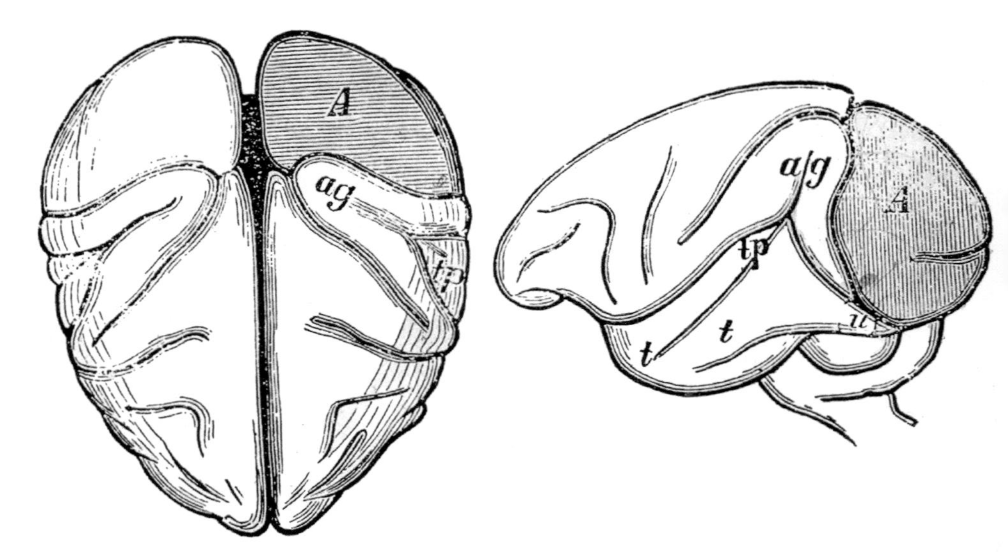

Figure 8.11
Picture from Munk's 1881 *Über die Funktionen der Grosshirndinde*. The lesion illustrated produced hemianopia in the monkey.

a small caliber bullet at a high velocity (figures 8.15 and 8.16). Unlike bullets that had caused most injuries in previous wars, these newer bullets often penetrated the skull at one point and then exited from another, making a straight path through the brain.

Inouye devised a three-dimensional coordinate system for recording the entry and exit wounds. He then calibrated the course of the bullet through the brain and estimated the extent of the damage it would have caused to the primary visual cortex or the optic radiations. Based on his study of visual field defects in twenty-nine patients, Inouye produced a schematic map of the representation of the visual fields on the cortex. The visual cortex is deeply folded, so it is hard to grasp the way in which the visual fields are projected onto it. The cortex of the calcarine fissure was drawn as if it were flattened. With the flattened map it was easier to visualize how the visual fields were represented on the cortical surface. The central fields were now placed correctly in the most caudal part of the striate cortex, with the peripheral visual fields represented anteriorly. The central visual fields occupied a much larger area of cortex than the peripheral fields. Inouye's map also made it clear that there is an over-representation of the central visual fields in the primary visual cortex.

Figure 8.12
Salomon Henschen (1847–1930) was a neurologist and neuropathologist who studied systematically the location of the brain lesion in patients who, in their lives, had suffered from hemianopia.

Later, the Anglo-Irish neurologist Gordon Holmes studied the visual field defects sustained by soldiers of the First World War. His work resulted in a more accurate and detailed map of the representation of the visual fields on the striate cortex. His map still forms the basis for interpreting partial visual loss in humans.

Confirming the Map through Electrical Stimulation and Recording

In the period between the First and Second World Wars, the basic arrangement of the visual fields was studied by using electrical stimulation of the brain in neurosurgical patients. Since the brain itself does not have

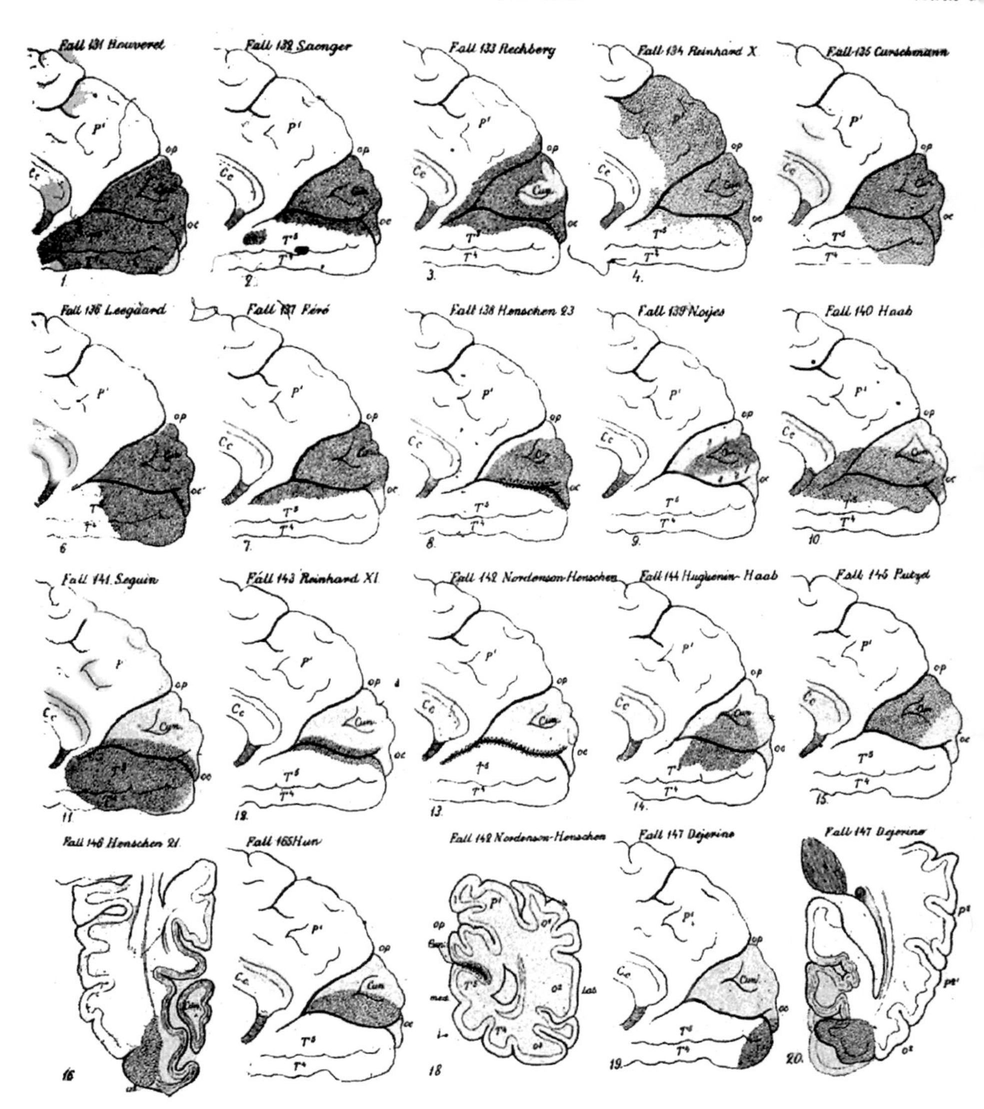

Figure 8.13
Location of the brain lesion in patients who had suffered hemianopia. The lesions were all centered in the calacarine fissure of the occipital lobe, coextensive with the region containing the stripe of Gennari.

Figure 8.14
Tatsuji Inouye (1881–1976) was a Japanese physician who studied visual loss in patients who had suffered brain wounds in the Russo-Japanese war. Courtesy Inouye Eye Hospital, Tokyo.

Figure 8.15
The Russo-Japanese war. The Russian soldiers (seen on the left) were armed with a rifle that, for the first time, used high-caliber bullets capable of penetrating the skull. Courtesy Heritage Images.

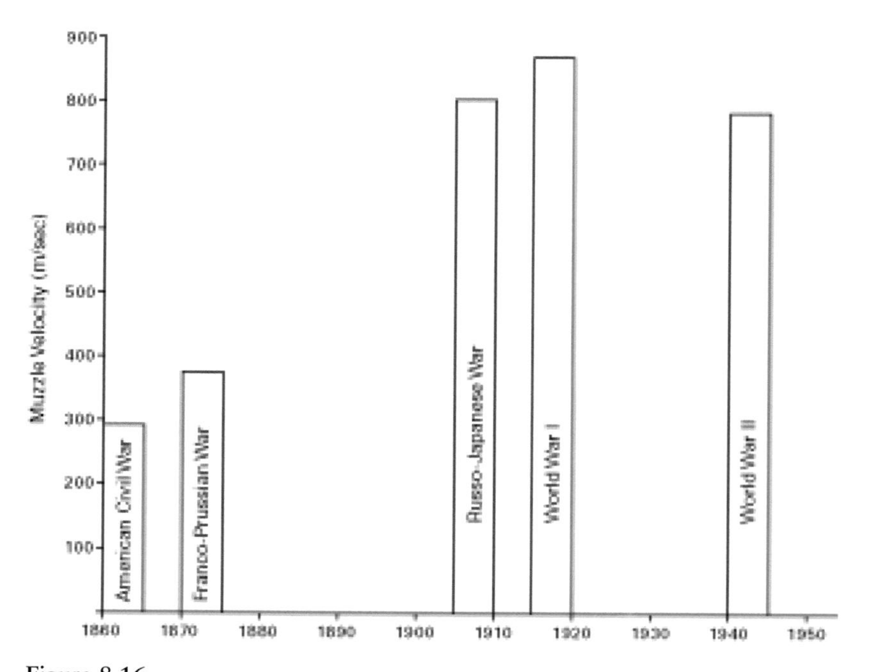

Figure 8.16
Muzzle velocities of rifles used in wars in the nineteenth and early twentieth centuries. Note the great change between the Franco-Prussian War and the Russo-Japanese War.

pain receptors, operations on it can be done under local anesthetics. The German neurologist and neurosurgeon Ottfried Foerster operated under local anesthesia on a patient who suffered from seizures caused by focal scarring of the brain. An electrical current was applied at a specific site on the cerebral cortex, and the resultant sensation was reported by the patient. Stimulation of the cortex at the occipital pole elicited a sensation of light that appeared to be centered in front of the patient. Stimulation of the cortex on the upper lip of the calcarine fissure five centimeters anterior to the occipital pole produced a phosphene that was centered in the lower visual field opposite to the side of the brain that had been stimulated.

Foerster's studies and those performed twenty years later by Wilder Penfield, a neurosurgeon at the Montreal Neurological Institute, confirmed the representation of the visual fields on the human striate cortex revealed previously by the Inouye and Holmes studies of the regions of blindness produced by focal lesions.

In the 1930s the American physiologist Philip Bard and his collaborators undertook a new kind of electrical measurement. They began to record the electrical activity that is evoked on the surface of the cerebral cortex of experimental animals by different points on the body surface. The electrodes at the time were too large to record the activity of individual neurons, but they were small enough to detect focal activity in a restricted group of cells. Bard and his colleagues found an orderly representation of the body surface on the primary somatosensory cortex, with neighboring points on the body represented at neighboring points on the brain.

One of Bard's colleagues, Wade Marshall, later collaborated with William Talbot in studying the activity evoked on the striate cortex of cats and Old-World monkeys. Talbot was a careful designer of optical research instruments at the Wilmer Eye Institute in Baltimore. Talbot and Marshall focused small spots of light on the retina of a monkey and marked the locus of maximal evoked activity on the cerebral cortex. Figure 8.17 is from their report published in 1941.

Talbot and Marshall's recordings were limited to the dorsolateral surface of the macaque cortex, comprising only roughly the central 10 degrees of the visual field. The work was extended by Peter Daniel and David Whitteridge (figure 8.18) in the early 1960s. They recorded more anterior cortical areas within the calcarine fissure of baboons, extending the mapping into the peripheral visual field.[6]

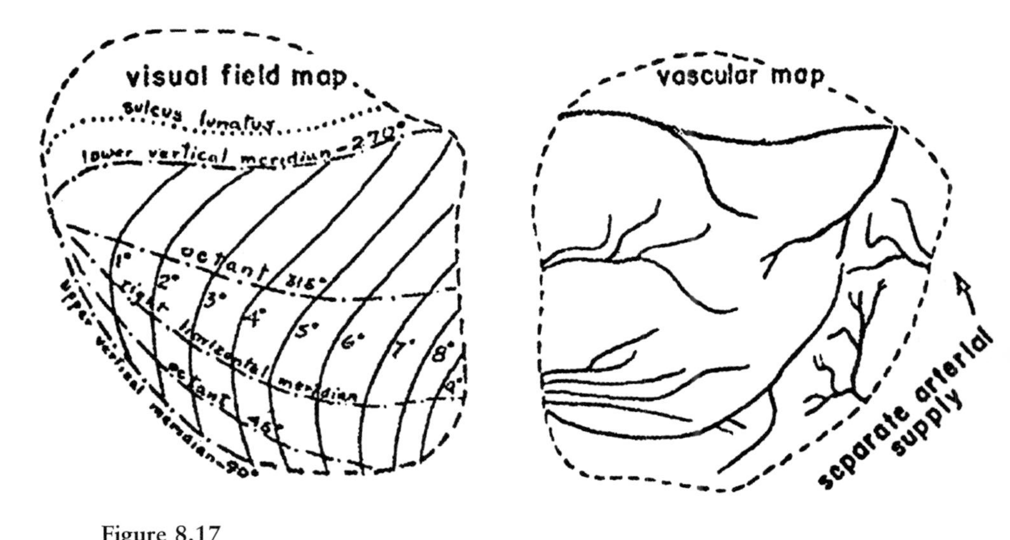

Figure 8.17
Talbot and Marshall mapped the visual field on the occipital lobe of a monkey. Courtesy Elsevier.

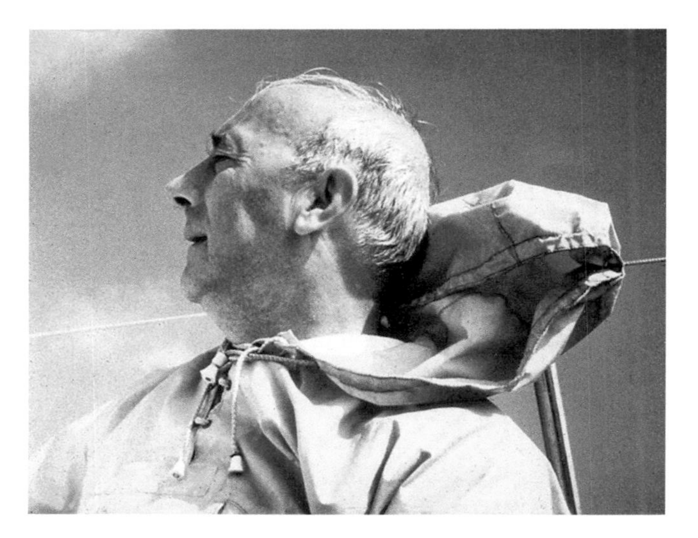

Figure 8.18
David Whitteridge (1912–1994), who, with his colleague Peter Daniel, extended Talbot and Marshall's map of the visual cortex in monkeys. They pioneered the technique of making a flat two-dimensional representation of a folded cortical surface.

The evidence from cortical lesions, electrical stimulation, and recording is all consistent. The visual fields are represented in an orderly way on the primary visual cortex.

Visual Areas outside the Primary Visual Cortex

Early researchers had suggested that regions outside the primary visual cortex also have a visual function. Hermann Munk, for example, labeled a region outside of the primary visual cortex in dogs as an area concerned with the storage of visual memories.

William Talbot in 1942 made a brief report to the Federated Society for Experimental Biology and Medicine that initiated modern study of the way in which the visual fields are represented beyond the primary visual cortex. Talbot recorded potentials evoked by vision from the surface of a cat brain. As expected, he found that the visual field is mapped in an orderly way on the primary visual cortex with neighboring points in the visual fields represented at neighboring points on the cortex. As Talbot continued to record lateral to the representation of the vertical meridian, he found that this region of cortex was also activated by focused spots of light but from increasingly peripheral regions of the visual field. Talbot had discovered a second visual area, later called "Visual 2," which is mapped on the cortex like a mirror image of the primary representation.

In subsequent years a number of areas of cortex were described in which visual stimuli evoke a gross potential. Although the pioneering work of electrical mapping was done in cats, the visual cortex in these animals is not "primary" in the sense that it is in monkeys and humans. In monkeys and humans the overwhelming majority of geniculocortical fibers terminate in the striate cortex. In cats the second visual area receives a direct and equally powerful input from the LGN.

In the late 1950s techniques were developed for isolating the activity of individual neurons. Single-unit recording has since become a standard method for studying visual processing by the brain. These studies confirmed that there are many visual areas beyond the striate cortex.

The Canadian David Hubel and the Swede Torsten Wiesel, working first at Johns Hopkins and later at Harvard University in the late 1950s, initiated a new and fruitful approach to study of cortical visual responses. During twenty-five years of research they mapped the primary cortical visual areas of cats and monkeys at the individual neuron level (figure 8.19). They discovered one of the fundamental principles of cortical

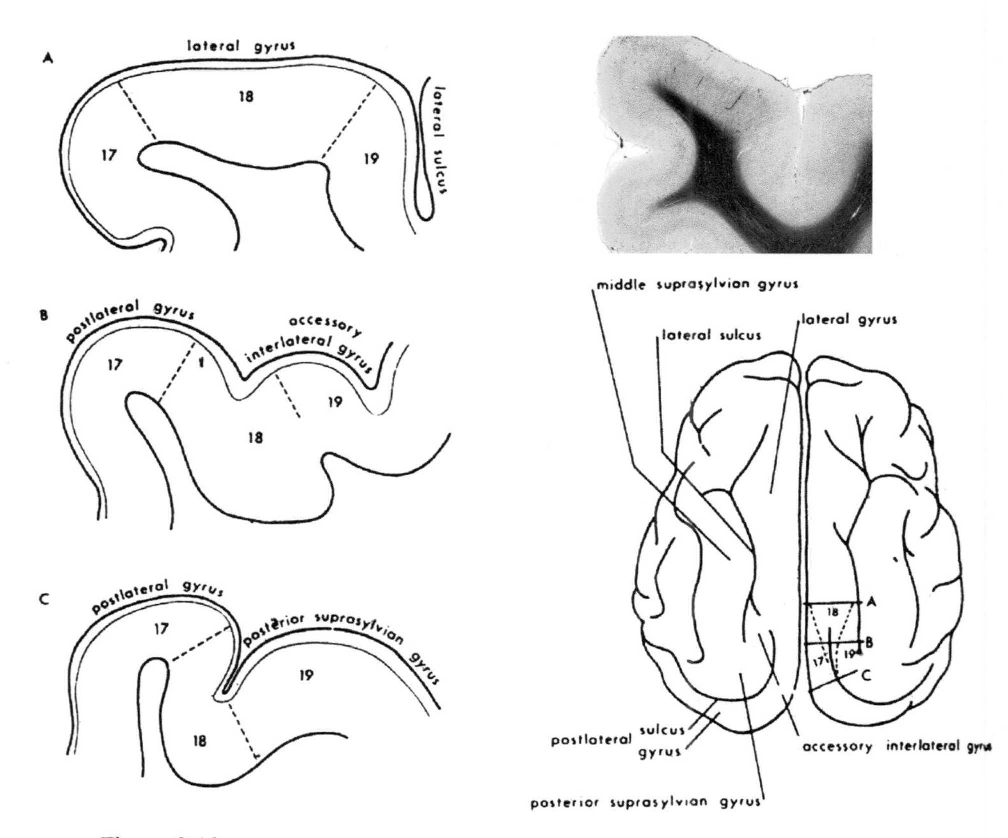

Figure 8.19
Upper right: Fiber-stained cat brain visual area. Left and below: Cat brain showing three visual areas on the cerebral cortex, from Hubel and Wiesel's 1965 paper. Courtesy American Physiological Society.

functions. Unlike the retinal and geniculate cells, neurons in the primary visual cortex were sensitive to *oriented* visual targets—that is, upright, horizontal, or at angles in between. They continued their experiments to the visual response properties on neurons in areas beyond the primary visual areas.

Following the work of Hubel and Wiesel, researchers on both sides of the Atlantic identified several more extrastriate visual areas, each of which appeared to be specialized for analyzing color, motion, or form. John Allman and Jon Kaas, then at the University of Wisconsin, studied the Owl monkey *Aotus*, and Semir Zeki at University College London studied Old-World macaques. By a count made in 1992, there are no less than thirty-two visual areas in the monkey brain. There are doubtless at least as many in the human cortex.

Early Behavioral Evidence for the Functions of Extrastriate Visual Areas

In humans and monkeys the striate area is virtually the sole cortical target of cells in the LGN. But in humans as well as monkeys the cortex adjacent to the primary visual cortex is also dominated by vision. An estimated one-third or more of the monkey and human cerebral cortex is devoted to visual processing. There are two great groupings of visual areas outside the primary visual cortex. In the early 1980s Leslie Ungerleider and Mort Mishkin, working at the National Institutes of Health in Bethesda, and Mitchell Glickstein and Jack May at Brown University, distinguished a medial group that is centered in the parietal lobe and a lateral group centered in the temporal lobe. There had been evidence for both from earlier behavioral studies.

In monkeys much of the parietal lobe cortex has direct or indirect input from the primary visual cortex. The angular gyrus of monkeys contains some of the extrastriate parietal visual areas. David Ferrier was a nineteenth century neurologist and physiologist working first in Yorkshire and later in London. When Ferrier stimulated the angular gyrus of the monkey brain electrically, he observed that the stimulation caused eye movements. When he ablated the region, the monkey appeared to be blind. Ferrier concluded, in error, that this region must be the primary visual cortex. Ferrier's protocols demonstrate that rather than blindness, his monkeys suffered a severe impairment in guiding their movements under visual control. Virtually identical symptoms were described by the Hungarian physician Rudolf Balint in 1909 in a patient who had suffered bilateral lesions of a similar region of the parietal lobes (figure 8.20) and by Gordon Holmes, studying casualties among British soldiers in the First World War.

Ferrier's monkeys, Balint's patient, and Holmes's soldiers were all unable to guide their movements accurately under visual control. The parietal lobe visual areas are principally concerned with spatial localization in the visual field and the visual guidance of movement.

In early studies of the effects of large temporal lobe lesions in monkeys, Sanger Brown and Albert Schäfer (in 1888) and Heinrich Klüver and Paul Bucy (in 1938) found that, in addition to the tameness that followed the operation, there were visual deficits. The specifically visual function

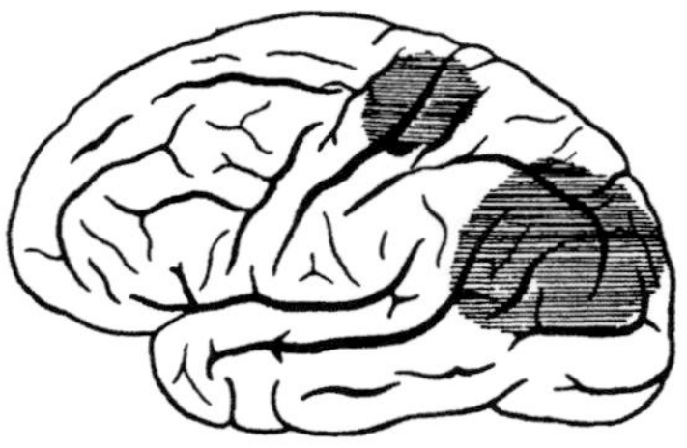

Schema I. Linke Hemisphäre.

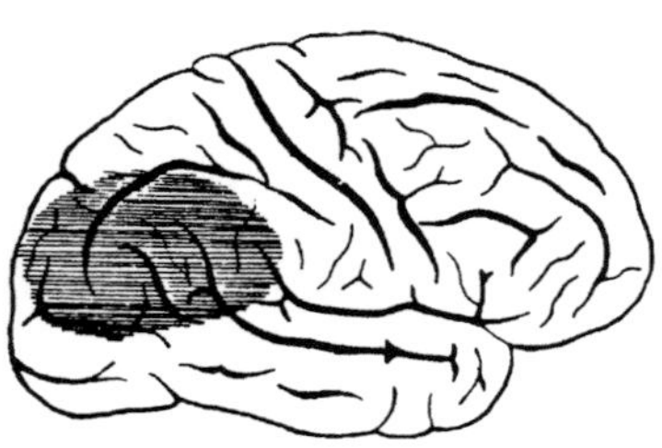

Schema II. Rechte Hemisphäre.

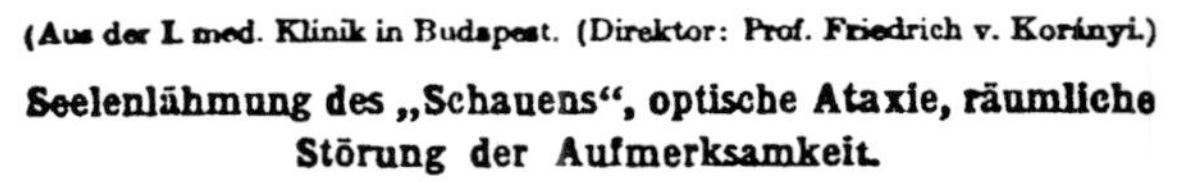

(Aus der I. med. Klinik in Budapest. (Direktor: Prof. Friedrich v. Korányi.)

Seelenlähmung des „Schauens", optische Ataxie, räumliche Störung der Aufmerksamkeit.

Von

Dr. RUDOLPH BÁLINT,

Assistenten der Klinik.

(Hierzu Tafel V—VIII.)

Diesen dreifachen Symptomenkomplex, dessen Komponenten ich unter obigem Titel zusammenzufassen versuche, habe ich bei einem auf der I. medizinischen Klinik längere Zeit hindurch beobachteten Patienten festgestellt.

Alle drei Komponenten des Symptomenkomplexes musste ich mit neuen Namen versehen, da in der Literatur keine ausführliche Beschreibung eines ähnlichen Symptomenkomplexes zu finden war, wenn auch den einzelnen Symptomen nahestehende Erscheinungen in der Literatur hie und da angedeutet vorzufinden sind.

Die Entwirrung und Deutung der an diesem Patienten beobachteten äusserst komplizierten Erscheinungen wurde mir

Figure 8.20
The brain of Rudolf Balint's patient and Balint's 1909 description of the visual deficits. The patient saw perfectly well but was unable to guide his hand under visual control after suffering bilateral parietal lobe lesion.

of the inferotemporal cortex was further clarified when animals were tested after smaller, more restricted lesions of the temporal lobes. In the 1950s K. L. Chow, working at the Yerkes Primate Laboratory in Florida, showed that lesions of the inferotemporal cortex cause a specific impairment in the acquisition and retention of visual discrimination learning. Fifteen years later Mortimer Mishkin demonstrated that the essential input to the inferotemporal cortex is by a series of corticocortical connections originating in the striate cortex. These studies demonstrated that temporal lobe lesions cause impairment in recognizing and remembering forms.

Our understanding of vision begins with a clear view of how images are formed on the retina. We know a good deal about both the nature of processing in the retina and the spatial organization of the visual pathways. Current interest focuses on the role of areas outside the primary visual cortex in processing color, form, and motion and the visual control of movement.

9

Audition

Nigh to the most intimate recess of this Den, a thin Membrane is placed, with a Circular Bone, fitted to the same, which wholly shuts up the Cavity of the Ear, and distinguishes the Interior Cloyster from the Exterior; so that the Impulse of the sound, shaking this Membrane like a Drum, delivers the Impression to the Sonorifick Particles planted beyond, and they being moved, affect the Fibres, with the Auditory or Hearing Nerve.

—Thomas Willis (figure 9.1), *Two discourses concerning the soul of brutes*

Very cursory observation often suffices to show that sounding bodies are in a state of vibration and that the phenomena of sound and vibration are closely connected. When a vibrating bell is touched by the finger, the sound ceases at the same moment that the vibration is damped. But in order to affect the sense of hearing, it is not enough to have a vibrating instrument; there must also be an uninterrupted communication between the instrument and the ear. A bell rung in vacuo, with proper precautions to prevent the communication of motion remain[s] inaudible. In the air of the atmosphere, however, sounds have a universal vehicle, capable of conveying them without break from the variously constituted sources to the recesses of the ear.

—Lord Rayleigh (figure 9.2), *Theory of Sound*

We live in a world of sounds. Human civilization is based in part on our ability to communicate with one another by speech. We are taught by speech, and we are persuaded to new points of view by speech. The ear is also our window to music—to the subtle interplay of melodies in a Bach Brandenburg concerto or to the titanic emotional power of the Verdi Requiem Mass. How does the brain code and process speech? What are the physical and biological mechanisms involved in our appreciation of music? Study of the ear and the auditory pathways can help us to begin to understand some of these issues. We must understand the way in which the ear processes sounds to understand the mechanisms that lead to hearing loss and deafness.

Figure 9.1
Thomas Willis (1621–1675) first described the arterial circle at the base of the brain that bears his name. He also was among the earliest authors to talk about the substance of brain, rather than the fluid in the ventricles, as being the important structure that controls functions. Courtesy Wellcome Library, London.

Figure 9.2
John William Strutt, 3rd Baron Rayleigh (1842–1919). His 1896 textbook, *The Theory of Sound*, is a classic of physics applied to analysis of a human sense. Courtesy Wellcome Library, London.

We know some important principles about how the ear codes different pitches and intensities of sounds. We know rather less about how sequential sounds like those of speech or music are put together by the brain. Let us begin our study of hearing by first reviewing some simple physical attributes of sound and then go on to describe what is known about how sounds are coded by the ear and transmitted to the brain.

The Physics of Sound

Sound is the effect on our ears of small rhythmical variations in the pressure of the air around us. Sound travels in waves through the air,

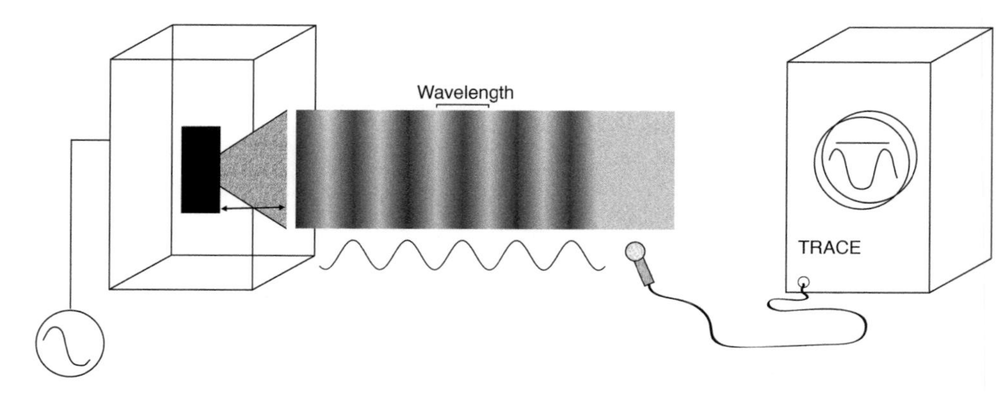

Figure 9.3
Diagram illustrating the nature of sound. The speaker on the left vibrates, alternately advancing and withdrawing, thus pushing air in front of it. The diagram illustrates the series of peaks and troughs of air pressure produced. A representation of the sound waves generated is illustrated on the face of the oscilloscope on the right.

very much like the waves in a pond that radiate outward from the point where a stone is thrown into the pond. Consider a tone, like that represented in figure 9.3.

The dots are a highly schematic representation of the air molecules that transmit the tone. If air is pushed forward rapidly by a loudspeaker in the direction shown by the arrow, it would momentarily compress the air just in front of the speaker. Now imagine that the speaker went back to its original position with the same speed as it had gone forward earlier. It would now leave a small region of thinner air just adjacent to the loudspeaker in which the air molecules were slightly less densely distributed than normally. Imagine that the speaker moves back and forth continuously. These two cycles, pressure and rarefaction, would be transmitted through the air adjacent to the speaker. If the speaker continued to vibrate, there would be an orderly sequence of increased pressure and rarefaction in the surrounding air, as shown in figure 9.3.

Suppose we were to put a probe somewhere in front of the speaker that could measure the changes in air pressure. (The probe might look very much like a microphone.) We could then convert these cyclical changes in air pressure into something we could see, such as a meter reading. But if we tried to use a mechanical meter, the excursions of the indicating needle would be so rapid that we would not be able to see them. Therefore, let us arrange for these cyclical changes in pressure to be first converted by a microphone into voltage changes that can be written out on an oscilloscope. Now if the loudspeaker in our example moved forward very suddenly, stayed there a short time, and then just as suddenly moved backward, and if this change were done three times in a row, we might see a trace on the oscilloscope face that looks like a square wave. Square waves are somewhat complex from both a physical and a mathematical standpoint. A more usual and easily generated signal would be one in which the speaker gradually moved outward to a maximum and then reversed itself like that shown on the oscilloscope.

The mathematical function that describes the tracing seen in figure 9.3 is a sine wave. One of the most useful mathematical discoveries ever made was that of the nineteenth-century French mathematician Jean Baptiste Fourier, who showed that repetitive waveforms can be analyzed into the sum of a set of sine waves. Fourier was a physicist and mathematician. The son of a tailor, he participated in the French Revolution. Although imprisoned during the subsequent Reign of Terror, he survived to continue scientific and mathematical study and contributions. From the physical and physiological as well as mathematical point of view, sine waves are fundamental.

In air at room temperature, sound waves are propagated at about 700 mi/h, or roughly 1,100 ft/s. Such a rate of propagation is very slow compared to the 186,000 mi/s speed of light. The speed of sound refers to the speed that the disturbance is propagated. Individual molecules of air do not move at 1,100 ft/s; rather, it is the sequence of pressure and rarefaction of millions of molecules of air that is propagated through the air at that speed. The relative slowness of sound propagation poses some challenges. Runners hear the starting gun a short time after it is fired. Organists in a large cathedral might hear the music they are playing well after the note is played.

Sine waves can be characterized by their frequency, amplitude, and phase. Frequency is measured by how many waves occur each second. The higher the frequency, the higher the pitch of the sound we hear. If our oscilloscope were set to record a sound wave for one-tenth of a second, and we found that during that time there were ten complete cycles occurring, the frequency of the sound would be 100 cycles per second. Modern auditory research workers use the unit *Hertz* (abbreviate Hz) to indicate the number of cycles per second. Two waves might have the same frequency, but if the height or *amplitude* of the two is different, all other things being equal, the higher amplitude wave would sound louder. Amplitude and frequency are almost enough to characterize a sine wave, but not quite enough.

Two waves can have the same amplitude and the same frequency but be out of *phase* with one another. Even though both have the same number of peaks/second and the same amplitude, they have a different *phase*. The peak of wave A occurs before the peak of wave B. We could characterize such phase differences numerically in angular measurement. If we think of one complete sine wave cycle as 360 degrees, then A differs from B by about 90 degrees. We say A leads B by 90 degrees. In some aspects of human hearing, phase is not an especially important attribute of the sound. But as we shall see when we study sound localization, we can use phase differences to determine the direction from which a sound comes.

We know the speed of sound in air is about 1,100 ft/s. We can use that fact to calculate the wavelength of some typical frequencies of sound that we can hear. An 1,100-Hz tone would have a wavelength of about 1 foot, a 110-Hz tone would have a wavelength of about 10 feet, and an 11,000-Hz tone would have a wavelength of about one-tenth of a foot.

Psychophysics of Hearing

What sound frequencies can we hear? How intense must a sound be for us to hear it? In order to understand the way in which auditory thresholds are measured, it is useful to understand the basic unit with which the physical intensity of sounds is measured—the decibel.

We could simply measure intensity of a sound in physical units. But people hear over a great range of sound intensities; hence it is useful to compress the scale. We do this by using *ratios* of sound intensity rather than their absolute values and then taking the logarithm of those ratios. Now the use of a ratio requires a reference intensity, and it is conventional to use an intensity close to the threshold for normal human hearing. Written out this threshold intensity is the tiny number 1/10,000,000,000,000,000 of a watt, or more simply 10^{-16} watt/cm^2.

Once we have such a ratio we can easily compress the scale for sound measurement by using as our measure the *bel,* which is the logarithm to the base 10 of the ratio of a measure of sound intensity over this reference sound. Thus if our measured sound intensity were 10^{-12} watt/cm^2, we would have

$$\log_{10} (10^{-12}/10^{-16}) = \log_{10} (10^4) = 4 \text{ bels}$$

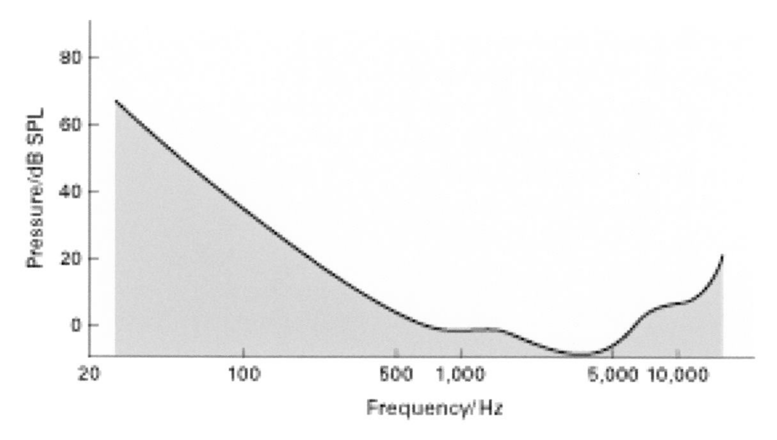

Figure 9.4
Human auditory frequency curve.

The bel turns out to be a rather large unit, so it is conventional to accept a smaller unit of one-tenth of a bel or decibel. In the above example the measured pressure of 10^{-12} watt/cm^2 would have a value of 40 decibels. Normal conversation in a quiet room is at about 50 to 55 decibels.

Human Frequency Response

Our eyes are sensitive only to the restricted ranges of wavelengths that we call light. In the case of sound we are also not uniformly sensitive to all frequencies. Human hearing has a range from about 20 Hz to 20,000 Hz, but we do not hear equally well at each frequency. Figure 9.4 shows a graph of human thresholds—the amount of sound intensity required for a subject to just detect a sound. As can be seen, people are best in the range of about 1,000–4,000 Hz, which is in the frequency range of the major components of human speech.

Hearing in Animals

There are great variations in the auditory capacity of different animals. Cats, for example, can hear tones of 40 kHz, about twice as high as the upper limit of human hearing. Some animals, including bats and desert rats, can hear tones as high as 100 kHz or greater. How could we measure the hearing ability of an animal subject?

William Stebbins, a psychologist working at the University of Michigan, trained monkeys in an apparatus with two keys. The animals were trained to press one of the keys to turn on a tone. When the tone was on the monkey was rewarded with a bit of food or fruit juice if it pressed the other key. When it switched to the second key the monkey was saying in effect, "I hear the tone." Once the basic behavior sequence was trained, weaker tones could be tested.

George Gourevitch, a psychologist at Hunter College in New York City, and his collaborators arranged a similar situation in which a rat was tested in an apparatus containing two levers. The animal was rewarded for pressing one of the levers only if there was no sound in the chamber. If the animal pressed the other lever B in the box, it would turn down an existing sound. With sufficient training, a rat would learn to press lever B until it no longer heard the sound. It would then press lever A, which would increase the intensity of the sound.

Stebbins and Gourevitch both achieved results on their sensory testing that are similar in precision to those that can be obtained from alert cooperative human subjects. Gourevitch then went on to use the method to evaluate the damaging effects of some antibiotic drugs on the ear.

The Structure of the Ear

Figure 9.5 is a diagram of the major structures of the human ear. The ear is divided into an outer portion or pinna with its associated outer ear canal—the meatus—leading to the eardrum. Sounds are conducted via the meatus to the eardrum, where the mechanical vibrations in the ear are converted to vibrations of the eardrum. The middle ear has as its principal constituents three tiny bones or ossicles—the malleus (or hammer), incus (or anvil), and the stapes (or stirrup)—which are connected in a sequential fashion. The final bone in the chain, the stapes, connects directly to the oval window, a membrane at one end of the cochlea.

The arrangement of the bones of the middle ear—the malleus, incus, and stapes—serves two important functions. The bones not only conduct a faithful reproduction of the sound waves to the inner ear but they also act as a lever, changing a relatively high-amplitude but weak vibration of the eardrum to a lower-amplitude but more powerful vibration in the inner ear. Transmission to the oval window acts to "gear down" the sound, much like shifting a car transmission into a lower gear.

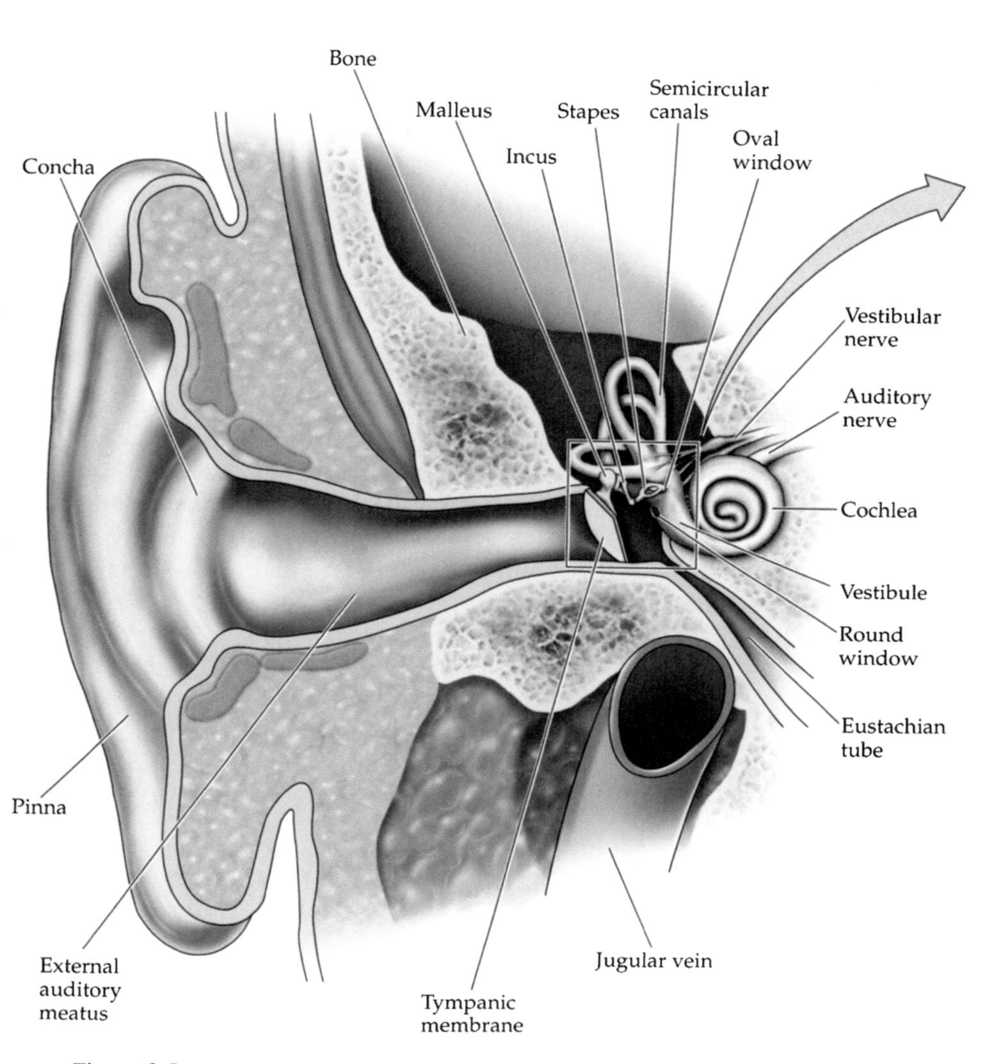

Figure 9.5
The structure of the ear. The major components of the ear are shown. Courtesy Dale Purves and Sinauer Associates.

The ear is partially protected from damage by reflexes that operate at high sound intensity to lessen the amount of possible damage to the inner ear. These protective reflexes are mediated by two tiny muscles—the tensor tympani and stapedius muscles. Both muscles contract at high sound levels, thereby making the coupling through the middle ear bones less effective.

The Structure of the Inner Ear

The organs of hearing lie within the cochlea (figure 9.6), which resembles a snail shell. The cochlea is essentially a long bony tube filled with fluids. Inside of the tube is a region that contains the sense organ for hearing. The rolled up tube has a cochlear partition occupying its center. Two fluid-filled regions are connected to one another at the end of the cochlea; their junction is called the helicotrema. Sound waves are carried to the inner ear where they produce vibrations that are transmitted via the perilymph throughout the cochlea.

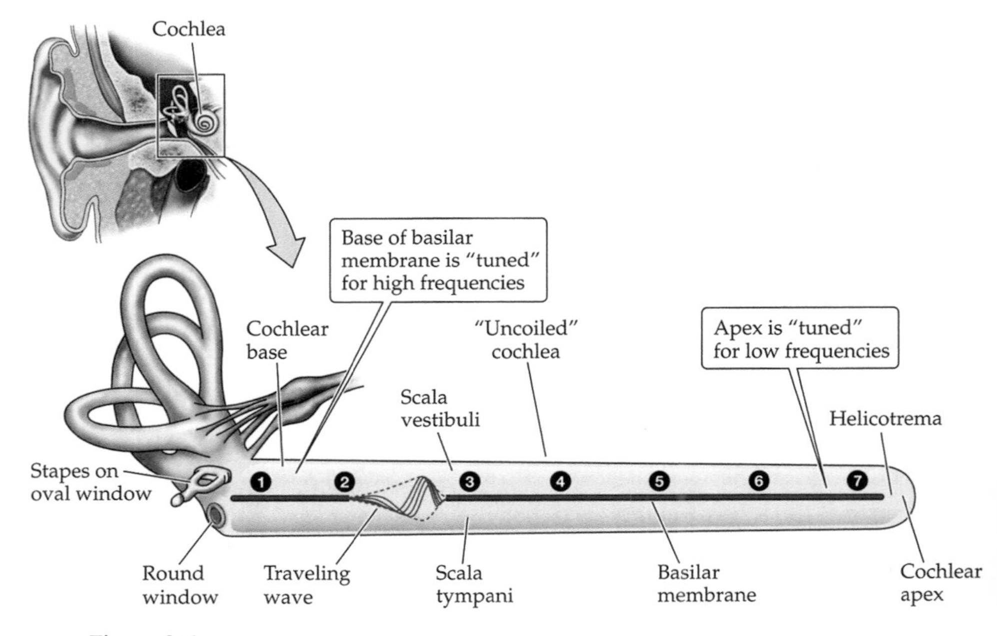

Figure 9.6
The cochlea is unrolled to show the relationship of position and frequency response along the cochlea. Courtesy Dale Purves and Sinauer Associates.

The analysis of how the ear responds to vibrations was brilliantly done by the Hungarian scientist Georg Von Békésy. Von Békésy started working as a sound engineer and was famous for making great changes in the efficiency of the Hungarian telegraph system. In his hearing experiments Von Békésy removed the ears of experimental animals and human cadavers. He placed a mechanical probe on the oval window and studied the response of the membranes of the inner ear to imposed vibrations.

When the internal fluid is set in motion, the most important structure within the ear responding to these vibrations is the basilar membrane. This membrane is thin and stiff at one end of the cochlea (near the round and oval windows) and wider and less stiff at the opposite end of the cochlea (near the helicotrema). When a mechanical stimulus is applied to the oval window, it initiates a traveling wave along the basilar membrane. The amplitude of this traveling wave varies as a function of the frequency of the tone. For low tones the basilar membrane vibrates at its greatest amplitude near the helicotrema. For high tones the vibration is greatest at the other end of the cochlea, near the round window.

Within the cochlea partition there are a large number of specialized hair cells that stretch between two membranes: the basilar membrane on one side and the tectorial membrane on the other. When the basilar membrane vibrates in response to sound, the ends of these hair cells are bent and deformed by a shearing action. Bending of the hair cells produces a generator potential in the auditory nerve fibers, which in turn causes action potentials in the auditory nerve fibers.

Response Properties of the Auditory Nerve Fibers

It is possible to record action potentials from a single auditory nerve fiber in an experimental animal. Tones can be presented to the ear of the animal, and the threshold of sound intensity necessary to actuate that fiber can be tested over a range of different frequencies. The resultant graph of threshold of response as a function of sound frequency is called a tuning curve.

Tuning curves are basic data of auditory physiology. As would be expected, auditory nerve fibers that connect to hair cells near the base of cochlea are maximally sensitive to high frequencies; auditory nerve fibers that are connected to hair cells near the helicotrema are maximally sensitive to low frequencies.

Place versus Frequency Theories of Hearing

Before modern recording methods were available there were two theories of how the ear codes different frequencies of sound. One of these theories is the *place theory* put forward by the great German physiologist Hermann von Helmholtz. Helmholtz suggested that within the cochlea there were a large set of resonators, each of which resonated approximately to a particular frequency of sound. The resonators were thought to be arranged to cover the entire range of pitches that could be discriminated. The resonators were thought to be set in action by sound-produced vibrations in the cochlea much as a cello string will be caused to resonate when a sound of appropriate frequency acts on it. Thus, different regions or places along the basilar membrane would be set in motion by different frequencies of sound. Helmholtz said we distinguish between two different pitches by distinguishing neural activity arising from nerve fibers connected to two different places on the basilar membrane.

We now know that Helmholtz's place theory is partially correct. Different regions of the basilar membrane are maximally activated by different frequencies. Place along the basilar membrane is a primary means of coding for pitch.

An alternative frequency theory held that the pitch of a sound was not coded by the place along the basilar membrane but by the vibration frequency itself. A 500-Hz tone was thought to drive some structure in the cochlea at five hundred times per second, and each excursion of this structure was thought to produce a single spike in an auditory nerve fiber.

In its simplest form this *frequency theory* of pitch cannot be correct because action potentials cannot be produced in a nerve fiber at a much greater rate than 1,000/s, and yet people can hear tones up to 20,000 Hz; some animals hear tones at frequencies over 100,000 Hz. Although this theory is wrong for high tones, there is some evidence for direct mechanical coupling from the basilar membrane to individual auditory nerve fibers that is related to the actual sound wave.

Central Auditory Pathways

There are a minimum of three synaptic relays from the first-order fibers of the auditory nerve to the cerebral cortex. Although we know something about the response properties of cells all along the pathway, research into auditory mechanisms has not yet reached the point where we can understand all of human hearing in cellular terms.

There is an orderly organization within the cochlea from high- to low-frequency response. If there were no crossing over of one fiber by another, then neighboring fibers within the auditory nerve would have similar "best" frequencies. An arrangement in which there is an orderly progression such that best frequencies change progressively from low to high is called *tonotopic organization*. Tonotopic organization is present in the lower relays of the auditory system, like the cochlea nucleus, and remains one of the basic features of organization at all levels in the auditory pathway. Although auditory response in the cerebral cortex is also organized tonotopically, it is weakly so.

The auditory system, like other senses, does not have a single connecting pathway. Even at the lowest level of the auditory pathways—the first relay in the cochlear nucleus—there are three independent subnuclei. Each of these three subnuclei—the dorsal, anterior ventral, and posterior ventral nuclei—receives an independent input from the auditory nerve, and each is tonotopically organized.

There is a core path from the inner ear to cortex responsible for fast high-fidelity conduction of simple auditory information. This fast system involves the anterior ventral cochlear nuclei. The posterior and dorsal cochlear nuclei, as well as other "accessory" auditory stations, up to and including cortex, contribute to descending fibers that modulate the afferent auditory information. These systems play a role in sharpening and tuning the system for selective sensitivity, such as listening to individual speakers at a cocktail party.

Sound Localization

There are other aspects of hearing that seem equally miraculous. People and animals use their ears not only to discriminate pitch and loudness but also *direction*. If you ask a friend to close her eyes, and then you snap your fingers, and ask her to point at the source of the sound, she will usually do this quite well. How does she know where the sound comes from? What are the physical features of the stimulus that can give the necessary cues for direction?

One possible cue is the relative intensity of the sound at the two ears. The head casts a sound shadow so that tones are louder on the side of the head on which they originate. Careful psychophysical measurement has confirmed that sound intensity differences are one important clue for sound localization. If two sounds of different intensity are delivered simultaneously to the two ears, we hear the sound as coming from the direction in which the sound was louder.

Compared to light, sound travels rather slowly. If a person's ears were 6 inches apart, a sound that originated 90 degrees from the right would arrive about one-half a millisecond earlier at the right ear. Half a millisecond would be the largest possible difference in time of arrival. If we changed the position of the sound, the difference in arrival time at the two ears would be less. Although the differences in arrival time at the two ears are very small, they are also cues for the direction of a sound. If clicks are presented to the two ears of a subject by earphones, say one-tenth of a millisecond apart, the sound appears to come from the direction of the earlier clicks.

How does the brain resolve such small differences in time of arrival and intensity? There is good evidence that a particular structure in the brain, the medial accessory olivary nucleus, plays a major role in processing auditory direction. Each neuron has a central cell body with dendrites emerging from it, projecting in two opposite directions. Studies of the input to the superior olivary nucleus from the cochlear nucleus found that fibers that arise from the left cochlear nucleus project to the dendrites on the left side of the olivary cells; fibers from the right cochlear nucleus project to the right side. The medial accessory superior olivary nucleus is a relay station with both a crossed and uncrossed input in the auditory pathway, and so it is in a position to compare intensity and time of arrival of a sound at the two ears.

Lesions of the medial accessory superior olivary nucleus cause a profound decrement in the ability of animals to localize sound, although animals with such lesions can still discriminate pitch or intensity of sounds.

Cochlear Microphonic, Auditory Brainstem Response, and Auditory Threshold

The experimental psychologist Glen Weaver discovered that if he placed an electrode on the round window, he could record an electrical potential with the same frequency as a sound with which he stimulated the ear. It was as if the inner ear were behaving like a microphone; hence, the phrase cochlear microphonic. This potential reflects generator potentials in the hair cells. The cochlear microphonic is a rough measure of the animal's hearing capacity.

More recently, noninvasive techniques have been developed to record the response of the auditory nerve from electrodes placed on the scalp and behind the ear. By using computer-based systems to average the synchronized response of the nerve to hundreds of tone pips at different frequencies, it is possible to define the minimal intensity of sound at different frequencies necessary to elicit a nerve response. The measure is called the auditory brainstem response or ABR. It maps the audiogram (behavioral auditory threshold function) quite well and is currently the measure of choice for screening for hearing loss in newborns.

Bats and Moths

Most bats fly and hunt at night. Although they do not see as poorly as proverbs relate, bats' vision is typically used only for crude recognition of terrain in the guidance of flight. Many small bats catch tiny flying insects, and such a demanding way of getting food requires good sensory mechanisms for detecting their prey—much better than are available by sight. Bats find and track flying insects by echolocation. While hunting, bats give off very high-frequency sounds that are normally well above our own threshold for hearing. If you could lower the frequency of a bat's calls so that they came within human hearing range, you would hear a hunting bat emitting a continuous stream of cries that serve as the stimulus for a kind of bat sonar. The cries are emitted as highly stable pulses, typically at a fixed frequency. The high-frequency calls are reflected from nearby objects and are detected by the bat.

Based on the difference in intensity and time of arrival at his two ears, the bat can detect the distance and direction of the target. And based on the shift in frequency (Doppler shift) of the echo, it can tell if the target is moving toward or away from it. Thus, bats make remarkable use of such high-pitched sounds to determine the location, direction, and even the flight path of their prey.

Donald Griffin, as a young zoologist at Harvard, studied how bats catch insects in a dark room. In one experiment he recorded the bat's cries, but he modulated the frequency so that the cries could be heard by the human ear. He also photographed the flight of the bat using specialized film. He found that as the bat came closer to its prey, the number of calls it emitted increased dramatically. From experiments like these, it became clear that the bat uses the echoes of its cries to find and follow the insects it feeds on.

Hunting bats use high-frequency sounds to catch moths. Some species of night-flying moths are prime targets for hunting bats. These moths have simple ears that detect the cries of bats. When the bat cries are first detected at a distance, the moth flies in a direction opposite to the source of the cries. It tries, as it were, to fly away as rapidly as possible. When the bat is closer and its cries are heard at a higher intensity, the moth takes direct evasive action. It may power dive downward toward the ground or fly with suddenly erratic changing of course like a fighter pilot in a World War II movie. These night moths have two simple and primitive ears, one on either side of the thorax, and each ear has only two auditory nerve fibers. The auditory nerve has different thresholds; the more sensitive of the two is activated by weak bat cries, and probably produces escape flight in a direction opposite from the source of the cries. When the higher-intensity auditory fiber is active, the moth takes more extreme forms of evasive action.

Cochlear Implants

Many of the features of the auditory system, particularly the encoding of sound by the inner ear, are reflected in a very practical way in the development of auditory neural prostheses—the cochlear implant. In the majority of cases of deafness, the primary pathology exists in the inner ear. Deafness or profound hearing loss typically is due to loss of sensory hair cells; some or perhaps much of the auditory nerve remains intact. The cochlear implant is a technology that bypasses the middle and inner ear to directly activate the nerves to the brain with encoded electrical signals. This device, now a series of electrodes mounted on a silicon substrate, is inserted into the inner ear, usually either through the round window or in a fenestra drilled in the bone adjacent to the round widow and then gently threaded up toward the helicotrema. Electrodes are distributed along the length of the implant. Because they are in a conductive fluid medium, they can more or less selectively stimulate groups of fibers that would normally be activated by hair cells tuned to different frequencies of stimulation.

The implanted electrodes are driven by current pulses from a device implanted in the mastoid bone behind the ear. This device is activated by radio frequency coupling to an external transmitter, which in turn is connected to a microphone in the ear canal. It looks like a hearing aid. Complex sounds arriving at the ear are analyzed by the microphone and analytic electronics into the sound's component frequencies, each of a specific amplitude. These are then applied as electrical pulses at a specific frequency and amplitude differentially to the electrodes along the cochlear spiral, with high frequencies activating basal electrodes and low-frequency information activating apical electrodes.

Although cochlear implants do not provide normal hearing, they provide speech discrimination, without lip reading, to the majority of profoundly deaf patients. For one thing they enable recipients to use the telephone. Patients who could previously hear report that after their implants everyone sounds like "Donald Duck," but this changes to normal-sounding speech in time. Infants born deaf and implanted within the first few years of life, some as young as six months, typically grow up with what appears to be normal hearing and speech. They enjoy normal reading skills, are mainstreamed in school with no special aids, and in many instances play musical instruments.

Psychophysical studies have clearly verified that loudness is dependent on amplitude of stimulation and the number of sites activated and that pitch is primarily encoded by place of activation and somewhat by rate of stimulation. The remarkable hearing abilities provided by cochlear implants is a testimony to the remarkable plasticity of the central auditory pathways and brain, which can extract useful information from a very crude signal and make it "normal."

10

Somesthesis and Vestibular Sense

This chapter deals with the sensations that arise from the surface of our body and the senses of position and movement. What receptors give rise to the sensations of touch, temperature, and pain? How are we provided with sensory information about the shape and surface of an object that we touch in the dark? How do we become aware of a pain in our foot if we step on a sharp object? How are we aware of the position of each of our limbs? What are the sensory inputs that assist in keeping us upright?

The chapter is about two kinds of senses. One kind signals touch, temperature, and pain from the surface of the body and the viscera. The other, the vestibular system, which is based in the inner ear, is critical for the control of posture and orientation in space.

The Skin Senses

Collectively the senses from our own body are called *somesthesis* or somatic sensation. There are several major attributes of somatic sensation that we hope to understand from a biological point of view.

1. Type: How many different kinds of somatic sensation can we experience?
2. Location: Where in the body is a sensation coming from?
3. Intensity: How strong is it?

Varieties of Skin Sensations

If you put your hand into a bowl of water that is below a certain temperature, it feels cold; a bowl of warm water feels warm. Similarly, we are aware of a gentle touch or a painful sting. Until the late nineteenth century the way in which these skin senses were coded was poorly

Figure 10.1
Magnus Blix (1849–1904) was a Swedish physiologist who studied the response of the skin to punctate stimuli. He found that sensations of cold, warmth, and touch are not uniformly distributed on the skin surface. Rather, small point-like spots are responsive to a single stimulus. Courtesy Germund Hesslow and Lund University.

understood. Was there some generalized receptor in the skin capable of reporting touch, pain, or temperature? Alternatively, were there specialized structures that were sensitive to a particular kind of sensation? In 1880 the Swedish physiologist Magnus Blix (figure 10.1), working at the University of Lund, did the simplest of experiments with rather unexpected results.

Blix tested the response of his own wrist and forearm to very gentle stimuli. His careful observations with very small probes uncovered a fundamental principle of sensory coding. Human skin is not uniformly sensitive. Rather, there are spots on the skin that are maximally sensitive to touch, warmth, cold, or pain.

Blix described the distribution of three classes of response to these weak stimuli. You can replicate Blix's observation on the skin of your own forearm by very light touch with the point of a pencil. It is, for example, usually easy to find small points that give the sensation of cold when touched. Cold spots are not distributed in the same places on the skin as spots that are sensitive to warmth. The two sensations are coded separately. Cold receptors may convey the sensation of cold even when they are activated by gentle prodding with a pencil. There are fewer warm receptors and more touch receptors. Blix seems to have spared himself more punishing stimuli that signal pain. As we shall see, there are a variety of specialized structures within the skin that help to tune it to different aspects of the stimuli. The so-called *nociceptive fibers* that signal pain typically end in the skin as free nerve endings with no special additional structures attached to them.

By 1900 the observations on the nature and distribution of receptors in the skin were replicated and widely accepted. The great physiologist Charles Sherrington (figure 10.2) took a strong interest in receptors. Sherrington, who was born in Islington, a suburb of London, and studied medicine at Cambridge, influenced and taught a whole generation of neurophysiologists. After a long and distinguished career, he was appointed Waynflete Professor of Physiology at Oxford University in 1913. Sherrington wrote the following description in Schäfer's *Text-book of Physiology* in 1900.

> Sense spots. The surface of the skin is found to be a mosaic of tiny sensorial areas. The elements of this mosaic are set not actually edge against edge as in the retina; between each element and its neighbours of like function extends a relatively wide interval, insentient when examined by stimuli of little above liminal intensity (Blix). The more locally limited and the nearer to the minimal the stimuli, the smaller appears each individual sensifacient field; ultimately by carrying the tests to further refinement, then individual fields may be reduced to mere "spots." Each of these "spots" is found to subserve a specific sense—touch, cold, warmth, or cutaneous pain. Each doubtless coincides with the site of some sensorial "end-organ," or with a tiny cluster of such. Rather, indeed, than to a mosaic may the skin be likened to a sheet of water wherein grow water-plants, some sunken and some floating. An object thrown upon the surface moves the foliage commensurately with the violence of the impact, its dimensions and their propinquity to its place of incidence. Where the foliage grows densely, not a pebble striking the surface but will meet some leaf; and beyond that or those directly struck, a number will be indirectly disturbed before equilibrium of the surface is re-established. Throughout almost all regions of the skin "touch spots," "cold spots," "warmth spots," and "pain spots" lie strewn [in] intermingled fashion. In some districts one variety predominates, in other districts another. On the whole, "pain spots" seem to be the most numerous, and certainly "warmth spots" are the least so.[1] (figure 10.3)

Touch (or pressure), warmth, cold, and pain are universally recognized as sensory categories, although each category may be subdivided further.

Figure 10.2
Charles Sherrington (1857–1952) was a major contributor to neurophysiology. His chapters in Schaefer's textbook of physiology are classics. Courtesy Wellcome Library, London.

In addition to sensations on the skin, many of the same sensations can be felt within our body—for example, pain, or a sense of warmth in the stomach or joints. In addition to skin and deep senses, we are also aware of the position of our body and limbs and of its motion in space.

The Structural Basis of Sensory Quality

The presence of spots on the skin that are especially sensitive to cold, warmth, touch, or pain led scientists to search for distinct types of receptors that could be recognized under the microscope and correlated with

Figure 10.3
Charles Sherrington's diagram of the response of the skin of the hand to gentle stimulation to show the punctate nature of the responses.

their function. In a typical experiment the heroic subject—usually the experimenter himself—was tested for the distribution of sensory spots on a small section of skin. The identified points would be marked with indelible ink and recorded. The patch of skin would then be anesthetized and surgically removed. The skin would now be sectioned for histological study in the hope of identifying a unique type of sensory ending that corresponded to each type of sensation. These brave experiments were often inconclusive; hence, they led some authorities to question whether there is any correlation between structure and function. But their doubts were premature. As we shall see, there is in fact good evidence for specialized functions for skin receptors with different morphology.

Much of our knowledge about the nature of sensory coding by the nervous system came from animal studies. In a typical experiment a recording would be made from a single sensory nerve cell or nerve fiber while the experimenter stimulated an area of skin. But using the results of these studies to interpret human sensation remained speculative. In the last half of the twentieth century it became possible to record directly from sensory nerves in human subjects. Tiny metal electrodes are driven through the skin and advanced into a sensory nerve.

In 1965 Åke Vallbo, then a young neurophysiologist, came to the University of Uppsala in Sweden to work with Karl-Erik Hagbarth, the head of the Clinical Neurophysiology Department. Initially they tested the recording method for safety, recording from their own nerves. It soon became possible for them to map out the region of skin that causes an individual sensory fiber to fire and to determine the kind of stimulus that is most effective.

The same nerve fiber could also be activated by a very low electrical current, and the subject was then asked to report the resultant sensation. These experiments cast new light on the relationship between different end-organs and the sensations that they code. It is likely that each of the specialized end-organs does, in fact, code for a particular type of sensation.

Sensory fibers that reach the skin and deep tissues usually branch repeatedly just before their ending. Some sensory fibers end in special encapsulated endings; some end as free nerve endings without any sort of specialized connective tissue covering the nerve terminal. There are a variety of specialized sensory encapsulations. The most striking of these is the Pacinian corpuscle, named for the Italian anatomist Filippo Pacini, whom we encountered in chapter 6. Pacinian corpuscles may be over a millimeter in diameter, about one hundred times the diameter of the nerve fiber that it covers. Pacinian corpuscles are exquisitely sensitive to slight mechanical stimuli. The attached nerve fiber responds to high-frequency mechanical stimulation. On the basis of its mechanical properties the Pacinian corpuscle clearly plays a role in the sense of vibration we feel when a vibrating object such as a tuning fork is placed against a joint.

Sensation type is correlated not only to specialized end-organs but also to the diameter of the nerve fiber. Mammalian sensory nerve fibers range in diameter from 20 μm down to 1 μm or less. The larger fibers are surrounded by a heavy myelin wrapping. Even the smallest, so-called unmyelinated fibers are bundled together so that several axons are covered by a single Schwann cell.

In addition to sensory input from the skin, there is also input from sense organs in muscle. Inside the muscles there are sense organs called *muscle spindles* (figure 10.4). Muscle spindles contain modified small muscle fibers that have a sensory nerve attached.

The largest-caliber fibers carry information from muscle spindles. Muscle spindles are directly involved in the stretch reflex (see chapter 11 of this volume) and help to signal the position and/or movement of limbs. The small myelinated fibers carry temperature information, and the larger fibers signal touch. The very finest unmyelinated fibers probably respond only to nociceptive stimuli: that is, stimuli that signal potential tissue damage and are associated with the sensation of pain (figure 10.5).

Organization of the Central Somatosensory Pathways

The cell bodies of the nerve fibers that innervate the skin and the internal receptors of the body are located in the *dorsal root ganglia* (figure 10.6) just outside the spinal cord.

The fiber of each dorsal root ganglion cell branches; one branch goes to the skin or deep tissue, and the other branch enters the spinal cord. Some of the axons[2] that enter the cord continue without synapsing all the way to the brain, and others synapse within the cord. In both cases the ascending somatosensory pathways are *somatotopically* organized. Neighboring points on the body surface are represented in neighboring points in the ascending spinal tracts and synapse on neighboring cells in the brain.

The great Spanish neuroanatomist Santiago Ramon y Cajal suggested a plausible way in which the mammalian system of sensory ganglia might have evolved (figure 10.7).

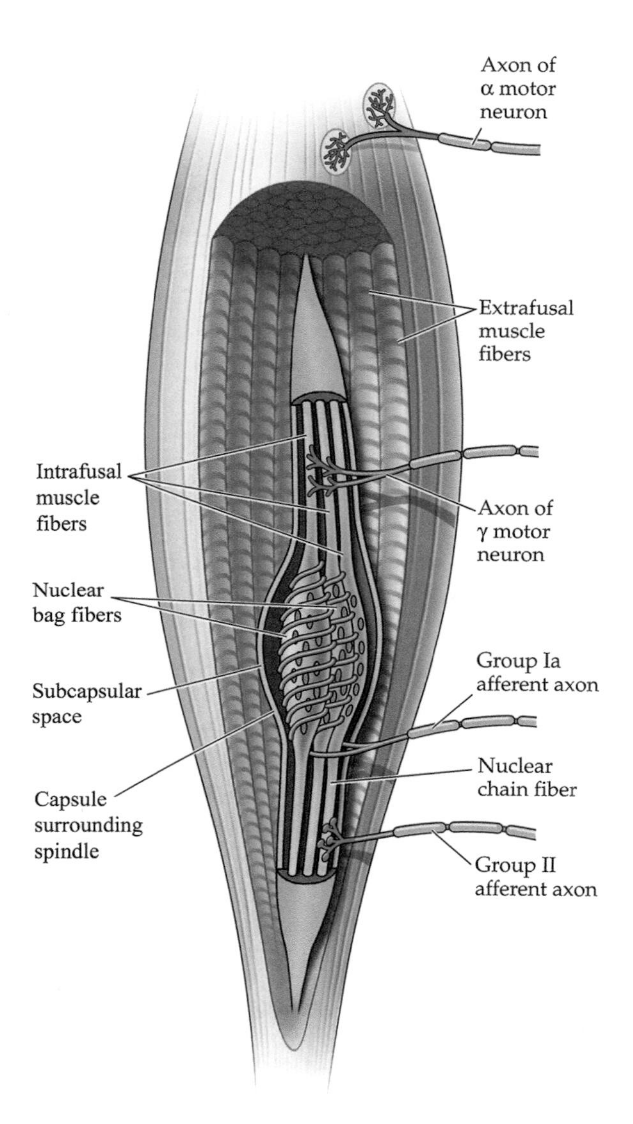

Figure 10.4
Embedded within muscles, muscle spindles are sensitive to the length of the muscle. They are important for reflex regulation of muscle length and awareness of a limb's position. Courtesy Dale Purves and Sinauer Associates.

Somatic Sensory Afferents that Link Receptors to the Central Nervous System				
SENSORY FUNCTION	**RECEPTOR TYPE**	**AFFERENT AXON TYPE**[a]	**AXON DIAMETER**	**CONDUCTION VELOCITY**
Proprioception	Muscle spindle	Ia, II (Axon, Myelin)	13–20 μm	80–120 m/s
Touch	Merkel, Meissner, Pacinian, and Ruffini cells	Aβ	6–12 μm	35–75 m/s
Pain, temperature	Free nerve endings	Aδ	1–5 μm	5–30 m/s
Pain, temperature, itch	Free nerve endings (unmyelinated)	C	0.2–1.5 μm	0.5–2 m/s

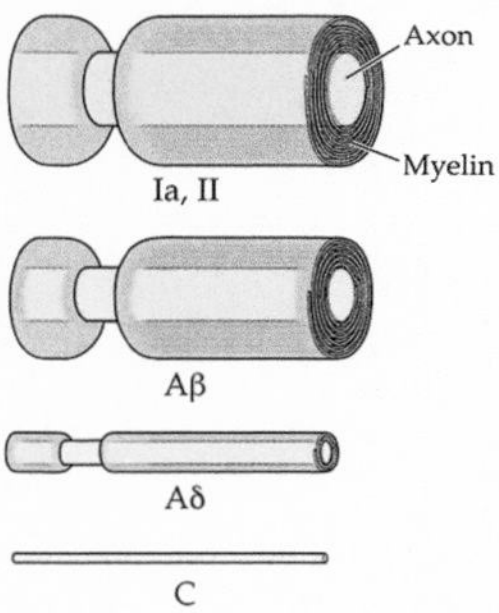

Figure 10.5
Functions of various caliber nerve fibers. Courtesy Dale Purves and Sinauer Associates.

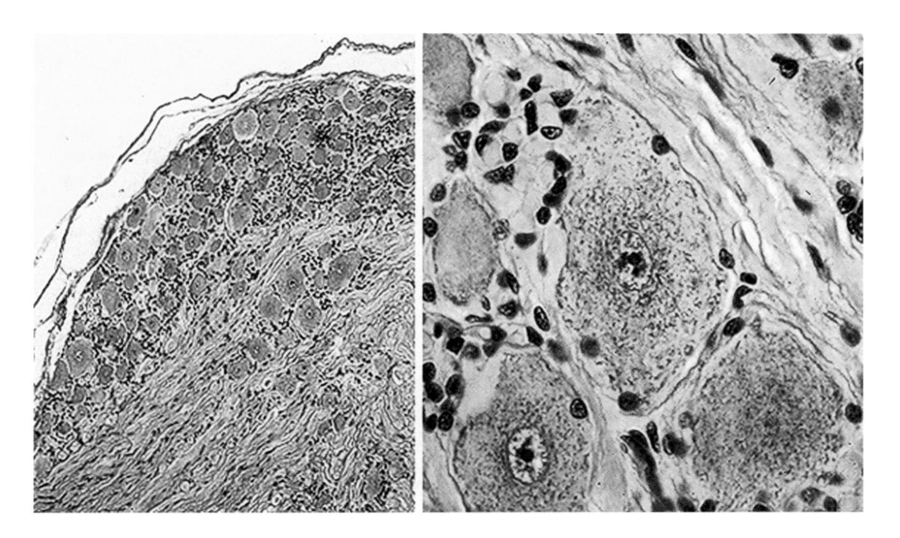

Figure 10.6
(Left) Low-power and (Right) higher-power Nissl stain. The dorsal root ganglion contains a group of sensory nerve cells that lie just outside the spinal cord. Its neurons give rise to a fiber that divides and sends one branch to the periphery; the other branch enters the spinal cord. Courtesy University of Washington.

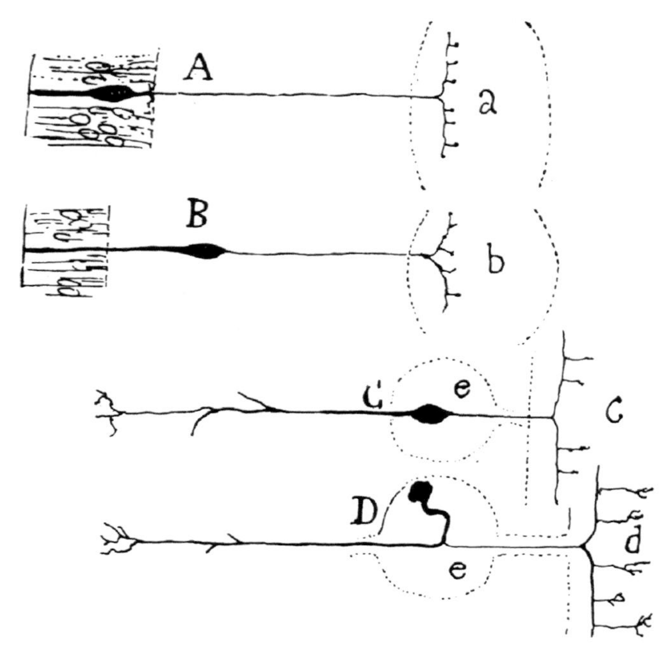

Figure 10.7
Cajal's scheme attempting to show how the anatomical arrangement of sensory fibers and their cell bodies might have come about. (A) Sensory cell body lying within the spinal cord. (B and C) Cell body migrates along the nerve fiber. (D) Sensory nerve cell bodies collect outside the cord as the dorsal root ganglion. Courtesy MIT Press.

Sensory Pathways

Two major pathways within the spinal cord bring sensory information from the skin and deep structures in the body to the brain (figure 10.8). The two pathways differ in the type of information they carry and their pattern of connection. One pathway, the dorsal columns, which includes the *gracile* and *cuneate* tracts, originates directly from dorsal root ganglion cells. Gracile and cuneate tracts are similar in function and signal light touch and position sense. Gracile tract fibers originate from the lower body, cuneate from the upper body. The cell bodies are in the dorsal root ganglia, and their axons ascend in the dorsal columns of the spinal cord. These fibers synapse directly on second-order sensory cells in the medulla. The other pathway consists of fibers that signal pain, temperature, and some touch. These fibers enter the cord and synapse on cells within the gray matter. Axons of these second-order cells in the

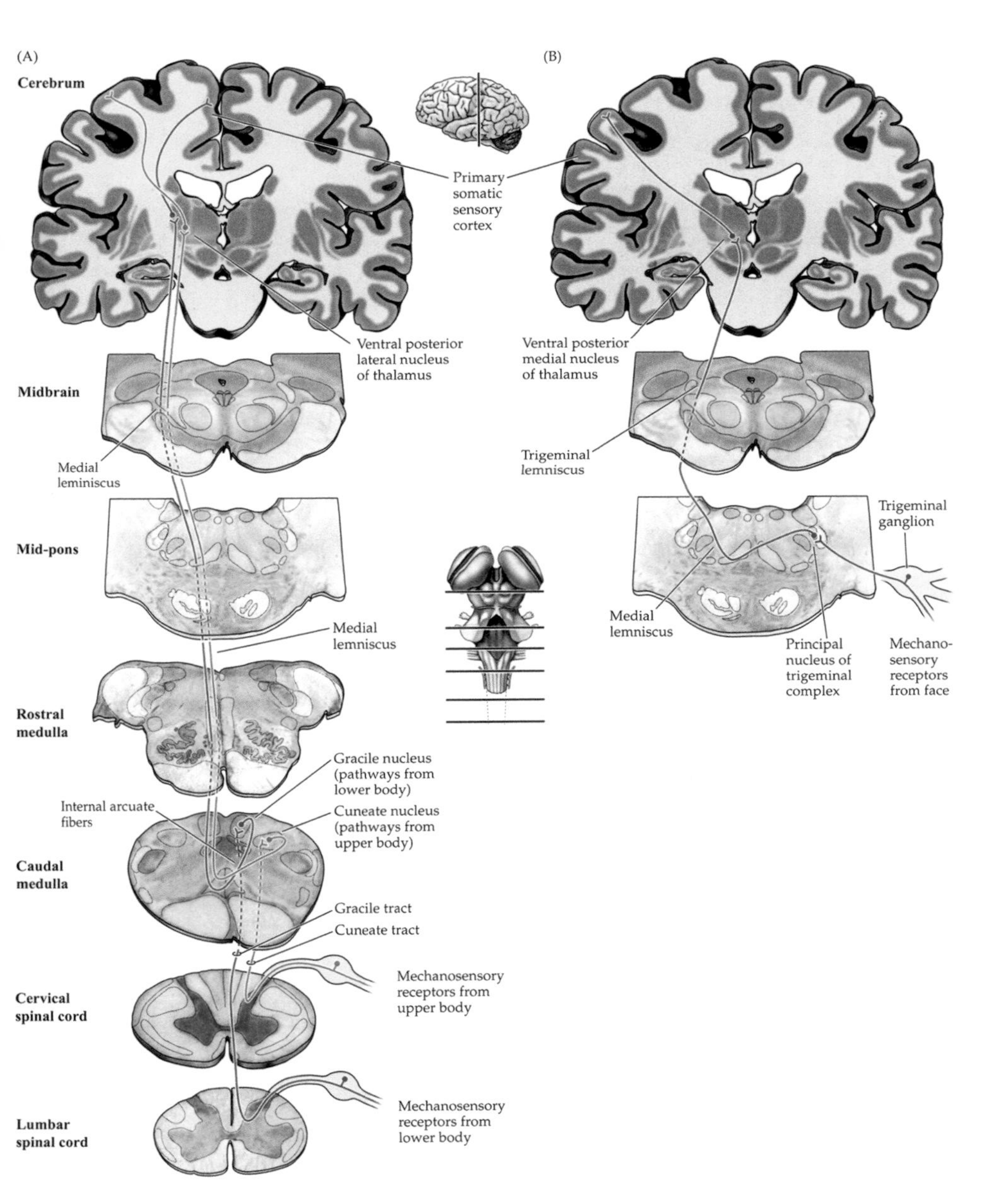

Figure 10.8
An outline of the main pathways conveying information from the body to the brain. Courtesy Dale Purves and Sinauer Associates.

gray matter give rise to the *spinothalamic* tract, either directly or after another synapse within the cord.

The thirty-two dorsal root ganglia distribute sensory fibers to the skin in an orderly pattern. The region of skin innervated by a single dorsal root ganglion is called a *dermatome.* The nineteenth-century German neurologist and neurosurgeon Ottfried Förster attempted to relieve spasticity by cutting dorsal root fibers in his patients. He was able to observe the distribution of sensory loss caused by those cuts. Each dorsal root ganglion serves a definite and restricted territory on the body surface, but the distribution of sensory nerve fibers from adjacent dorsal root ganglia overlaps with that from its neighbors. As a consequence, even if one entire dorsal root ganglion is destroyed or its fibers cut before they enter the spinal cord, no patch of skin would be completely without sensation. Neighboring patches of skin above and below would still have sensory nerve endings in them. In order to remove sensation completely from a given patch of skin, at least three adjoining dorsal root ganglia would have to be removed. Such a lesion would leave a patch of skin without sensation—the middlemost portion of the central one of the three dermatomes served by the three dorsal root ganglia.

The arrangement of the dermatomes is orderly, but in an upright human their arrangement can seem a bit puzzling. Their pattern is best understood if we imagine a person not upright but walking on all fours (figure 10.9). The dermatomes then show a regular pattern beginning at the back end with each successively higher dorsal root ganglion connecting to a region of skin closer to the head.

The Dorsal Column System

Some sensory fibers enter the spinal cord and continue in the long white columns on the dorsal side of the cord without synapsing. The fibers arising from the trunk and below are called the *gracile* tract. Fibers originating above this level are grouped together in the *cuneate* tract.

Fibers of the cuneate and gracile tracts ascend to the level of the medulla, where they synapse in two nuclei of the same name—the cuneate and gracile nuclei (also called the dorsal column nuclei). In a tall person a neuron might have a cell body in a dorsal root ganglion of the lumbar spinal cord, a sensory fiber ending in the toe, and the other process ascending in the gracile tract all the way to the level of the neck. The combined length might be almost 2 m.

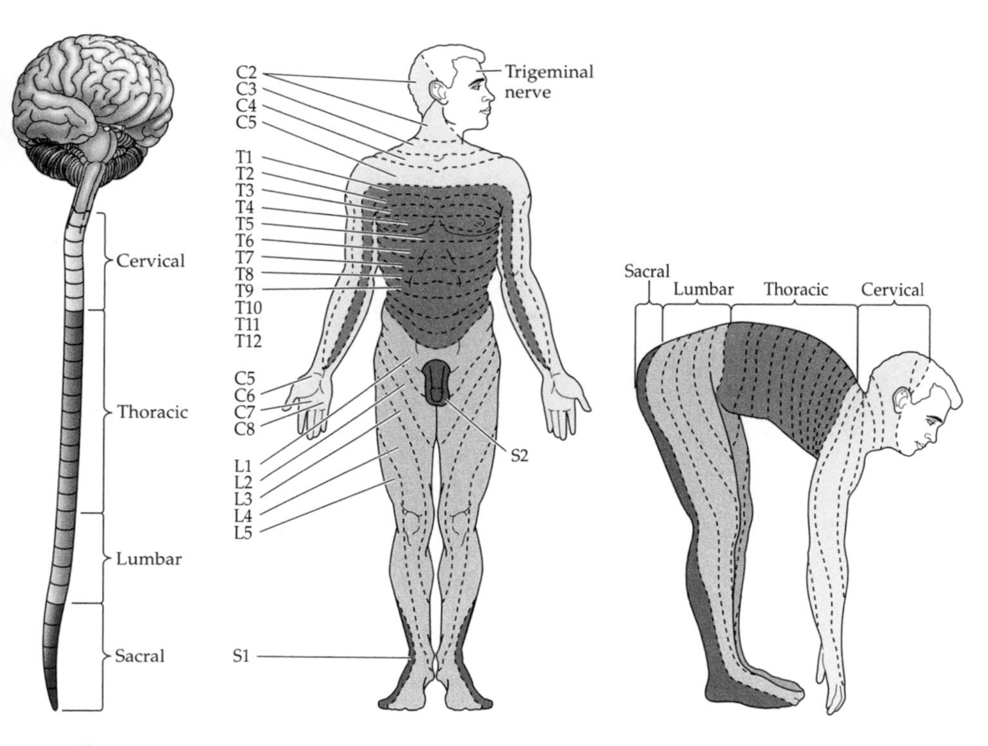

Figure 10.9
The sensory territory on the body surface of the successive dorsal roots. Courtesy Dale Purves and Sinauer Associates.

The gracile and cuneate nuclei are situated at the lowest level of the brainstem just above its junction with the spinal cord. Cells of the gracile and cuneate nuclei in turn send their axons in a fiber tract called the *medial lemniscus* that crosses the midline and ascends through the brainstem to the level of the thalamus where it ends in the ventral posterolateral (VPL) nucleus of the thalamus (figure 10.10). Fibers from the VPL in turn project to the cerebral cortex where they end in the somatic sensory cortex, just behind the central sulcus (figure 10.11). Thus, from the body surface to the sensory cortex there are three successive neurons in the pathway. The first is a cell in the dorsal root ganglion, the second a cell in the cuneate or gracile nuclei, the third in the thalamus. The dorsal column system principally relays information about touch or pressure on the skin surface—"fine touch" and position sense.

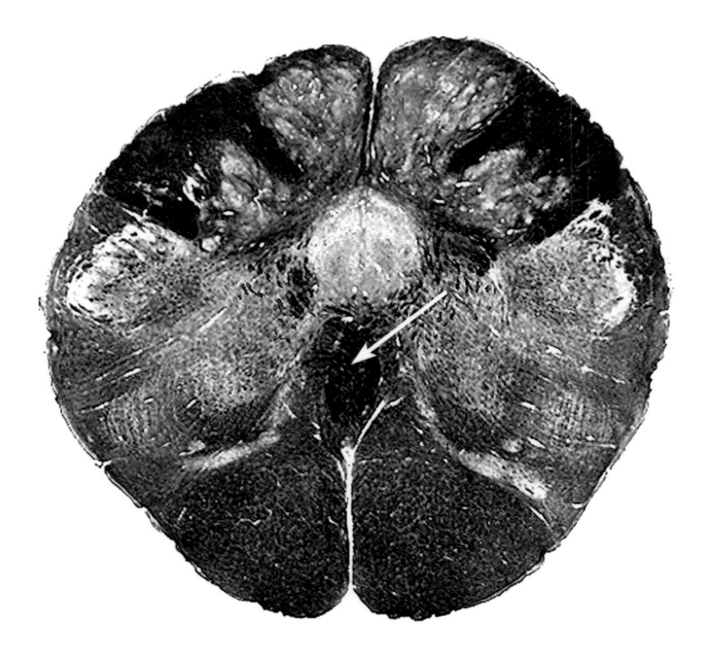

Figure 10.10
A section through the human medulla to show the crossing of the sensory fibers on their way to the thalamus. Courtesy University of Washington.

Somatotopic Organization within the Ascending Tracts in the Spinal Cord and Brain

The representation of the body surface within the dorsal columns is orderly. Caudal parts of the body are represented medially, and uppermost parts of the body are represented laterally in the dorsal columns. Spatial organization is maintained such that neighboring points on the body surface are represented at adjacent fibers within the tract. Spatial ordering of this sort is called somatotopic organization, and it is a prominent feature of sensory organization.

How does somatotopic organization come about? The first fibers to enter the dorsal columns arise from a region of the body surface that is very far caudal—the skin of the buttocks and the back of the legs. Fibers from these regions enter the cord at the low sacral level and ascend in the dorsal columns. Fibers from successive dermatomes enter the dorsal columns and ascend along with these original fibers. If fibers from successively higher dermatomes do not cross the axons that are already present, then successively higher parts of the body would be represented

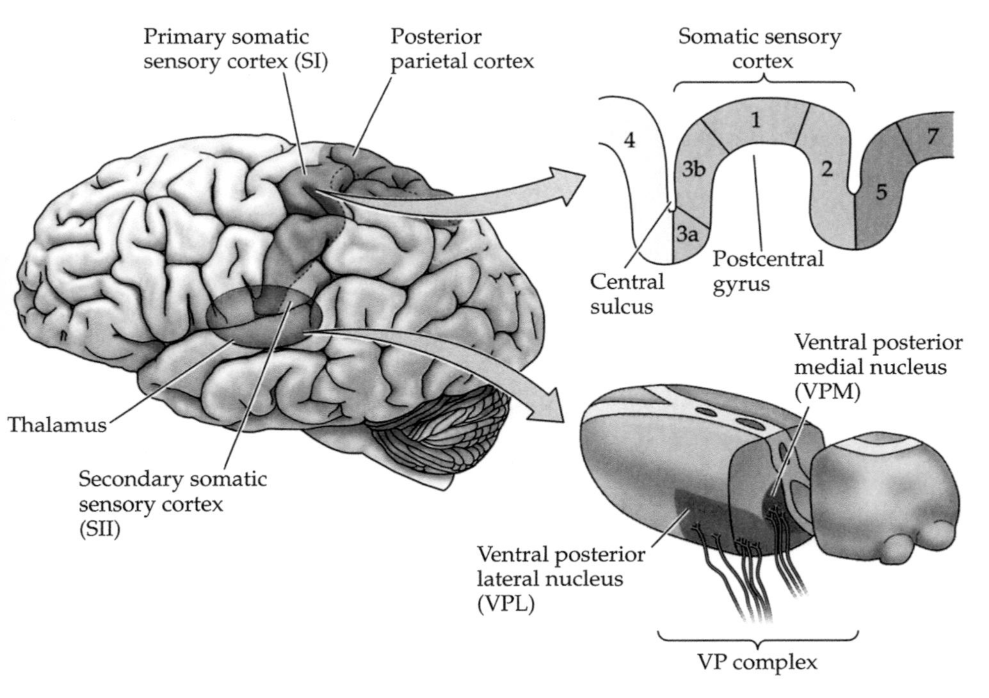

Figure 10.11
A diagram to show the way in which the sensory fibers from the body are represented on the surface of the cerebral cortex. Courtesy Dale Purves and Sinauer Associates.

at successively lateral portions of the dorsal columns. At the level of the upper cervical spinal cord, the entire body, from the neck down, is represented in the dorsal columns. Fibers from the neck region occupy the most lateral part of the dorsal columns, and fibers from buttocks and legs occupy the most medial. You can visualize the arrangement by analogy to a map of the eastern United States. Suppose that telegraph wires carried a message from southern Florida to Maine. As the wires extended north they would be joined by fibers from Georgia, South Carolina, and so forth, finally ending in Maine. If successive additions to the cable did not cross the wires already present, they would maintain a representation of a map of the eastern states.

The Spinothalamic Tract

The second major sensory pathway carrying sensory information from the skin and deep structures in the body to the brain is the spinothalamic system. Like the dorsal columns, the spinothalamic tracts begin with neurons whose cell bodies are in the dorsal root ganglia. One branch of the dorsal root ganglion cell innervates the skin and deep tissues, and the other enters the cord. Unlike the dorsal columns, these axons ascend only a short distance within the cord before they synapse onto cells in the dorsal horn of the spinal cord. Cells in the dorsal horn, either directly or by way of relays within the cord, give rise to the spinothalamic tract. Spinothalamic fibers cross the midline of the cord and ascend to the level of the thalamus. Whereas the dorsal columns signal light touch and body position, the spinothalamic system signals pain and temperature and some crude touch sensations.

The pattern of crossing or decussation of sensory tracts has important consequences for interpreting the effects of lesions to the spinal cord.

The Brown-Sequard Syndrome

The impairment of touch and position sense on one side of the body and pain and temperature on the other after hemisection of the cord is called the Brown-Sequard syndrome after the physician who first described it.

Charles Edward Brown-Sequard (figure 10.12) was born on the island of Mauritius to an American father and French mother. He took a degree in medicine in Paris and practiced as a neurologist and neurophysiologist during the second half of the nineteenth century. Brown-Sequard studied the effect of partial lesions of the spinal cord in experimental animals. His work showed that the pattern of crossing or decussation of sensory tracts within the spinal cord has important consequences for interpreting the effects of lesions to the spinal cord in humans.

In his initial experiments he saw that when the left or right half of the spinal cord is cut through completely, the opposite side of the body appeared to be lacking in sensation. He later found that sensory input on the same side as the cut is also affected, but with a different type of sensory loss. The crossed spinothalamic fibers signal pain and temperature as well as crude touch. The dorsal columns on the same side signal fine touch and position sense.

Brown-Sequard's experiments helped show the resulting sensory losses if someone were to be stabbed with the knife entering the spinal cord between two vertebrae such that the right half of the cord is cut completely through. Because of the arrangement of the fibers, the right dorsal columns would be severed, hence blocking the sensory information

Figure 10.12
Charles Edward Brown-Sequard (1817–1894) was a neurologist and scientist who studied the response to experimental lesions of the sensory pathways in the spinal cord. The pattern of loss after lesions in the center of the spinal cord that interrupt crossing fibers is still called the Brown-Sequard syndrome. Courtesy Wellcome Library, London.

carried by those tracts from all segments of the cord below the cut. Because the dorsal columns are not crossed, the impairment of touch and position sense would be on the same side of the body as the cut. The hemisection of the cord would also interrupt spinothalamic fibers, but these would be largely *crossed* fibers. Sensory input carried by the spinothalamic tracts—principally pain and temperature sensation—would be affected on the *opposite* side of the body. The impairment of fine touch and position sense on one side of the body and pain and temperature on the other after hemisection of the cord is called the Brown-Sequard syndrome.

Sensory Input from the Head

The pathways that we have considered so far arise from sensory fibers in the body below the neck. A similar system of sensory fibers innervates the head. There is a large sensory ganglion called the trigeminal ganglion, which is located just outside the skull at the level of the brainstem. The trigeminal ganglion contains the cell bodies of primary afferent fibers from the head. Fibers from the trigeminal ganglion enter the brain. Some of these fibers relay in the main sensory nucleus of the trigeminal nerve, which is the counterpart of the dorsal column system. Some fibers end in the spinal nucleus of the trigeminal nerve, which is the trigeminal counterpart of the dorsal root cells, with axons that contribute to the spinothalamic system. Fibers arising from the trigeminal nerve relay sensory information from the head and cross to synapse in the ventral posterior medial nucleus (VPM) of the thalamus on the opposite side.

It is as if the head now becomes tacked on to the rest of the body. The head representation lies medially, in VPM and the rest of the body laterally, in VPL. VPM and VPL together relay sensory information to the sensory cortex so that it contains a complete representation of the body surface relayed via the thalamus to the sensory area of the cortex.

Fibers from VPM and VPL relay sensory information to the cerebral cortex with the result that neighboring points on the body surface are represented at neighboring points on the cortex. But the body is not represented "democratically." Just as there are differences in the number of axons arising from central and peripheral retina there are great distortions in the territory in the brain devoted to different body regions. The skin of the back and trunk for example, although they occupy a vast expanse of body surface, are represented in a very small region of the brain. The sensory input from the fingertips and mouth has a much larger area of cortex devoted to their representation.

The mapping from body surface to the cortex shows interesting differences among species. Humans, monkeys, and raccoons have a large percentage of their sensory cortex devoted to their fingers. Rats and mice have a relatively large input from their whiskers. The orderliness of the sensory map is reflected in the way in which whiskers are represented. If the rat cortex is cut tangential to its surface and appropriately stained and magnified, definite circular regions become apparent in the region of sensory cortex that receives input from the whiskers (figure 10.13).

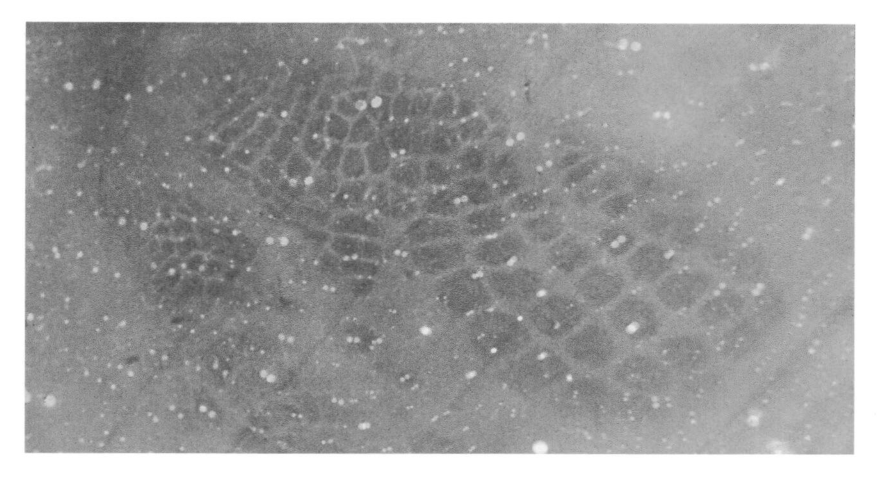

Figure 10.13
Cytochrome oxidase preparation of rat sensory cortex. The large squares are the region where sensory fibers from the whiskers and body surface terminate on the cerebral cortex. When viewed in three dimensions they seem to resemble barrels.

These circular regions extend through the middle layer of cortex in a roughly cylindrical shape and are called "barrels." The barrels are not of equal size. A collection of very large barrels appears toward the back end of the "barrel fields."

Mice and rats have a set of unusually large whiskers, called the *mystacial vibrassae.* Each of the especially large barrels receives its sensory input from a single whisker. If in a newborn mouse a single whisker is removed, a single barrel degenerates. The barrels represent a processing unit in which information arising from a single whisker is represented in cortex isolated from the representation of neighboring whiskers. Cells in the cortex just above and just below the cells of layer IV in the barrel fields integrate information from neighboring whiskers.

Mice and rats are dependent on their whiskers as a major source of sensory information. The whiskers are actively mobile; they are constantly being "whisked," and they are used to guide movement. Mice and rats will often walk along walls with the whiskers just touching the walls. It is clear that the large surface area of the sensory cortex of mice and rats devoted to whisker representation reflects the degree to which whisker information is used for guiding movements by the animal.

The Sense of Position and of Movement

We now consider the vestibular system—sense organs in muscles and joints and the remarkable sense organ in the inner ear that monitors position and movement of the head.

If we move a limb or if it is passively moved, we perceive its new position. We are constantly aware of the position of our body in relation to the ground underneath us. How do we know where our limbs are? How do we perceive it when they are moved? How do we distinguish whether our body is upright or lying down?

In order to try to answer these questions, it is useful to talk first about the sorts of messages that are carried to the brain. Physiologists distinguish between *afferent* nerve fibers and *sensory* fibers. Afferent fibers are *all* of those that carry information toward the central nervous system. Sensory fibers are one class of afferent fibers—those that give rise to conscious sensation.

We are not necessarily aware of activity in all afferent fibers. Increased pressure in the carotid artery activates afferent fibers that act reflexively to lower the blood pressure. The machinery works without our being aware of its operation. In some cases the distinction between afferent and sensory fibers may become blurred; it is not always certain whether the output of certain classes of afferent fibers contributes to conscious sensation.

The afferent messages that deal with position and movement of the body are among the most important we have. They are so much a part of our normal experience that we fail to appreciate these mechanisms. We do not think it remarkable that we can stand even though the constant pull of gravity requires that we have constant monitoring and adjustment of the contraction of muscles to maintain the body's position. Without precise regulation of limb and body position, humans and animals would be reduced to a helpless mass, unable to stand upright and unable to move. In certain neurological and muscular diseases we can see the drastic consequences of impaired position senses.

The sense of position can be broken down into two broad categories. The sense of position arising directly from receptors in the body is mediated principally through receptors in muscles, tendons, and joints. Another sense, arising from the bony labyrinth in the inner ear, signals the movement and position of the head.

Position Sense of the Body and Limbs

Ask a friend to close his eyes and extend his arm to the side with the index finger pointing. Now move his limb so that it points at some object on the wall. The object should be close enough so that you can tell accurately where the finger is pointing. Now, ask the subject to drop the arm back to his side and then to reposition it exactly in the same place where it was before. People can do this very accurately. Under optimal conditions a person can point to within an inch or two of the original target.

In a similar experiment ask the subject to close her eyes while you passively move a limb. For example, you grasp the forearm with one hand and the upper arm with the other and slowly extend the arm at the elbow. How slowly do you have to go before movement at the elbow is undetectable to a blindfolded subject? At some joints, especially the shoulder and hip, people are extremely sensitive to such passive movement; they can detect a rate of movement of as little as 1 degree/s of the shoulder or hip joints. Our sense of position and movement is better for proximal joints like the shoulder and hip than for distal joints like the fingers and toes. Thus, sensitivity to position and movement is different from that for resolving where we are touched. We are far more sensitive to resolving the precise location of where we are touched on the fingertips than at the shoulder.

How do we detect the position of a limb and its movement? For some years the muscle spindles were thought to be involved only in reflex regulation of the length of the muscle. The spindle was thought to act as part of a sort of power-steering mechanism for voluntary movement. But subsequent experiments have demonstrated that along with receptors in the joints they are also involved in the sense of position or movement.

Vestibular System

In addition to the cochlea, which contains the organ for hearing, the inner ear also has a set of specialized receptors that are principally involved in the senses of balance and posture and in the reflex control of eye movements (figure 10.14). These so-called vestibular senses are divided into two broad groups. One group comprises a pair of structures called the *utricle* and *saccule*. These structures contain a group of sensory hairs with crystals attached to them. The crystals are embedded in a

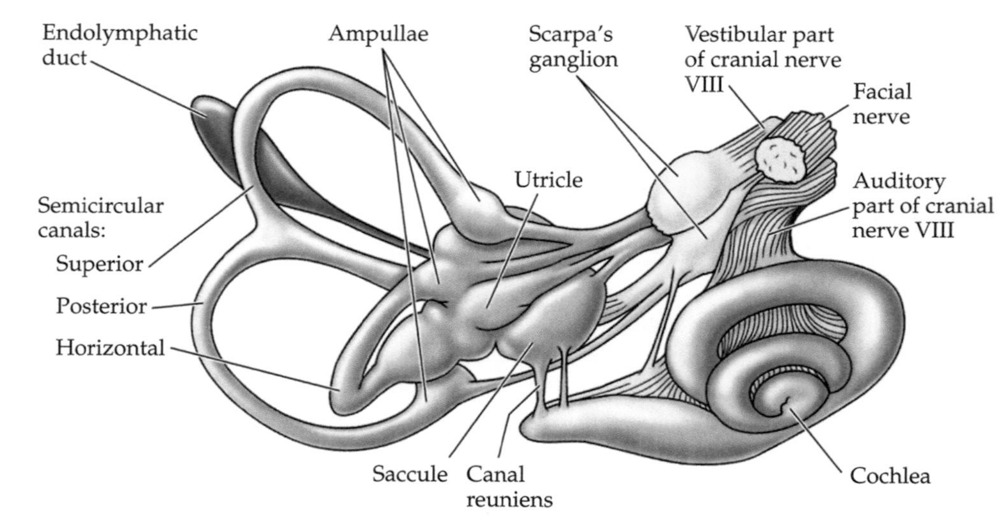

Figure 10.14
The main structures sensing balance and hearing. The three semicircular canals are all at right angles to one another and signal angular movement of the head. The cochlea is the sensory apparatus for hearing. Courtesy Dale Purves and Sinauer Associates.

gelatinous mass. If the head is tipped away from the upright position these crystals act as weights; they deflect the sensory hairs and signal the postural change.

A second group consists of three *semicircular canals* that, like the utricle and saccule, can detect changes in the angular movement of the head. The three canals are arranged at right angles to one another, and so they can detect movement of the head in any direction. Mostly we are unaware of the output of the vestibular system unless things go wrong. Seasickness is caused by excessive stimulation of the vestibular system. A most unpleasant disorder in the inner ear is called Ménière syndrome. People with Ménière syndrome typically become dizzy and disoriented. In extreme cases they may be unable to remain upright.

A pioneer in the study of the vestibular system in the inner ear was the Viennese physician Josef Breuer. Breuer was born in Vienna in 1842 and studied medicine at the University of Vienna. As a young physician he collaborated with Heinrich Ewald Hering, studying the reflex control of the vagus nerve in respiration. When the daughter of one of his patients (Anna O) developed hysterical symptoms after the death of her father, Breuer collaborated with a younger colleague, Sigmund Freud, in studying its mechanisms. After a time, Breuer left further work with the "talking cure" to Freud.

Although in private medical practice in Vienna, Breuer managed to study the anatomy and functions of the vestibular apparatus of the inner ear and its role in the sense of position and movement. One of the more important functions of the semicircular canals can be seen by a simple test. The vestibular canals are reflexly connected to the eye muscles. If you change the position of your head, your eyes have a tendency to move reflexively in an equal and opposite direction. If you are rotated in a barber's chair, the movement of the head induces activity in the semicircular canal. The semicircular canals by reflex connections cause the eye to move in a direction opposite to the rotation. The net effect of this reflex is to hold the visual world steady even when the head is moving.

The vestibuloocular reflex (VOR) is not there for barbers' chairs. The reflex serves to stabilize the visual world automatically in the absence of any conscious effort on your part, and it allows you to use your eyes under conditions when your head and body may be moving rapidly. To experience this, read this page aloud while moving your head back and forth as fast as you can from side to side. Now move the book back and forth at the same speed as your head was moving but keep your head perfectly still as you move the book. You probably will be able to read the text while your head is moving but not when the book moves.

If you replicate this experiment on a friend, you will be able to see why. As the head moves, the VOR stabilizes the position of the eyes so that they can remain on the page. No such stabilizer is working when the book is being moved. The VOR is like a gun-control apparatus on the deck of a ship. One necessary correction for accurate aim is a device that corrects the altitude of the guns for wave action. The elevation of the guns must be changed to compensate for the rolling of the boat caused by the motion of the sea.

Sometimes it is useful to be able to override the reflex. For example, imagine being rotated back and forth in a barber's chair. Now suppose that you want to read while you are being rotated. The VOR would now work *against* you, but you can consciously overcome it if you choose to hold your eyes still. The VOR used to be considered a pure example of a hard-wired and immutable reflex connection. But it is highly modifiable; that modifiability is one of the best examples of plasticity in the nervous system and is discussed more fully in chapter 16 of this volume.

11

Chemical, Heat, and Electrical Senses

Most sense organs respond to *physical* stimuli in the world around us, such as pressure, light, or sound. Some animals have specialized senses that respond to heat or electricity. Smell and taste are unique in that they are *chemical* senses. Odorants are substances carried in the air that we breathe. We taste flavors by the chemical properties of substances in the foods we eat or the liquids we drink. Smell and taste are independent senses in humans. They are not as distinct, however, in some animals. Some fish, for example, have taste-like chemical sensors located on the surface of their bodies; these sensors enable perception of substances dissolved in the water in which the fish swims. Such taste receptors function in a way similar to the mammalian sense of smell.

The Sense of Smell

The sense of smell is not essential for humans. People who are completely unable to smell can get on well enough in the world. For some animals, however, especially certain insects, the sense of smell is particularly important. Pacific salmon, for example, are born in fresh water rivers and streams and migrate to the sea in the first year of their life. The salmon spend most of their lives as free-swimming animals in the ocean. In the fourth year of life, they return to spawn in the very place from which they were first hatched. Their sense of smell guides the homing salmon back through the many branches and tributaries of the rivers to the place where they were born. Salmon remember each turn and junction by its distinctive smell on the outward journey. They read this remembered "smell-map" on their way home.

Powerful effects of the sense of smell can also be seen in the reproduction of rodents. Pregnant female mice can distinguish the distinctive odor of urine of the father of the litter they are bearing from that of all other males. If a pregnant mouse smells the urine of a strange male, she aborts. If the odor is that of the father of the litter, the pregnancy is unaffected.

The receptor cells for odors in mammals have fine processes that send one end into a mucous layer. The other end of the olfactory cell has an axon that connects the cells to the brain. In man, olfactory sensory cells are found in a small mucus-covered surface at the very top of the nasal passage, about the size of a postage stamp. Humans have about five million such olfactory sensory cells. Rabbits have a much greater area containing olfactory receptors with an estimated hundred million or so receptors. Bloodhounds can have as many as two hundred million or more olfactory receptors.

Given that the human eye has three distinctly different cone receptors, could there be a similar fundamental set of receptors for smell? Early scientists attempted to identify a basic set of odors that could be used to classify all possible odors. Linnaeus, the great Swedish botanist who had published the first systematic classification of all plants, attempted to do the same for smells. In 1756 he listed seven basic odors, which seemed to be related to their degree of pleasantness/unpleasantness, from fragrant to repulsive. Linnaeus proposed seven odoriferous qualities: aromatic, fragrant, musky, garlicky, goaty, repulsive, and nauseous.

Linnnaeus's scheme was simply a list of odors with different properties. An early attempt at a more systematic classificatory scheme was the 1916 smell prism of Hans Henning, with six corners labeled putrid, ethereal, resinous, spicy, fragrant, and burnt. All smells were thought to be combinations of these six elementary odors.

Although the prism was a convenient scheme for classifying odors, it is somewhat arbitrary and far from being universally accepted. Several alternative schemes have been proposed, but there is no universal scheme whereby odorants can be classified. The nature of quality coding—how people distinguish one odor from another—remains only partially understood.

One promising attempt toward developing a classifying scheme for odors is to look for people who are not able to smell certain substances. When tested for their ability to detect a particular odorant, these people demonstrate a threshold much higher than that for the rest of the population. One such smell is the smell of sweat; some people usually do not smell it. Such people when tested can detect sweaty odors only at very high concentrations. By analogy with deficiencies of color vision, studies of specific anosmias provide clues to finding a scheme for classifying olfactory receptors. Although the problem of how we code odor quality is not solved, there is abundant anecdotal and laboratory evidence for great sensitivity in the smell of some substances. Humans are so sensitive to the smell of ozone, for example, that only a few molecules may be enough to activate a single olfactory receptor.

Olfactory Sensitivity in Insects and the Power of Pheromones

The word *pheromone* comes from a Greek stem: *phero*, to carry, and *(hor)mone,* to excite or arouse. Hormones are chemical substances produced in one organ and transported to another organ where they produce a specific effect. Pheromones are hormones carried between two individuals, usually members of the same species.

In the nineteenth century the brilliant French naturalist Henri Fabre found a female moth in an oak leaf and brought her into his home. He placed her in a cloth cage with an open window and found to his astonishment that within a short time sixty male moths had collected around the cage. He repeated the experiment with another species of moth and found the same results. Dozens of males were attracted to the female. Fabre did some simple experiments to try to determine how the males of these moth species, which were normally very scarce, were attracted to the captive female. He found that if the female were in a sealed cage, even though visible, the males did not appear.

Fabre's experiments, although brilliant, led him to the wrong conclusion. He argued that it was unlikely that the female could be releasing some sort of odorant to attract the male moths. He believed that the concentration of the odorant released must be so weak it would be completely undetectable by the male a short distance away. Fabre felt "that one might as well expect to tint a lake with a drop of carmine dye." Fabre suggested that perhaps some sort of vibration from the female was the critical stimulus for attracting the males.

Contrary to Fabre's reasoning, it is now certain that female moths attract the males by an olfactory cue. The female silkworm moth *Bombyx* releases a pheromone so powerful that it may be detected by the male of her genus many miles away. The scent of the pheromone serves as a cue that guides the male toward her. Pheromones have been found in many species, and they can influence a variety of behavioral responses.

Adolf Butenandt, a leading German biochemist, discovered the pheromone in 1959 and named it bombykol. Bombykol is powerful stuff. The effective concentration of bombykol on the moths' antennae may be as few as a hundred molecules per cubic centimeter of antenna surface to produce an effect. But the receptors occupy only a small percentage of the surface of the antenna. It seems likely, therefore, that only a very few molecules of bombykol are sufficient to attract the male. The moth can then home in on this scent and follow it to the female.

Other species of moth use pheromones in a similar way. Although the chemical structure is similar, it is distinct enough that pheromones only work on male members of the same species. Pheromones can be remarkably specific chemicals. If only one atom in the molecule is in the wrong place, the chemical does not attract male moths. One of the useful applications of this sort of research is in the use of attractants for insect control. If a sex attractant lures all of the males of the species, fields can sometimes be cleared of insect pests in a few years.

Sense of Taste

Taste, like smell, is a chemical sense. The tongue contains many taste buds that are located in specialized structures called papillae. The papillae are identifiable as dark red spots on the tongue. Each taste bud contains several taste receptor cells. A nerve links the taste receptors to the brain.

The primary receptors for taste are somewhat better understood than for those for smell. There is good evidence for a basic set of taste qualities—sweet, salty, sour, and bitter. The existence of distinct taste qualities is suggested by the specific action of certain chemicals. For example, the leaf of the plant *Gymnema sylvestre* contains a substance that temporarily abolishes sweet taste. If you rinse your mouth in a weak solution made from the plant, an astonishing thing happens; you lose the sense of sweetness. Sugar becomes tasteless and feels like sand. The other tastes—salty, bitter, and sour—are unaffected. The loss of sweet taste lasts for an hour or so and then recedes. One interpretation of this effect is that there is a specific receptor for sweetness that is blocked competitively by the molecules of gymnemic acid.

Another substance has a totally different effect on the perception of sweetness. A plant called "miracle fruit" (*synsepalum dulcificum*) changes the sense of taste in an even more remarkable way. For a period of time after miracle fruit is chewed, sour substances taste sweet as well as sour. A lemon eaten after miracle fruit tastes like sweetened lemonade, both sweet and sour. Miracle fruit alters the sweetness receptors in such a way that for a time they can be activated by sour substances as well.

Some tastes seem to provoke a reflex-like response in the facial muscles. There is a Facebook website showing babies tasting lemon. After a single drop the infant makes an instantly recognizable "sour face." The same unmistakable sour face response can be elicited in newborn babies, just a few minutes old.

These observations are strong evidence for a small underlying set of fundamental taste qualities. But just as cones in the retina, although maximally sensitive to a single wavelength, also respond to light of nearby wavelengths, so taste receptors may respond to more than a single taste.

Taste in Flies

Flies like the taste of sugar. In his wonderful little book *To Know a Fly*, the entomologist Vincent Dethier described the way in which a fly chooses its diet of sugar and how to fool it. Dethier had received his PhD from Harvard in 1939, and was in the U.S. Army Air Corps throughout the Second World War. Some of his research on flies was conducted while he was in the Army's Chemical Corps. The common house fly has two independently located sugar detectors: a powerful one on its foot and a second on its proboscis. If the fly's foot is immersed in a sugar solution, the fly pushes out its drinking apparatus. If the sugar detector on its drinking apparatus tastes sugar, then the fly drinks until its little stomach is full. Dethier describes how if the foot is kept immersed in the sugar solution, the fly will continue to drink even if the other receptor encounters a bitter solution that it would normally reject. The input from the receptor on the foot is stronger.

Senses We Do Not Have: Thermoreception in Snakes and Electric Location by Fish

In 1937 G. K. Noble and Arthur Schmidt were working at the Museum of Natural History in New York City. They described a series of fascinating experiments with two kinds of snake—rattlesnakes and boa constrictors. For example, they placed a rattlesnake on a platform on which it was free to move and to strike at objects in front of it. They then confronted the snake with a pair of light bulbs. Neither light bulb was lit at the time, but one had just been turned off and so was slightly warmer than the other. The rattlesnake would strike repeatedly at the warmer of the two light bulbs. The snake did not have to see or smell the light bulb to strike at it. Even if blindfolded and even with forked tongue removed prior to the test, the animal struck accurately at the warmer of the two light bulbs as the bulbs moved slowly in front of them. The heat radiating from the light bulb was the essential stimulus for guiding the strike. All other possible senses were excluded. Essentially similar results were obtained with boa constrictors.

What is the sense organ that snakes use for detecting the warm bulb? In the boa constrictor there is a row of temperature-sensitive nerve endings that run roughly parallel to the lips. In the rattler there is a specialized organ like a small pit (hence the name the pit-vipers) that is heavily supplied with nerve endings (figure 11.1).

Some years later, in 1952, Professors T. H. Bullock and R. B. Cowles of UCLA recorded electrical activity of the nerve endings that end in the pit organ of the rattlesnake. They showed that tiny amounts of heat markedly affect the firing rate of nerve endings in the pit. If a warm object were held in the region of the pit, the average firing rate went up. If a cold object were held near the pit organ, the firing rate went down. The sensitivity of the snake's facial pit is remarkable. The human hand,

Figure 11.1
Face of a western diamondback rattlesnake. Between the mouth and the eyes are two prominent pits, which contain highly sensitive heat detectors. Courtesy Dylan Cebulske.

when held at a distance of more than one foot, was enough to activate receptors in the pit organ. They calculated that the sensitivity of individual receptors within this specialized pit was a small fraction of a degree of temperature difference between an object and its surroundings. The usefulness of such a paired organ in an animal whose diet consists of small, warm-blooded mammals is obvious. Even at night the snake would only have to line up its head so that the input from the two pit organs was equal and then strike.

The special senses of animals may be extreme examples or extreme modifications of senses that humans have. *Bombyx* can smell a few molecules and use it as a sexual attractant. The rattlesnake can strike at a mouse in total darkness. Although we do not directly experience these senses, we too smell odors and feel warmth. A much harder sense for us to understand is the electric sense.

Certain species of fish produce an electrical discharge. The electric eel, which produces a powerful electrical discharge of several hundred volts that can stun or kill prey, has been known since ancient Greek times. There are other fish species that use electricity in a subtler way, guiding locomotion and detecting objects in space. One of these species, the African knife fish *Gymnarchus niloticus*, was studied by Hans Werner Lissmann, who discovered the electric sense in fish while working at Cambridge University. Before the Second World War Lissmann had spent many years analyzing the way in which animals move, from snails and toads to snakes. While studying the swimming mechanism of the fish *Gymnarchus,* he established that the creature uses a weak electric signal to detect prey and predators in its vicinity. Lissmann concluded, "I am now convinced that *Gymnarchus* lives in a world totally alien to man. Its most important sense is an electric one different from any we possess."

Some of the fish with weak electric properties can be bought in an aquarium store. The elephant nose fish (figure 11.2) produces an electric discharge from an electrical organ in its tail. When the electric organ discharges, a current flows through the surrounding water. In calm water with no objects around it, there will be a symmetric flow of current from the tail to the head. Most of the current will go through the water immediately along the fish's side from tail to the head. If an object whose electrical resistance is different from that of water is in the vicinity of the fish, the electric field would be distorted. If the object had a lower resistance than the water around it, more current would flow through the object. If the object had a higher resistance than the water around it, less current would flow through it.

Figure 11.2
The elephant nose fish is in a class of weakly electric fishes. A 6-volt signal is constantly generated from an electric organ near the fish's tail, and there are detectors along the side of its body that act something like a radar screen, detecting objects in the vicinity of the fish. Courtesy Jennifer H. Smith.

The fish has along its body surface a number of specialized electric receptive organs. A complex set of nerve tracts puts this information together to provide the fish with an electrical view of the world. Perhaps the closest analogy for such a sense is a radar screen. Most mammalian sense organs are associated with a highly developed and elaborate development of the cerebral cortex. Curiously, in the case of the electrical fishes, the organ responsible for receiving these responses is the cerebellum—a structure vital in human movement (see chapter 13 of this volume).

12

Reflexes

We and other animals have an array of several senses and a complex array of muscles. One of the fundamental jobs of the nervous system is to link sensory input to appropriate muscle output. The simplest of such arrangements is the reflex.

Most of the body mass of humans and other mammals is made up of muscle tissue. Muscles are the flesh of poets and playwrights; the steaks and chops of butchers and waiters. The body has a large bony frame, the skeleton, with joints that allow a part of the body to be moved relative to all of the other parts. Muscles are attached across each joint and provide the machinery to move that joint. Consider a relatively simple joint (figure 12.1). Two bones meet at a joint; two muscles attach across the joint by means of tendons inserted into the bones.

Contraction of one of the muscles will make the angle between the two bones smaller. Contraction of the other muscle will increase the angle of the joint. Such a relatively simple arrangement of bones and muscles in which there is only a single direction of movement is present at only a few places in the body, such as the elbow or distal parts of the fingers. A more complex set of muscles and attachments is required to produce the greater range of movement at joints such as the shoulder and the wrist.

Muscles, when activated, contract. It is the orderly contraction of skeletal muscles that produces the full range of human movement from the relatively simple movements at the hips and knee when we walk to the precise and delicate sequencing and synchronization between the fingers of the left hand and the wrist and shoulders of the right arm, which is required for playing the cello. Muscles of mammals, when activated, can only contract; typically, they pull the bones on which they act in a single direction. In order to produce a range of movement around even the simplest joint, there must be at least two muscles arranged on

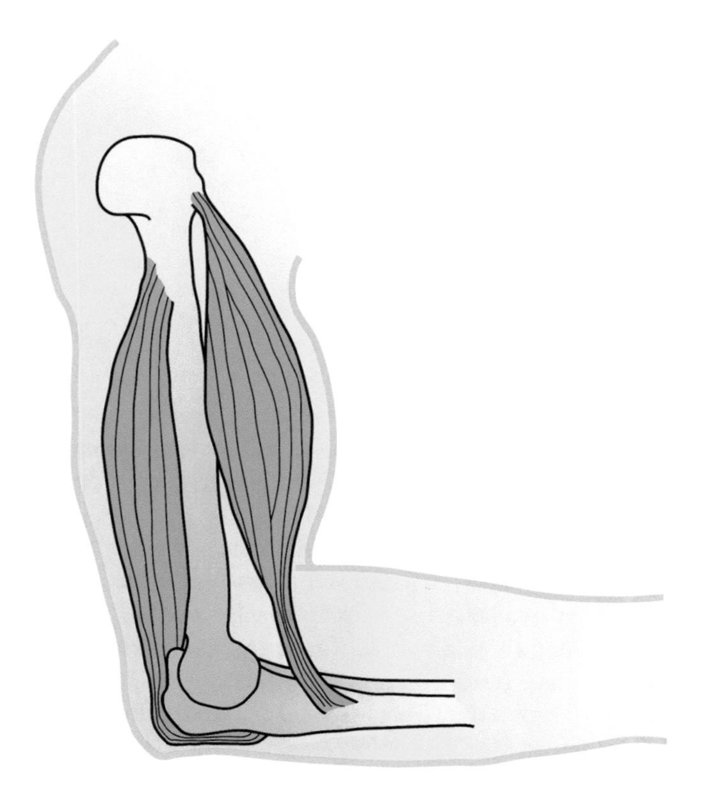

Figure 12.1
In this simplified diagram a single flexor and a single extensor muscle are illustrated. The elbow, and all other joints in the body, actually have more muscles and more complex arrangements of their insertions.

opposite sides of that joint. When the biceps muscle of the arm contracts, it decreases the angle at the elbow, thus flexing it. The triceps muscle, located on the opposite side of the upper arm, has a reverse effect and extends the arm at the elbow. If a muscle is principally responsible for a given movement, we call the muscle that produces the movement an *agonist* for that movement. The biceps muscle is an agonist muscle for elbow flexion. The muscle that opposes the movement is called the *antagonist*. Triceps is an antagonist for flexion at the elbow and an agonist for extension at the elbow.

The human body has three sorts of muscles: striate, smooth, and cardiac. Striate or striped muscle is so called because of its appearance under the microscope. Biceps, triceps, and all of the muscles that span joints, move the eyes, and vary our facial expressions are striated muscle.

Figure 12.2
The figure shows a motor axon dividing as it approaches a target muscle and ending in a characteristic structure on the muscle, the motor end plate.

Smooth muscle forms the major part of structures like the stomach and gut. The cardiac muscle of the heart is in a class of its own. Under the microscope it looks like striate muscle, but unlike striate muscle elsewhere it can function autonomously. The heart can beat without a nerve supply. Since hearts can beat without a nerve supply, they can be transplanted.

All voluntary muscles are of the striated kind and are under direct excitatory control of the nervous system. In normal life those muscles require an input from a motor nerve to contract (figure 12.2). The smooth muscles, such as those in the gut, and the cardiac muscles are involuntary. Involuntary muscles also have an input from the nervous system, but they do not require that the nerve be active for the muscle to contract.

The spinal cord is encased within the bony vertebrae. There are thirty-two pairs of nerves emerging from its dorsal and ventral sides. Although it had been clear for many years that these nerves are necessary for sensation as well as movement, there was no understanding about how they were organized. A critical advance was the recognition that the dorsal spinal roots are sensory and the ventral roots motor. It was the Frenchman François Magendie (figure 12.3) who produced the firm evidence. Magendie was a physician and scientist trained in Paris, and he remained there for the remainder of his life. In a paper published in 1822 he

Figure 12.3
Despite Charles Bell's claims, it was François Magendie (1783–1855) who first showed conclusively the sensory nature of dorsal roots and the motor function of fibers in the ventral root of the spinal cord. Courtesy Wellcome Library, London.

described an experiment on a young dog in which he cut the dorsal roots going to one hind limb. The animal lost sensation in that limb, although it still could still move. Although it was more difficult to cut the ventral roots, Magendie managed to do it, sparing the dorsal root. The limb became immobile, but sensation was preserved. He wrote:

> I repeated these experiments on several species of animals; the results that I have just announced were confirmed in the most complete manner. Both for the anterior and the posterior limbs . . . it is enough for me to announce today as positive, that the anterior and posterior roots of the nerves that arise from the spinal cord have different functions; that the posterior appear more particularly related to sensation, and the anterior to movement.[1]

Charles Bell (figure 12.4), an anatomist living in London, claimed credit for the discovery; hence it is known as the Bell-Magendie law, which states that the dorsal spinal roots are sensory in function; ventral

Figure 12.4
Charles Bell (1774–1842) was one of the investigators who studied the functions of the dorsal and ventral spinal roots. Courtesy Wellcome Library, London.

roots are motor. Although Bell claimed to have made the same discovery ten years earlier, his claim was based on a letter he had written to his brother, a route to fame and recognition that seems to have become less commonly used. The evidence strongly supports Magendie's priority for the discovery of the functions of the dorsal and ventral roots.

The spinal cord in cross section is shaped rather like a butterfly, with prominent dorsal and ventral divisions, called dorsal and ventral horns (figure 12.5). Motor neurons are located in the ventral horn of the spinal cord and in the motor nuclei of the brainstem.

The axons of motor neurons in the spinal cord exit by way of the ventral roots of the cord to end on muscles. The axons of motor neurons in the brainstem exit from the brain by way of cranial nerves to supply

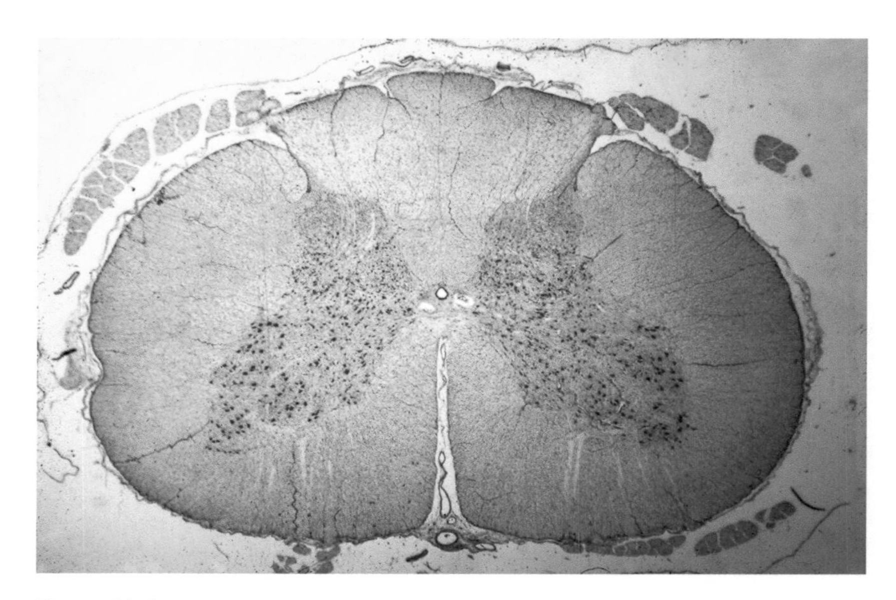

Figure 12.5
A cross section of the cervical spinal cord stained for cells: dorsal on top; ventral below. The nerve cells are contained within the butterfly-shaped region at the center of the cord. The paler-staining areas outside contain the fiber tracts.

muscles in the head. There is a rough correspondence between the *dermatomes*, the region of the body that is supplied by a single sensory root, and the *myotomes,* the set of muscles supplied by the motor fibers from the same segment of the cord.

When the axons of a motor neuron reach the muscle, they terminate at specialized junctions called motor end plates. Just before the axon terminates, it breaks up into many smaller branches with each branch forming its own motor end plate. A single motor axon may in this way connect to hundreds of muscle fibers. The number of muscle fibers supplied by a single axon is called the *innervation ratio*. The innervation ratio is related to the delicacy of control that can be exerted by a muscle. In the large muscles of the trunk and thigh, a single motor axon typically connects to hundreds of muscle fibers. In muscles that control the position of the eyes or the fingers, the innervation ratio is very low, associated with a much greater precision of movement.

Because the motor neuron and the axon that arises from it are the last step in the pathway by which the nervous system controls muscles, the English neurophysiologist Sir Charles Sherrington called the motor neuron and its axon the "final common path." When a motor neuron fires, the action potential is conducted down its axons and activates all of the muscle fibers to which it is connected, causing contraction of the muscle.

What causes a motor neuron to fire? There are two broad classes of inputs to motor neurons. Some connections involve local circuits within the spinal cord and serve reflex functions. The cell bodies of the neurons that provide such local connections to the motor neurons are either in the spinal cord or they are sensory neurons whose cell bodies are in the dorsal root ganglia. Motor neurons are also activated directly or indirectly from the brain by way of long descending pathways. We can appreciate these two kinds of connections by considering the case of a person or animal whose spinal cord is cut—unfortunately, a relatively common event during wartime, or a tragic consequence of a motorcycle accident. The Assyrian kings were lion hunters, and their deeds were sometimes recorded by artists of the court. The figure of the injured lion in chapter 1 shows an early but accurate view of the effect of spinal transection.

Immediately after the cord is cut there is a complete paralysis of the muscles of the body that are controlled by spinal motor neurons below the cut. If the spinal cord is cut in the midthoracic region, voluntary movements of the legs would cease instantly and never return. The motor neurons in the lumbar spinal cord that connect to and control the activity of the leg muscles may not be damaged, but the pathways by which the brain activates those neurons are permanently severed.

If a patient with a cut spinal cord is given adequate nursing care, some movement may return. There would be an eventual restitution of some reflexes in response to local sensory stimulation in the body below the level of the cut spinal cord. Within a few weeks the leg might be withdrawn in response to noxious stimulation of the foot. In a few months stretch reflexes might return; the leg might extend briskly at the knee if the quadriceps muscle is stretched by tapping the patellar tendon that inserts in the shin just below the knee cap. The "knee jerk" may now appear in an exaggerated form. Because the descending motor pathways from the brain to the spinal cord do not regrow, it is clear that some neuronal circuits are left, since reflex movement survives. Much of our knowledge about reflex control comes from the study of men and animals in which the spinal cord had been cut.

Spinal Reflexes

The French philosopher and scientist Renee Descartes recognized that there must be a link that carries the sensory input to the spinal cord or brain and then back to the muscle.

Inspired by the elaborate fountains of France, Descartes proposed a hydraulic linkage between sensory input and motor output. A painful input would be communicated by a hydraulic channel to the brain, and the muscles would withdraw the leg (figure 12.6). There was at the time some confusion about whether muscles get bigger or smaller when they contract. It was often assumed that muscles must expand when activated. The English scientist Jonathan Goddard addressed the question experimentally. He read a paper at the Royal Society in 1669 on his experiment and its observations:

> A case was made of latten capacious and convenient to receive immers'd in water the Arme of a man, so as the larger orifice or entrance into it might be stopped close by the part of the arme next the shoulder, with a small glass pipe cemented to it—towards the other end, opening into the cavity. Upon putting in of water first a little warmed, and afterwards of the arme, so as it closed the wide orifice of it, and the water did rise in the small glass canale, first it was visible that the water rose upon every pulsation of the artery, and subsided upon every intermission: and then the person being ordered to make a contraction of clutching his fist of both arms, that within the case and that without at the same time; upon every such contraction the water in the glasse canale did descend much more than upon the intermission of the pulse before mentioned.[2]

Goddard's experiment makes it clear that muscles do not expand in volume when they contract.

Reflex and Voluntary Movement

Although Descartes did not make an explicit distinction between voluntary and reflex withdrawal of the limb, the idea of a reflex is implied in his writing. The idea of an involuntary sensory-induced movement that might take place automatically was put forward explicitly by the Scottish physician and scientist Robert Whytt (figure 12.7).

Whytt dismissed the idea that involuntary reactions could be organized entirely within the muscle itself. He made a detailed study of the pupillary response to light demonstrating that the response must involve a link from optic nerve via the brain back to the muscles controlling the size of the pupil.

Whytt listed a number of such involuntary reactions. Although the term *reflex* to describe these involuntary reactions was not yet invented, most of Whytt's involuntary reactions would be recognized today as reflexes. Anatomical evidence was poor, so Whytt was necessarily somewhat vague on the exact pathways involved in reflexes. Whytt was a

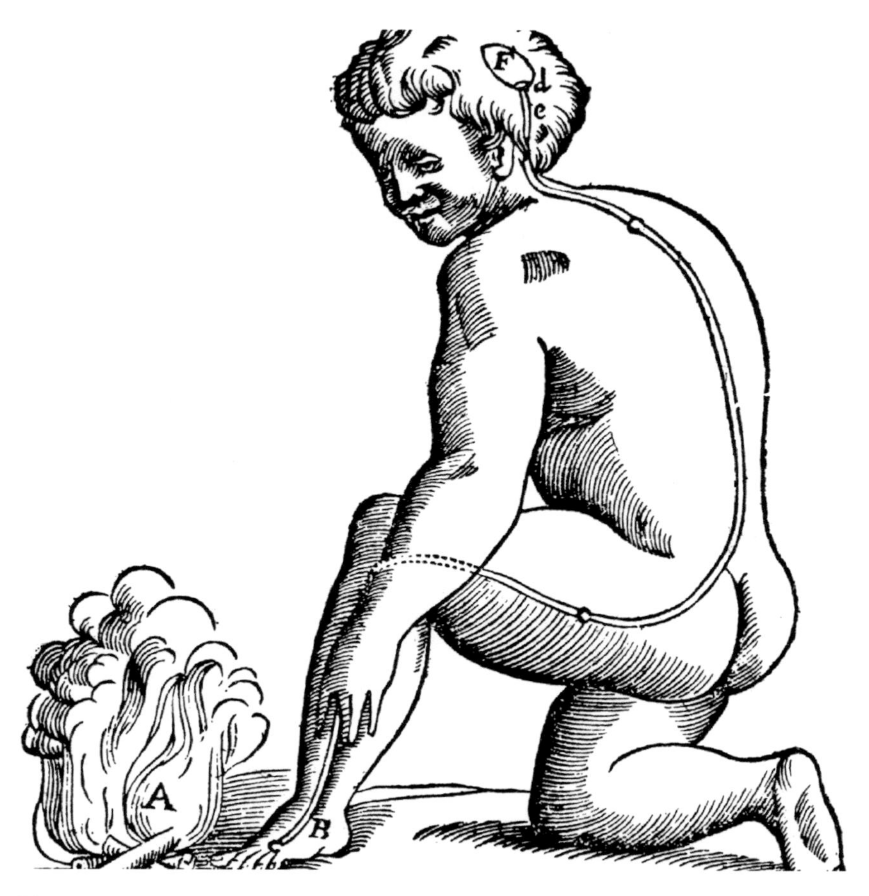

Figure 12.6
An early, hydraulic view of reflex functioning. A sense organ in the foot relays the information to the brain, which in turn sends fluids to inflate the muscle, withdrawing the limb. The figure is from René Descartes' book *Traité de l'homme (Treatise of man)* published in 1664.

AN

ESSAY

ON THE

VITAL and other INVOLUNTARY

MOTIONS of ANIMALS.

By ROBERT WHYTT, M.D. F.R.S.
Phyſician to his MAJESTY,
Fellow of the Royal College of PHYSICIANS,
AND
Profeſſor of Medicine in the Univerſity of *Edinburgh*.

Inanimum eſt omne quod pulſu agitatur externo; quod autem eſt animal, id motu cietur interiore et ſuo. Nam hæc eſt propria natura animi atque vis. ——Quæ ſit illa vis, et unde ſit intellegendum puto. Non eſt certe nec cordis, nec ſanguinis, nec cerebri, nec atomorum.

CICERO. Diſput. Tuſcul. lib. I.

The ſecond Edition, with Corrections and Additions.

EDINBURGH
Printed for JOHN BALFOUR.
M,DCC,LXIII.

Figure 12.7
Robert Whytt (1714–1766) was a Scottish physician and scientist who had the idea of a reflex, although he was not the one who coined the term. Portrait courtesy Wellcome Library, London.

Scot, which perhaps accounts for one of his few errors. He suggested that the sound of bagpipes leads to a reflex tendency to make urine.

Whytt clearly had the idea of a reflex, but he did not coin the term. That was left to Marshall Hall (figure 12.8) in the mid-nineteenth century. Hall was born in Nottingham, England, and studied medicine at Edinburgh. However, he practiced medicine and did his studies of reflex mechanisms in London.

The term reflex was based on the concept of a mirror-like arrangement, whereby the input would be reflected from the spinal cord or brain, much as a mirror reflects light. Hall also coined the term "reflex arc" to emphasize the dependence of reflexes on a connection to and from spinal cord or brain.

Hall showed that many responses must be organized at the level of the spinal cord since they could still be elicited when the cord was isolated from brain, or even in segments of an animal cut off from the rest of the body.

Figure 12.8
Marshall Hall (1790–1857) coined the term "reflex." He showed that reflexes are based on connections within the spinal cord and that they can even be elicited from an isolated segment of the cord. Courtesy Wellcome Library, London.

As we noted, severing the spinal cord causes immediate, complete, and irreversible paralysis in those regions of the body whose muscles are controlled by motor neurons that are located below the cut. If the cut occurs very high in the cervical cord, it is fatal. Without the descending tracts that control the muscles that expand the chest, we cannot breathe. The motor neurons would now no longer control the orderly increase and decrease in the volume of the chest, which is breathing. If the cut were lower in the cord, the motor neurons controlling the muscles in the chest would retain their descending control from the brain. In the case of a midthoracic cut, breathing and use of the arms and fingers would be preserved. Leg motor neurons would escape direct damage, as they are located below the cut, but they would immediately lose all control from descending pathways. An initial state of spinal shock would follow in which reflex functions of the cord would be depressed. The neurons of the lower cord, deprived of tonic facilitation from descending tracts, would be in a state of lowered excitability. Reflexes that are controlled by local circuits within the spinal cord would at first be difficult to elicit. But in humans and other mammals the spinal cord can recover local reflex functions. Reflexes would survive without influence from descending pathways from the brain. In the case of a person or animal whose spinal cord was cut in the midthoracic region, stretch and flexion reflexes would recover.

The Stretch Reflex and Timing at the Synapse

Study of stretch reflexes illustrates how the nervous system controls the simplest sort of motor functions. The clinical use of reflexes as diagnostic tools for determining the integrity of spinal cord and brain circuits followed. In papers published in the same year, 1875, Wilhelm Heinrich Erb (figure 12.9) and Karl Friedrich Otto Westphal, both German neurologists, described the knee jerk and its manifestation in neurological disease and in healthy subjects.

The knee jerk is based on a stretch reflex. An examiner strikes a subject's knee just below the kneecap with a rubber hammer, and the leg kicks upward. What is the nature of the stimulus, and what muscles produce the response? The rubber hammer striking the knee pushes down on the patellar tendon, which puts a sudden stretch on the quadriceps muscle. The response is a brisk contraction of the same muscle. The act of tapping the knee delivers a sudden stretch to the tendon and its attached muscle.

Erb and Westphal disagreed in their interpretation of the basis of the knee jerk. Their argument boiled down to a question of latency, the time that elapses between the stimulus and the response. In a controversy that lasted for many years, Westphal argued that the response of the leg to stretch of the muscle happened far too quickly for the message to get from the knee to the spinal cord and back. Many respectable physicians and scientists shared Westphal's view that the knee jerk is a purely local response of the muscle to stretch (figure 12.10).

Latency, the timing of nerve and muscle action, has always been a major factor in the analysis of reflex circuits. It was the analysis by Charles Sherrington and his colleagues that demonstrated the reflex nature of the response to stretch. If either the dorsal (sensory) roots of

Figure 12.9
Wilhelm Heinrich Erb (1840–1921), a German physician, was among the first to describe the knee jerk, the extension of the foot when the knee is tapped. Erb affirmed that it is based on a reflex, as opposed to direct mechanical effect on the muscle. Courtesy Wellcome Library, London.

Is the Knee-jerk a Reflex?—The most interesting question in this connection is whether the jerk is a true reflex act or is due to a direct mechanical stimulation of the muscle. Opinions are divided upon this point. Those who believe that the jerk is a reflex lay emphasis upon the undoubted fact that the integrity of the reflex arc is absolutely essential to the response. The quadriceps receives its motor and sensory fibers through the anterior crural nerve, and pathological lesions upon man as well as direct

Figure 12.10
Although it had been clearly described by Erb in 1875, the question of whether the knee jerk is reflex was not settled for fifty years. This discussion of the issue is from a 1905 medical textbook.

the spinal cord or the ventral (motor) roots are cut, stretch reflexes are abolished. Sherrington concluded that the knee jerk requires an afferent message from the leg to the spinal cord and an efferent message to the muscle from the spinal cord. The efferent message from the muscle is relayed within the spinal cord so that it activates motor neurons that control the quadriceps femoris muscle. Sherrington raised the questions: How does the muscle signal stretch? And how is that signal input channeled within the spinal cord? What sort of connections are there within the spinal cord from the entering sensory fibers to the exiting motor neurons? These questions were explored and answered by analysis of stretch reflexes.

The Stretch Reflex in Decerebrate Cats and Timing at the Synapse

Sherrington, working in the early twentieth century, made an intensive study of stretch reflexes in cats in which all of the brain from the midbrain upward had been removed—the so-called *decerebrate cat.* He studied both the acute initial response to sudden stretch and the response to chronic stretch. When the muscles of a decerebrate cat are stretched, they show a prompt and powerful tendency to shorten. He arranged the experiment so that one end of the muscle is fixed and the other could be stretched over a measured distance. Because the muscle was held rigidly it could not shorten, but when its motor neurons were activated it exerted a force that could be measured. Sherrington and his junior colleague Edward Liddell analyzed the tension that a muscle develops in response to stretch.

Suppose that the muscle were a purely passive structure like a rubber band. Figure 12.11 develops the idea of the relationship between length and tension in an intuitive way. Suppose that we suspend a weight from a spring. We could plot a graph for the length of the rubber band as a function of the weight that is connected to it.

Alternatively we could plot the same data the other way round. How much weight do we have to add to achieve a given length of the rubber band?

Like a spring, muscles develop additional passive tension when they are stretched. But the tension that a muscle develops is much greater than the purely passive response. Liddell and Sherrington compared the response of the stretched muscle before and after its sensory or motor nerve had been cut. When these nerves were cut, all reflexes were abolished and the muscle behaved passively; it developed no more tension

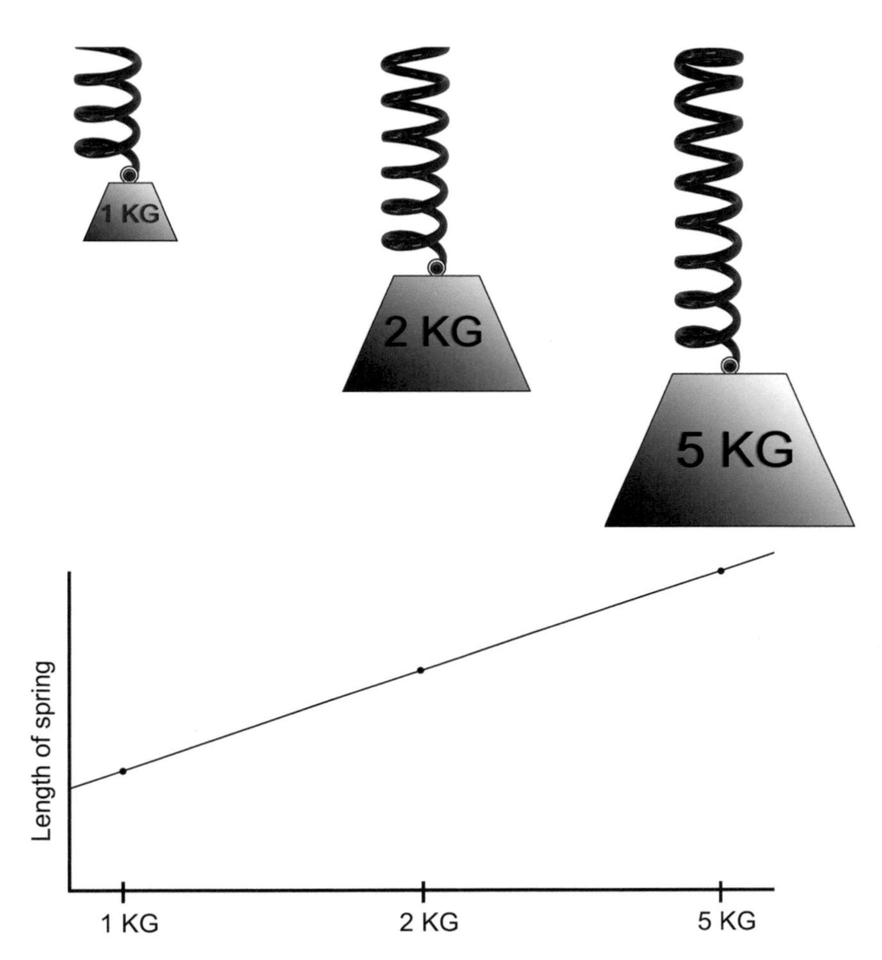

Figure 12.11
If a weight is suspended by a spring, the heavier the weight, the longer the spring becomes. The same would be true of a muscle unconnected to the spinal cord. An essential step in understanding stretch reflexes was to disentangle the purely passive elastic properties of muscle from reflex response to the stretch (see figure 12.12).

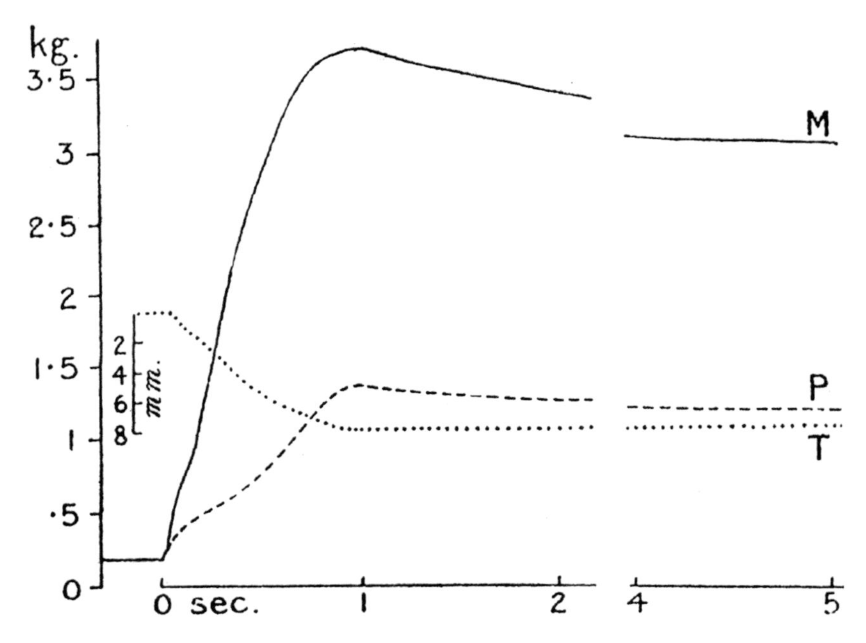

Figure 12.12
From Liddell and Sherrington's 1924 paper on myotatic reflexes. A graded amount of stretch on a muscle, with a measure of the force of contraction elicited. T tracks the length of the muscle stretched to an additional 8 mm in length. M reflects the reflex contraction caused by the stretch. P shows the purely passive, spring-like response in a muscle in which the attached nerve has been cut. Courtesy of the Royal Society.

than it would have in a dead animal. The difference in response between a muscle that is unconnected to its nerve supply and a muscle with its motor and sensory connections intact is a measure of the strength of the reflex response to stretch. Most of the tension that was produced when the muscle was stretched was not passive tension but was caused by active reflex shortening of the muscle. Sherrington published their findings on the stretch reflex in 1924 (figure 12.12).

The Receptor for the Stretch Reflex

Sherrington proceeded to analyze which receptor is activated to produce the stretch reflex. There were, initially, three candidates: skin sensors, tendon organs, and muscle spindles. Removing the sensory afferents from skin or tendon left the stretch reflex intact. Because neither skin nor tendon organs are necessary for the stretch reflex, Sherrington concluded that the crucial input for the stretch reflex must arise from sense organs or receptors within the muscle itself. This line of evidence suggested strongly that the important receptor for the stretch reflex was likely to be the muscle spindle, a specialized receptor within the muscle.

Muscle Spindles

Most skeletal muscles have within them a specialized sensory structure called the muscle spindle (see figure 10.4), a group of six or so specialized muscle cells contained within a spindle-shaped capsule of connective tissue. There are two sorts of sensory endings in the muscle spindle. One of the sensory fibers is called the *annulospiral* or primary ending. The annulospiral ending is a sensory fiber whose terminal wraps itself in a spiral fashion around the middle of a modified muscle cell within the spindle. Sensory fibers that form annulospiral endings are large, as much as 18 μm in diameter for a cat muscle spindle. The secondary ending in the muscle spindle has a smaller-caliber axon and terminates in a different portion of the muscle spindle.

When a muscle is stretched, the muscle spindles that are contained within that muscle are stretched along with it. Stretch on the muscle spindle pulls on the sensory endings and lengthens them, much as a pull on the two ends of a spring lengthens it. Stretch is the adequate stimulus for activating both the primary (annulospiral) and the secondary endings in the spindle. The muscle spindle signals stretch of the muscle in which it is embedded. The primary, annulospiral ending is especially sensitive to the change of length. It fires briskly just after the muscle is pulled. The secondary ending responds primarily to the static length of the spindle.

When a muscle is stretched, the sensory fibers from the spindle produce an afferent volley—a burst of action potentials that is conducted to the spinal cord by the sensory fibers. The primary sensory ending in the muscle spindle is the terminal part of a sensory fiber that has its cell body located in a dorsal root ganglion, just outside the spinal cord. One branch of the sensory fiber goes out to end in the muscle spindle; the other branch enters the spinal cord through the dorsal root.

We now have a set of necessary facts for asking further questions about the neural organization of stretch reflexes. There is an afferent signal when the muscle is stretched. This afferent signal must somehow connect to motor neurons of the same muscle since stretch of a given muscle leads to prompt contraction of the same muscle. How does the connection made from the sensory input to the motor neuron produce the stretch reflex?

Timing at the Synapse and the Response to Stretch: The *Myotatic Reflex*

First, the conclusion, and then, the evidence. One of the major connections from the spindle afferent fiber is to the motor neurons of the same muscle in which the spindle is located. Stretching a muscle produces a burst of afferent activity in the sensory ending of the spindles contained within the muscle. The afferent fibers so activated connect directly, that is, *monosynaptically* onto motor neurons of the same muscle. This synaptic activation of the motor neuron causes it to increase its firing rate, thus activating the muscle and shortening it. Sudden or sustained stretch on a muscle causes it to shorten. The stretch reflex is one of the basic "hard-wired" elements of the vertebrate nervous system.

How do we know that the stretch reflex is monosynaptic, and *why* is there a stretch reflex anyway? In his textbook, *Histologia del sistema nervioso*, Santiago Ramon y Cajal, the Spanish neuroanatomist, illustrated some of the sensory fibers that enter the spinal cord directly contacting motor neurons of the ventral horn (figure 12.13). The idea of a monosynaptic connection from sensory fibers onto motor neurons was plausible on a purely anatomic basis, but Cajal's methods could not specify the source of the sensory input or the sort of information that these incoming fibers might carry. A first question to be answered was whether *any* afferent fibers make functional monosynaptic connections with any motor neurons. This question was answered by the research of Birdsey Renshaw, an American neurophysiologist. Renshaw stimulated the sensory fibers entering the spinal cord and recorded a response from the motor fibers emerging from the cord (figure 12.14). He established that the minimum synaptic delay was about 0.6 ms. Renshaw was considered one of the most promising young physiologists of his day, but he died tragically of polio in 1948 at the age of thirty-seven.

David Lloyd, a physiologist working at the Rockefeller Institute in New York in the 1950s, went on to prove that the stretch reflex is monosynaptic. Lloyd imposed a sudden stretch on the muscle. When the muscle was subjected to a brisk, brief pull, an afferent volley was initiated in the sensory endings of the spindles in the stretched muscle. This afferent volley could be recorded from nerve fibers just before they entered the spinal cord. A short time later a reflex discharge of motor axons was recorded. When conduction time through the cord was subtracted from the total time, the resultant time allowed only enough time to permit passage across only a single synapse. The earliest

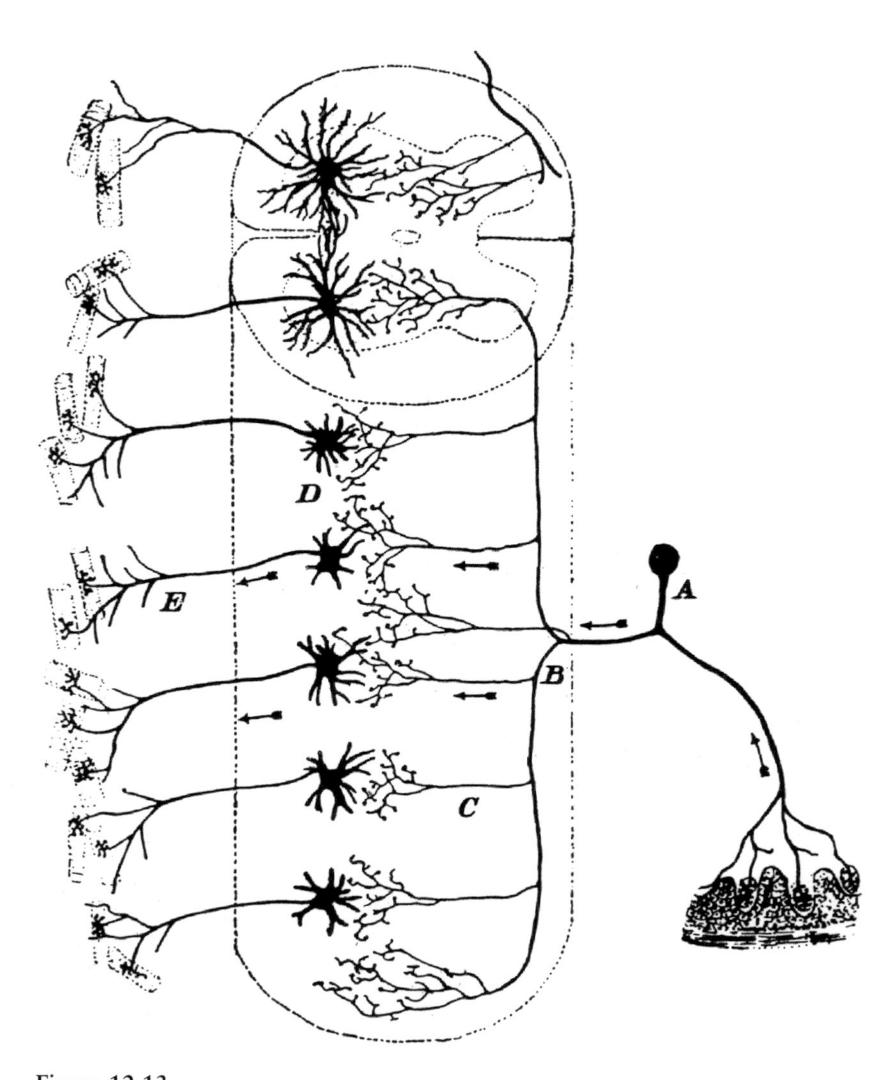

Figure 12.13
Based on his studies using the Golgi stain, Cajal showed that some incoming sensory fibers appear to terminate directly on motor neurons. Courtesy Cajal Legacy, Instituto Cajal (CSIC), Madrid, Spain.

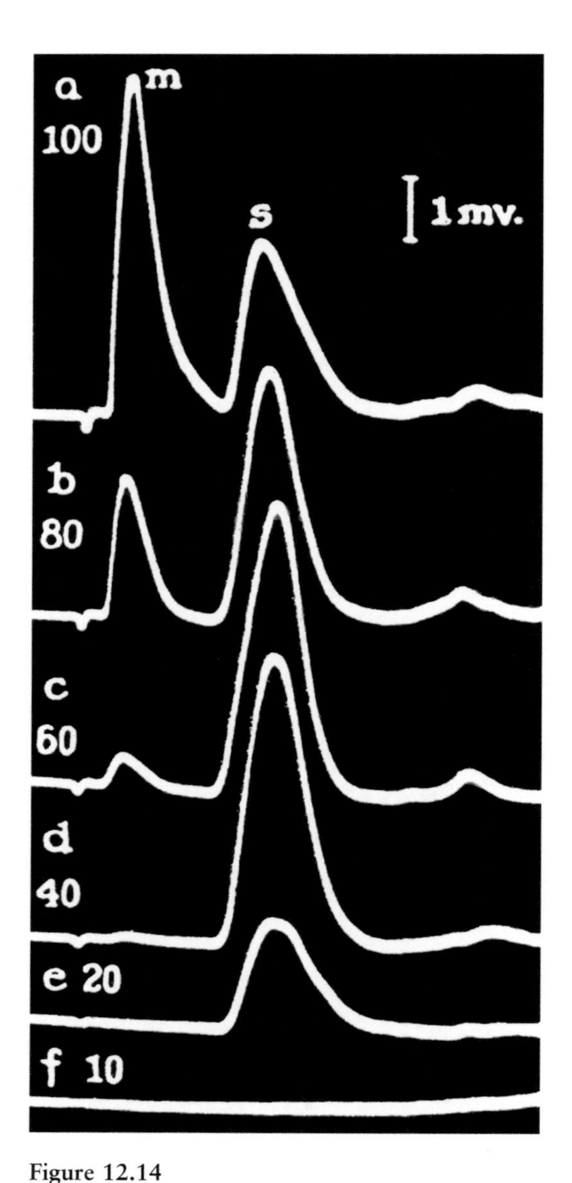

Figure 12.14
From Renshaw's 1940 paper. The traces record the electrical response of motor fibers from the spinal cord. Reading from the bottom, a weak electrical stimulus to an electrode within the dorsal region of the spinal cord elicits a reflex response (*s*) by activating sensory nerves that end on motor neurons. With successively more intense stimuli, an earlier response (*m*) appears that is caused by the stronger current causing direct stimulation of the motor neurons. Renshaw found that there is an irreducible time of 0.6 ms between the two waves, which reflects the delay at the synapse. Courtesy American Physiological Society.

stretch reflex is monosynaptic, although there are also later, polysynaptic connections.

But why is there a stretch reflex? The knee jerk is really a very atypical situation. Muscles are typically subjected to continuous and graded stretch. In humans the stretch reflex is particularly prominent in quadriceps muscle, a muscle that has powerful antigravity function, assisting us in standing. The hallmark of stretch reflexes is their prominence in muscles involved in posture and standing.

The Flexion Reflex

Like the stretch reflex, the flexion reflex requires an appropriate afferent input and results in a definite motor output. But the flexion reflex is organized differently. The sensory endings that initiate the reflex and the synaptic connections within the spinal cord are unlike those of the myotatic reflex.

The flexion protects the limbs from damage. If your hand brushes against a hot stove, it is instantly withdrawn. Flexion reflexes may still be present in an animal or person with the spinal cord cut. But although flexion reflexes are mediated locally within the cord, the pattern of their synaptic connections differs from that of the stretch reflex. Stretch reflexes principally activate motor neurons of the same muscle that was stretched. In contrast, painful stimulus to one finger can produce brisk withdrawal of the entire arm. Flexion reflexes thus typically involve activation of all the flexor muscles in the limb. Because the muscles of a single limb are controlled by motor neurons in several segments of the spinal cord, the flexion reflex is more widespread in the distribution of its connections within the spinal cord than are the stretch reflexes. The essential pathway for the stretch reflex is monosynaptic; there is a direct connection from the afferent terminal of the muscle spindle onto motor neurons of the same muscle. In contrast, even the shortest pathway for the flexion reflex is disynaptic. Sensory endings that give rise to flexion reflexes do not terminate directly on flexor motor neurons but, rather, on interneurons, which in turn activate the flexor motor neurons.

The flexion reflex illustrates a major point about the organization of motor control. The noxious stimulus that initiates the stretch reflex not only excites flexor motor neurons in the stimulated limb, but at the same time it also inhibits extensor motor neurons of the same limb. Even in this relatively simple motor act there is an orderly balance of inputs to agonist and antagonist muscles. If the noxious stimulus causes a simultaneous activation of extensor as well as flexor motor neurons, the limb would be made rigid and hence could not withdraw. Flexion reflexes must involve not only activation of the agonist, flexor muscles but also inhibition of the motor neurons that control the antagonist, extensor muscles. The result is a smooth withdrawal of the limb from the noxious stimulus.

The Crossed Extensor Reflex

So far we have considered reflex activation and contraction of motor neurons and muscles on the *same* side of the cord to which the sensory input was delivered. When the right quadriceps femoris muscle is stretched, there is a contraction of the same muscle. A painful stimulus to the right toe produces a reflex withdrawal of the entire right leg. In parallel with flexion reflexes on the same side, there are also reflex connections to motor neurons on the opposite side of the cord. An important example is the crossed extensor reflex. At the same time as a painful stimulus elicits flexion on the side of the sensory input, there is a reflex activation of extensor motor neurons on the opposite side of the cord. A painful stimulus to the right produces not only withdrawal of the right leg but also reflex extension of the left leg. Physiologists have interpreted such crossed-extension reflexes functionally as producing increased support from a limb in contact with the ground to compensate for loss of support from the flexed limb.

Long Spinal Reflexes

Reflexes may be local and restricted to a single spinal segment in their distribution like the stretch reflex, or they may involve several segments of the spinal cord. The flexion reflex acts over several segments, and the crossed extension reflex acts on the opposite side of the cord. Scratch reflexes are organized over nearly the entire spinal cord. Rubbing, tickling, or tapping of a point on the skin of the back of an animal will elicit rhythmic flexion and relaxation, a scratching movement of the hind limb. Although the scratch reflex appears purposive and "voluntary," it can be elicited in an animal in which the spinal cord has been cut; hence, the brain is not required to mediate the reflex.

Reflexes are important for the insights they provide into the normal functioning of the nervous system. They are also valuable diagnostic tools. *Stedman's medical dictionary* lists over 300 reflexes, most associated with a neurological lesion. The French Neurologist Joseph Babinski (figure 12.15) described and analyzed one of these in detail in 1903.

Figure 12.15
One of the outstanding neurologists of his day, Joseph Babinski (1857–1932) described the reflex response to stroking the foot. Courtesy Wellcome Library, London.

Joseph and his older brother Henri Babinski were the sons of Polish emigres Alexandre and Henriette Babinski. The name Babinski was widely known at the time, not because of Joseph, but because of Henri who was one of the great chefs of Paris.

If the sole of the foot is stimulated by drawing a blunt instrument across it, in most cases there is either no response or the toes may curl inward. In some cases, however, there is an extension of the toes, most prominent in the big toe (figure 12.16). Babinski (modestly) called the abnormal reflex "the phenomenon of the toe." What we now call the Babinski sign is indicative of absence of pyramidal tract influence on the motor neurons controlling the toe. Because the pyramidal tract is unmyelinated in a newborn infant, the Babinki sign is usually present.

The Babinski reflex and its appearance in patients with damage to descending motor pathway introduces us to the next chapter, which deals with brain control of movement.

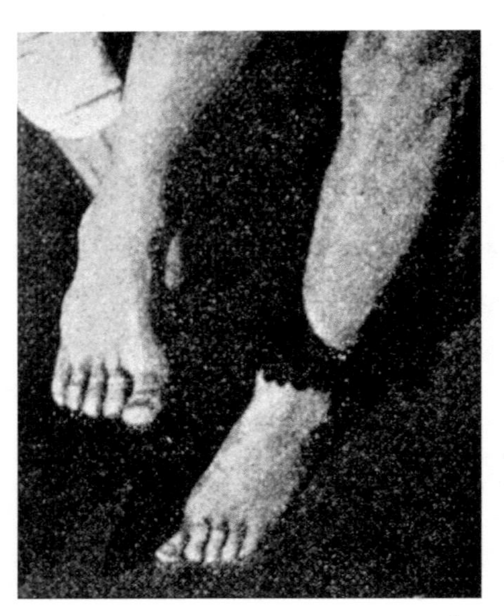

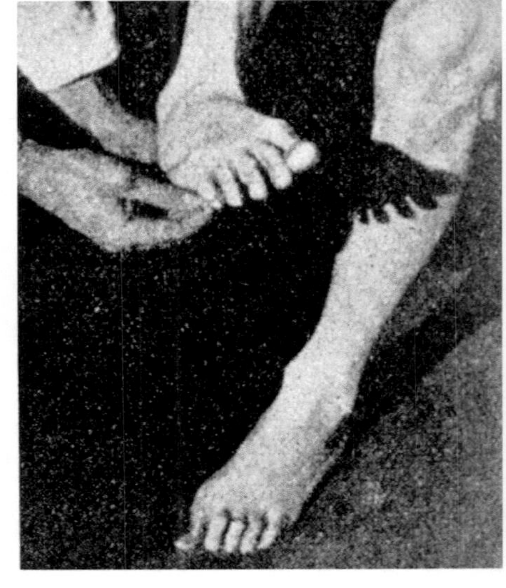

PARAPLÉGIE SPASMODIQUE.

FIG. 1. — Pied au repos.

FIG. 2. — Pied au moment de l'excitation Abduction des orteils, d'une intensité moyen

Figure 12.16
Babinski showed that in the absence of a functioning pyramidal tract there is a characteristic backward movement of the toes when the sole of the foot is stroked.

13

Brain Control of Movement

Much of our normal posture and movement relies on local reflex circuitry within the cord. Sensory messages are automatically relayed to the appropriate motor neurons or interneurons that are prewired to produce appropriate muscular response. Voluntary movement is different; it requires control from structures higher in the nervous system. A person with a complete spinal transection cannot execute any voluntary movement of those muscles whose motor neurons are located below the level of the cut. The brain and its descending pathways must be present. How does the brain control movement?

Experiment of Fritsch and Hitzig

By 1870 the cerebral cortex was recognized as being necessary for normal sensory, motor, and intellectual functions. But there was very little experimental or clinical evidence about whether the cortex is divisible into different parts, each with different functions. In 1861 Pierre Paul Broca had shown that the patient "Tan," who had suffered from profound aphasia during his life, had a localized lesion in the left hemisphere of the brain. The functions of other areas of the cortex were still poorly understood.

The most important single contribution to an understanding of the cortical localization of movement was the work of two young German physiologists, Gustav Fritsch and Eduard Hitzig, published in 1870. Hitzig had been a military surgeon in the Franco-Prussian War. While treating soldiers with severe head injuries he had, on occasion, applied weak electrical stimulation to the back of the head of some of the soldiers in his care, and produced eye movements. Later, back in Berlin, he collaborated with Gustav Fritsch to study brain localization in experimental animals. Working at home on Frau Hitzig's dressing table, they exposed

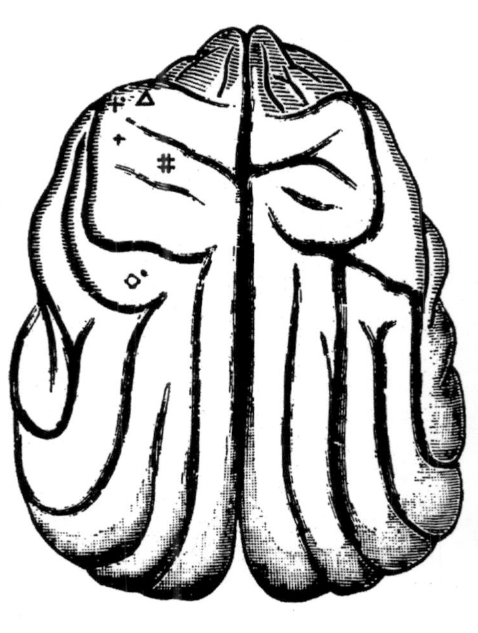

Figure 13.1
Eduard Hitzig (1838–1907; seated) and Gustav Fritsch (1838–1927). Right: A figure of a dog brain from their 1870 paper showing the sites at which electrical stimulation of the cerebral cortex produced movement. Portrait courtesy Wellcome Library, London.

the brain of a dog and demonstrated that electrical stimulation of the cortex could produce isolated movement of individual small groups of muscles on the side of the body opposite to the side of the brain that was stimulated. There is a region in the frontal lobes with a particularly low threshold of electrical excitability for producing such movements (figure 13.1). They wrote:

> A part of the convexity of the hemisphere of the brain of the dog is motor (this used in the sense of Schiff), another part is not motor. The motor part, in general, is more in front, the non-motor part more behind. By electrical stimulation of the motor part, one obtains combined muscular contractions of the opposite side of the body.[1]

Fritsch and Hitzig further showed that if the same region of cortex that caused movement were ablated, the dogs could still use the affected limbs, but the animals were now clumsy in their use of that limb.

Fritsch and Hitzig's demonstration that a part of the cortex is especially concerned with the control of movement was soon accepted and extended to other species. David Ferrier, a physiologist and neurologist working in England, showed in 1876 that electrical stimulation of the cortex on the banks of the central sulcus produced movement in the macaque monkey. Such movements are most readily elicited by stimulation of the cortex just in front of the central sulcus.

Shortly after Fritsch and Hitzig's paper appeared, Vladimir Betz, an anatomist from the Ukraine, described a characteristic cell type found in a deep cortical lamina, which he called the fourth (but is now generally accepted as the fifth) layer of cortex. He wrote:

> Now the area which we have designated as the anterior one contains cell formations which, until now, have never been described. They appear in the abovementioned sites and are the largest pyramidal nerve cells of the entire nervous system, I should like to call them "giant pyramids." . . . The cells of this anterior cortical region do not form a continuous layer, but rather are embedded as nests of one, two, three, or more cells.
>
> Such consistency in the region where these cells can be found manifested as a very definite cortical layer as well as in a specific convolution prompted me to devote my attention to that particular part of the animal brain, namely the dog's, in which Fritsch and Hitzig achieved such brilliant physiological results, i.e., the lobe which borders the cruciate sulcus. I now found such cells of the same shape and in exactly the same position in nests in the dog as well as in man.[2]

Motor Cortex

After these discoveries were made, neurosurgeons operating on the exposed brain of patients found that when the human brain is electrically stimulated by low currents, movements of the opposite side of the body are produced. The motor cortex of humans, like that of other mammals studied, is organized somatopically—that is, neighboring points in the brain, when stimulated, produce movements of neighboring regions of the body. Thus, the whole body is laid out like a map on the surface of the brain. A figure can be drawn, called a *homunculus,* or little man, that illustrates the way in which each region of the body is controlled by a given locus on the motor cortex. The homunculus in the human brain is organized as if a little person were standing with his or her feet between the hemispheres and the body bent round over the dorsal surface of the brain.

The motor map is very similar to the somatotopic map for sensory input to the brain. The two maps face one another across the central sulcus. The homunculus reflects the fact that the motor system, like the sensory pathways, has unequal representation for different parts of the body. There is a disproportionately large representation for those muscles that are capable of the most delicate movements. In humans and raccoons a large portion of the motor cortex is devoted to control of the fingers. Rats devote much of the motor cortex to control of the whiskers. In humans only a relatively small portion of the motor cortex is connected to muscles that control the trunk.

Initially, only a single motor area was known. In subsequent years it became apparent that other cortical areas are involved in the control of movement. A second *premotor area* in the frontal lobe is just in front of the classical motor cortex. There is also a *supplementary motor area* in man and monkeys with another representation of the entire body surface, which is located entirely between the hemispheres.

One current problem is to sort out the differential functions of these motor areas. There is evidence, for example, that activity in the supplementary motor area may precede that in the motor cortex. This area may play a critical role in planning a movement that is then carried out by way of the motor cortex, although all of the cortical motor areas also contribute to the descending motor tract.

What does electrical stimulation that produces movements feel like? We can ask people whose motor cortex was stimulated during an operation under a local anesthetic. Or we can ask subjects in whom movements were produced by induction from a magnetic stimulator placed on the surface of the head; the stimulator can electrically excite underlying neurons by inducing a current in the brain. Subjects report a feeling that the movement is automatic and not under their own control. The subject says that he or she did not choose to move the arm; it just moved.

Descending Motor Pathways: The Pyramidal Tract

Motor neurons are part of the final common path for movement production. Voluntary commands must be transmitted over pathways that descend from the brain down the spinal cord where they end either directly or by way of interneurons on motor neurons. One of the earliest of the descending tracts to be recognized is the pyramidal or corticospinal tract. This is a long, unbroken pathway that has its cell of origin in the cortex and descends through the brainstem to the spinal cord.

Figure 13.2 shows the location at which the pyramidal tract crosses over to descend in the spinal cord. The cells of origin are in the cortex. The axons of these cells descend through the *internal capsule*, a mass of

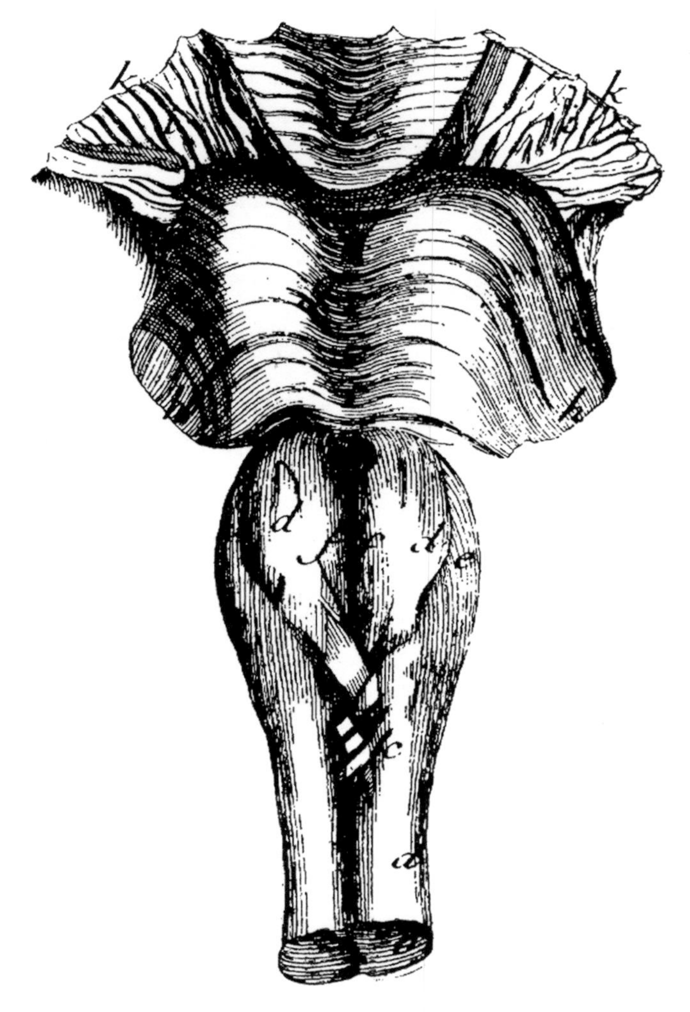

Figure 13.2
The pyramidal tracts are seen descending along the medulla and decussating (crossing over) to descend in the opposite spinal cord. Courtesy University of Washington.

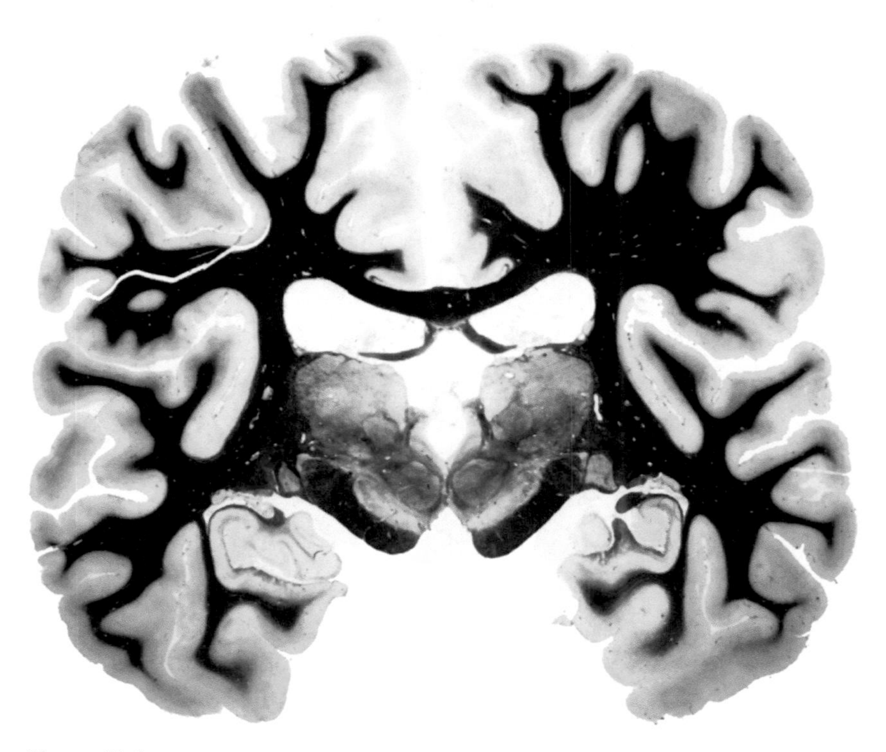

Figure 13.3
Cross section of a human brain; fiber stained. The massive tracts at the base of the midbrain are the basis pedunculi or crura cerebri. The fiber tracts form the base of the cerebral peduncles.

white matter that underlies the cerebral cortex. At the level of the midbrain, the pyramidal tract fibers are a part of the *cerebral peduncles* (peduncle: Latin, "a stalk"), a great bundle of nerve fibers at the base of the midbrain composed principally of fibers whose cell bodies are in the cerebral cortex (figure 13.3).

At the level of the pons, the pyramidal tract fibers split up into many fascicles, groups of nerve fibers that course directly through the pontine nuclei to reemerge at the base of the brain at the level of the medulla oblongata (figure 13.4).

The pyramidal tracts from the left and right side lie very close to one another near the midline on the ventral surface of the medulla. At the point where the medulla merges into the spinal cord, most of the fibers of the pyramidal tracts *decussate*. They cross and take up a new position in the lateral columns of the cord (figures 13.5 and 13.6). A few of the

Figure 13.4
The pyramidal tract (indicated by the arrow) is seen at the base of the brain with the inferior olive above. Courtesy University of Washington.

fibers remain uncrossed and descend in the dorsal white columns to form an uncrossed pyramidal tract. Some of these uncrossed fibers cross at a lower, spinal level.

In the higher primates including man a small percentage of the pyramidal tract's fibers end directly on motor neurons, especially on those controlling distal muscles. Other pyramidal tract fibers end on interneurons in the cord. These interneurons in turn connect to motor neurons. There is a direct connection from cells in the motor cortex to motor neurons that control the fingers. When stimulated in primates, these pyramidal tract fibers make monosynaptic excitatory connections with finger motor neurons.

In strokes a region of the brain is affected by loss of its blood supply, either by a burst artery or an artery blocked by a blood clot. A frequent location for strokes is in a branch of the middle cerebral artery that supplies the internal capsule—the broad stretch of white matter that contains the pyramidal tract. One of the characteristic symptoms of such a lesion is paralysis of the opposite side of the body. Physicians formerly believed that the motor defects that often follow such strokes were all due to the

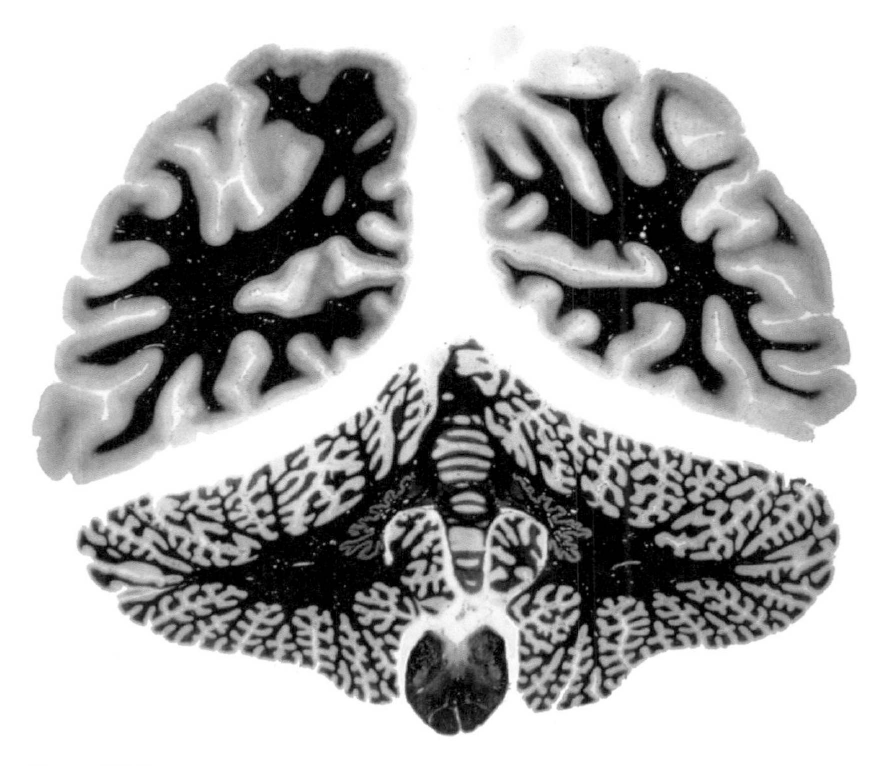

Figure 13.5
Human brain fiber stained. Pyramidal tracts are the small triangular structures at the base of the brainstem.

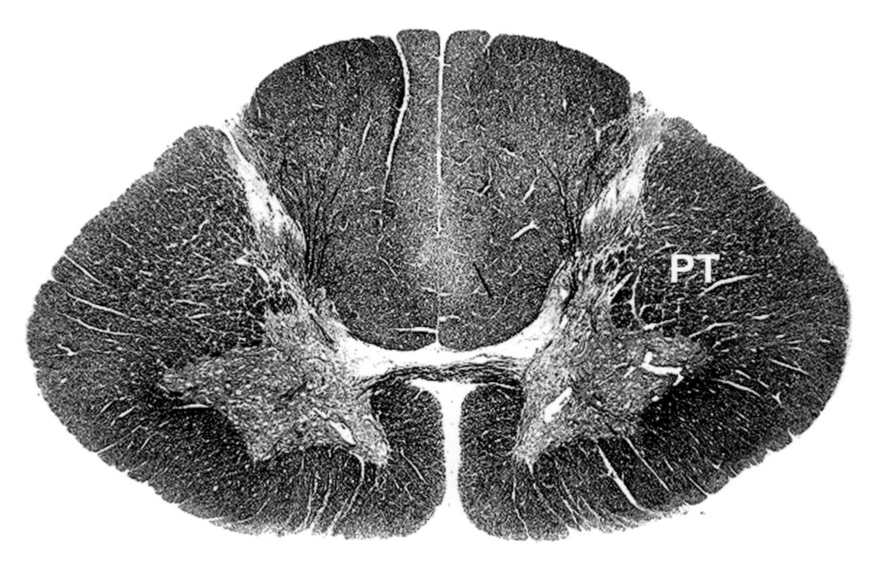

Figure 13.6
Cross section of spinal cord to show position of pyramidal tracts (PT) in lateral columns. Courtesy University of Washington.

lesion interrupting pyramidal tract fibers in their course through the internal capsule. But although the pyramidal tract is an important tract, it is certainly not the only descending motor pathway. If a person or an animal has a cut pyramidal tract without other damage, there are definite symptoms of motor loss but not a complete paralysis.

Donald Lawrence and Hans Kuypers, two anatomists working at Western Reserve University in Cleveland, Ohio, in the 1960s, described in detail the effects of pyramidal tract lesion in monkeys in which the tracts had been cut without damage to other pathways. The animals were not paralyzed. They could still climb and walk normally, and they could reach and grasp accurately. But after a pyramidal tract lesion, the monkeys lost completely the ability to move the fingers independently. They could still pick up a small piece of food but they did so by closing all the fingers simultaneously.

Some pyramidal tract fibers terminate directly on motor neurons that control the fingers. The cells of origin of the pyramidal tract are active prior to movements. The pyramidal tract acts preferentially on motor neurons that connect to the muscles of fingers and toes.

Chronic recording techniques have allowed experimenters to observe the activity of cells in the brain of awake monkeys while they are performing a voluntary act. Monkeys are trained to perform a simple movement, and the activity of cells in the motor cortex is recorded during performance of the task. By electrically stimulating the pyramidal tract fibers in the medulla, one can also establish whether a given cell from which one is recording has an axon in the pyramidal tract. Detailed study of such pyramidal tract cells during a simple repetitive movement confirms that pyramidal tract cells (PT cells) fire prior to movement.

In newborn infants the pyramidal tract is present, but its axons are not fully myelinated until the infant is about one year old. Late myelinization is correlated with the slow development in babies of certain reflexes, which are dependent on the presence of an intact and functioning pyramidal tract.

Other Descending Motor Pathways: The Rubrospinal, Vestibulospinal, Reticulospinal, and Tectospinal Tracts

In the past all descending motor pathways other than the pyramidal tract tended to be lumped together as "the extrapyramidal system." But later research into the anatomy and functions of these pathways revealed a more appropriate alternative classifying scheme. The vertebrate spinal cord has three broad columns of fibers: dorsal, lateral, and ventral. In monkeys and humans the dorsal columns are largely sensory. The descending tracts are contained in the lateral and ventral white columns.

Working at Western Reserve University in the United States, later at Erasmus University in the Netherlands, and finally at Cambridge University in England, Hans Kuypers showed that there are two broad classifications of descending tracts. Those that control distal muscles occupy a lateral position in the spinal cord; those that control proximal muscles descend in the ventral columns.

The *rubrospinal* tract, involved in the control of distal muscles, originates in the red nucleus of the midbrain. The tract crosses the midline and descends in the spinal cord. The pyramidal tract and the rubrospinal tract jointly share responsibility for the control of distal muscles, like the wrist and fingers and the ankle and toes.

The other major descending pathways constitute a medial group: the vestibulospinal tract, which has its cells of origin in one division of the vestibular nuclei in the medulla, and the reticulospinal tract, which has its cells of origin in the reticular formation in the brainstem. These two tracts are concerned with control of proximal muscles, that is, muscles that are close to the center of the body such as the muscles of the trunk, the abdomen, and the shoulder. A third tract, the tectospinal tract, originates in the superior colliculus and descends in the ventral columns for a few segments within the cord.

Brain Structures that Regulate Movement: The Basal Ganglia and the Cerebellum

We know that there is a region of the cerebral cortex especially devoted to the control of movement. We also know that the pyramidal tract is not the only pathway whereby the motor cortex can influence spinal motor neurons. Where are the other structures in the brain involved with the control of movement and how are they linked together?

Traditionally, the pyramidal tract was thought to be necessary for all normal voluntary movement. This view arose from the fact that the severe paralysis that often followed strokes destroyed the fibers of the internal capsule, the cerebral white matter that contains the pyramidal tract fibers (figure 13.7). Such severe strokes usually destroy more than just pyramidal tract fibers. Because the pyramidal tract is so prominent and organized in the medulla, postmortem examination would reveal the pyramidal tract degenerated in sections cut below the level of the injury.

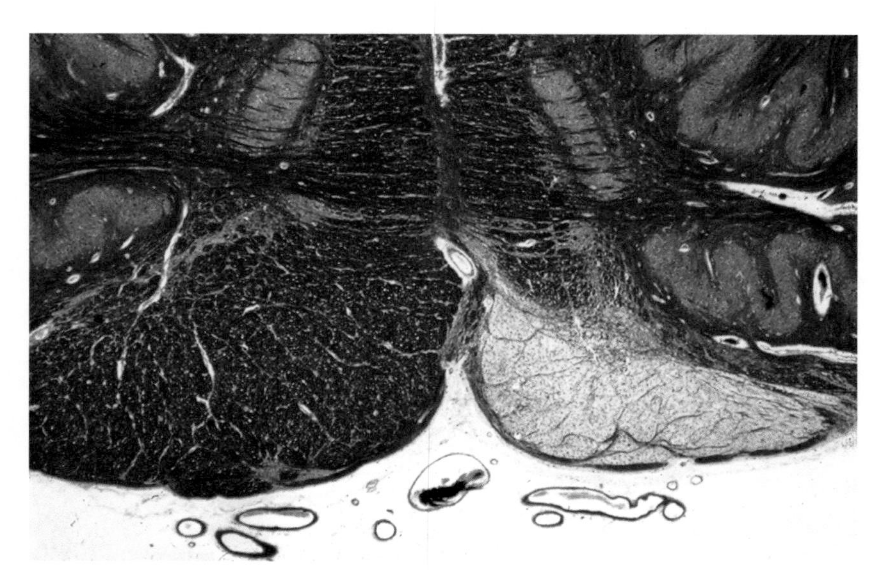

Figure 13.7
A fiber-stained section through the human medulla. The pyramidal tract on the right is completely degenerated due to an earlier stroke. Normal pyramidal tract on the left.

If the pyramidal tract is cut in monkeys or chimpanzees, there is only a modest disability in movement, not a crippling paralysis. Such experiments have brought about a change in the view of the functions of the pyramidal tract. The nomenclature for motor pathways was traditionally based on an assumed preeminence of the pyramidal tract. Hence, all motor pathways were grouped as pyramidal or extrapyramidal systems, with the added implication that the pyramidal tract was cortical in origin and the extrapyramidal system originated from subcortical structures. The dichotomy of pyramidal versus extrapyramidal pathways is misleading because it does not help us understand the origin and connections of the other motor pathways.

The Basal Ganglia

Situated at the anterior end of the brain is a very prominent set of nuclei, the basal ganglia. The basal ganglia comprise three structures: the *caudate* nucleus, the *putamen*, and the *globus pallidus*. The *caudate* nucleus has a long tail-like extension into the temporal lobe and takes its name from the Latin for "tail" (cauda). The *putamen* is from the Latin for "shell."

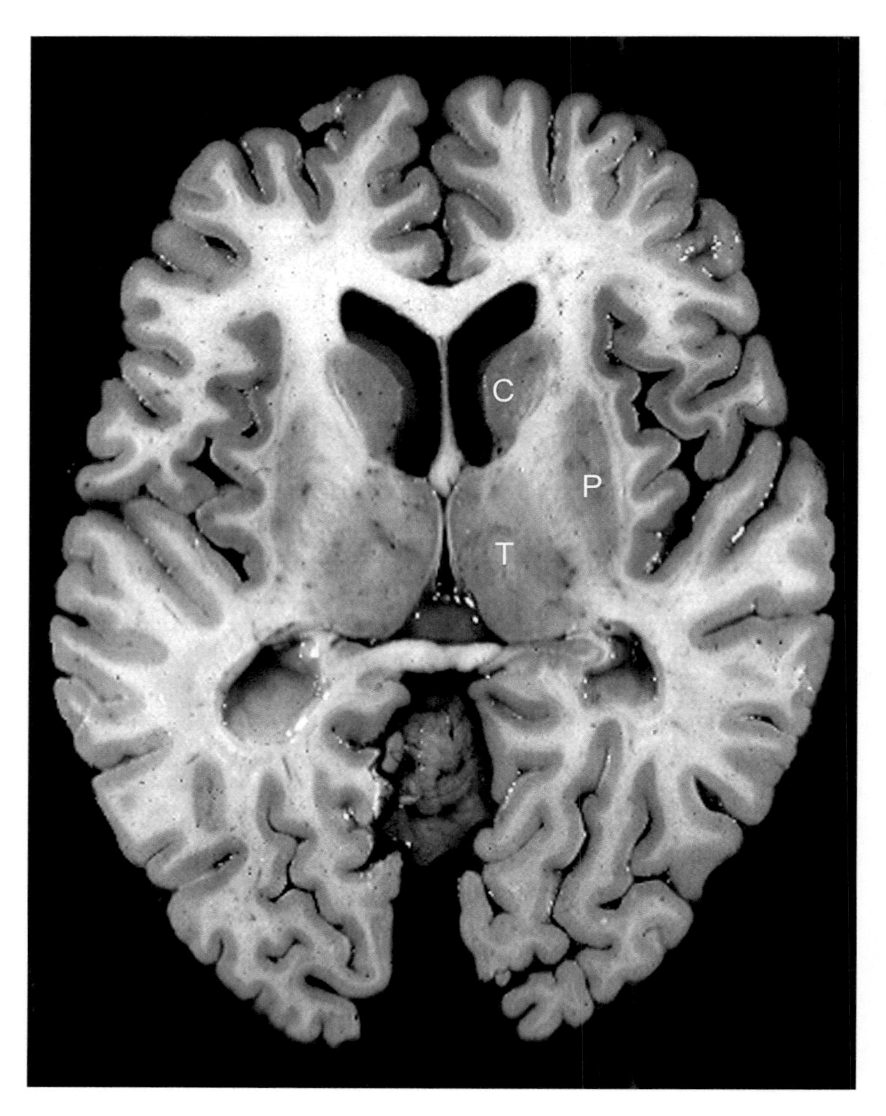

Figure 13.8
MRI horizontal section through the human brain; caudate nucleus (C), putamen (P), and thalamus (T) are illustrated. Courtesy University of Washington.

The *globus pallidus* (Latin for the "pale globe") is so called because the many myelinated fibers that pass through it give it a whitish appearance in contrast to the gray appearance of adjoining cell masses.

Farther back in the brain, lying just above the cerebral peduncles, is another cell mass. This is the *substantia nigra* (Latin for "black substance"), so called because in unstained brains one layer of its cells have a conspicuous black speckling. The basal ganglia and substantia nigra are critically involved in the regulation of some classes of movement, and they are often the principal site of degeneration or injury in patients with severe disorders of movement. In Parkinson disease, for example, patients typically have a resting steady tremor and lose much of the control of certain muscle groups (see chapter 17 of this volume). The lack of control of the muscles of facial expression may give the faces of patients suffering from Parkinson disease a characteristic mask-like appearance. In chorea or athetosis, the movement disorder can be even more distressing, producing constant and uncontrollable movements of writhing or dance-like twisting.

The Connections of the Basal Ganglia

Caudate and putamen are really like one large nucleus. The continuity between caudate and putamen is masked because the fibers of the internal capsule divide them into separate nuclei. The principal input to both caudate and putamen comes from the cerebral cortex. The caudate and putamen in turn have their principal efferent connections to the globus pallidus, which also receives a major input from the substantia nigra. The medial division of the globus pallidus has as one of its single most important targets the ventral lateral nucleus of the thalamus, which projects to motor areas of the cerebral cortex.

We have a sort of paradox. One of the principal circuits in this great so-called extrapyramidal system of the brain ends on a region of the motor cortex, some of whose output to the cord travels by way of the pyramidal tract! Many severe disorders of movement are associated with lesions or degenerative changes in the basal ganglia. Replacement of one of the precursors of the transmitters that are liberated by the fibers of this tract, L-DOPA, can, for a time, reverse the motor symptoms of Parkinson disease. The identification of the lesion in Parkinson disease and its management by L-DOPA illustrate one of the major successes in the application of basic neural science to the treatment of a severe human disorder (see chapter 17 of this volume).

The Cerebellar Pathway

In addition to basal ganglia, another major circuit related to movement involves the cerebellum. The cerebellum, literally the little brain, is attached to the brain by three prominent pairs of peduncles or stalks. Like the cerebral cortex, the surface of the cerebellum is deeply folded, but even more so. If we look at the surface of the human brain, we see only about a third of it; the rest is contained within the fissures and sulci. In the cerebellum we see only one-tenth. The human cerebellar cortex, if unfolded and laid flat, would have about half of the surface area of the cerebral cortex (figure 13.9).

The cerebellum is histologically much simpler than the cerebral cortex. The cerebral cortex is conventionally divided into six layers; the cerebellar cortex has three layers (figure 13.10).

In the middle of the cerebellar cortex there is a single layer of very prominent large neurons, the Purkinje cells. They are named for the great Czech physician and scientist Johannes Evangelista Purkinje, who first described them early in the nineteenth century. Just below the Purkinje cells there is an array of a very densely packed small *granule* cells. In addition, the cerebellum contains three other neuronal types: the basket cells, stellate cells, and Golgi cells.

The Purkinje cell is the only output from the cerebellar cortex. There are two sorts of afferent inputs to the cerebellar cortex. One of the afferent fibers, the climbing fiber, terminates directly on the dendrites of the Purkinje cell. The other type of input, the mossy fiber, contacts granule cells. Granule cells in turn send their axons outward toward the cerebellar cortex and end as long parallel fibers in contact with the dendrites of the Purkinje cells. These two inputs, the mossy fiber and climbing fiber, are the only afferent connections to the cerebellar cortex. The other cell types in the cerebellar cortex—the stellate, basket, and Golgi cells—receive their inputs directly or indirectly either on the Purkinje cell directly or onto the mossy fibers as they terminate.

The Purkinje cell is the only source of axons that leave the cerebellar cortex. Where do its inputs come from? The overwhelming majority are from the granule cell. Granule cells receive a massive mossy-fiber input. Some mossy fibers arise directly from cells in the spinal cord, the so-called spincocerebellar tracts. But by far, most mossy fibers come to the cerebellum from the pontine nuclei, a prominent group of cells on the ventral surface of the brainstem.

The single largest source of afferent fibers to the cerebellum is from the pons. Most of the fibers that project to the pons come from the

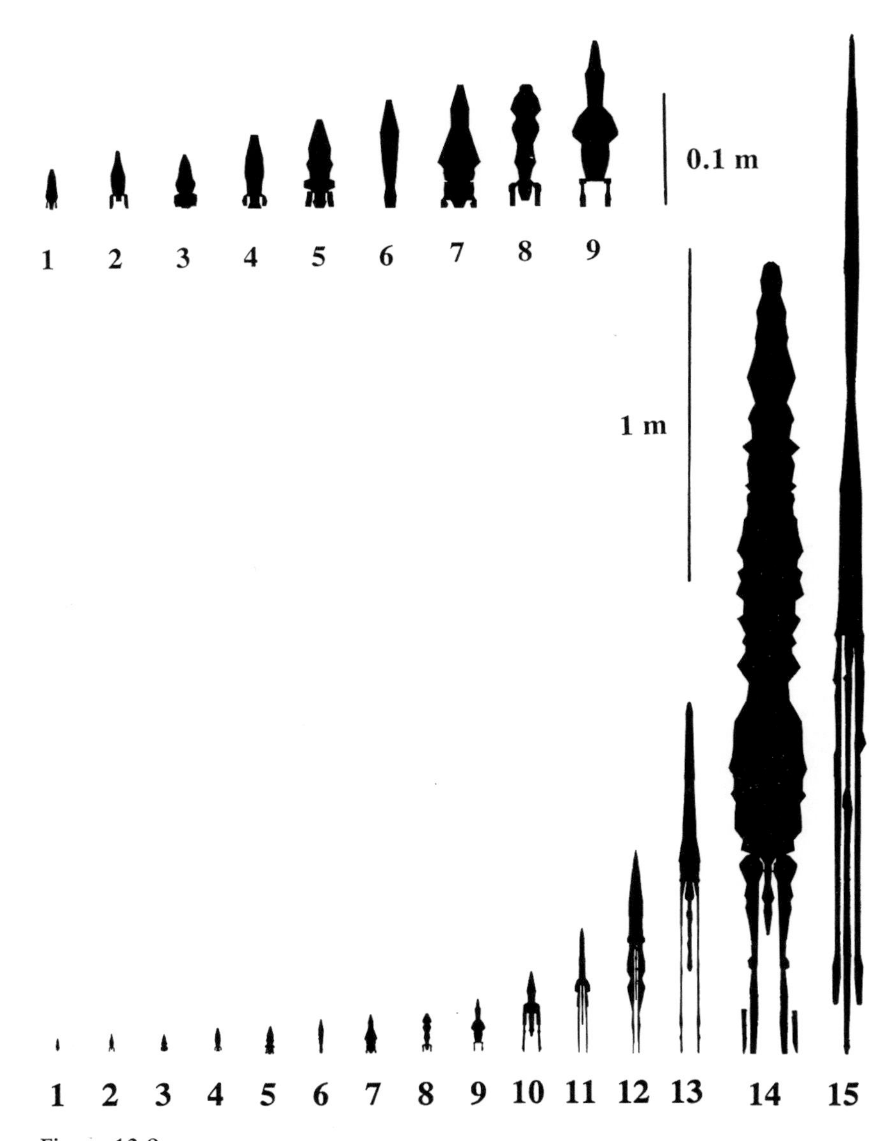

Figure 13.9
In life the cerebellum is deeply folded. The true dimensions of its surface are revealed if it is unrolled. The figure, from a study by Fahad Sultan and Valentino Braitenberg, shows the unrolled dimensions of several mammalian species. The human brain is number 14. Number 15 shows the unrolled cerebellum of a cow.

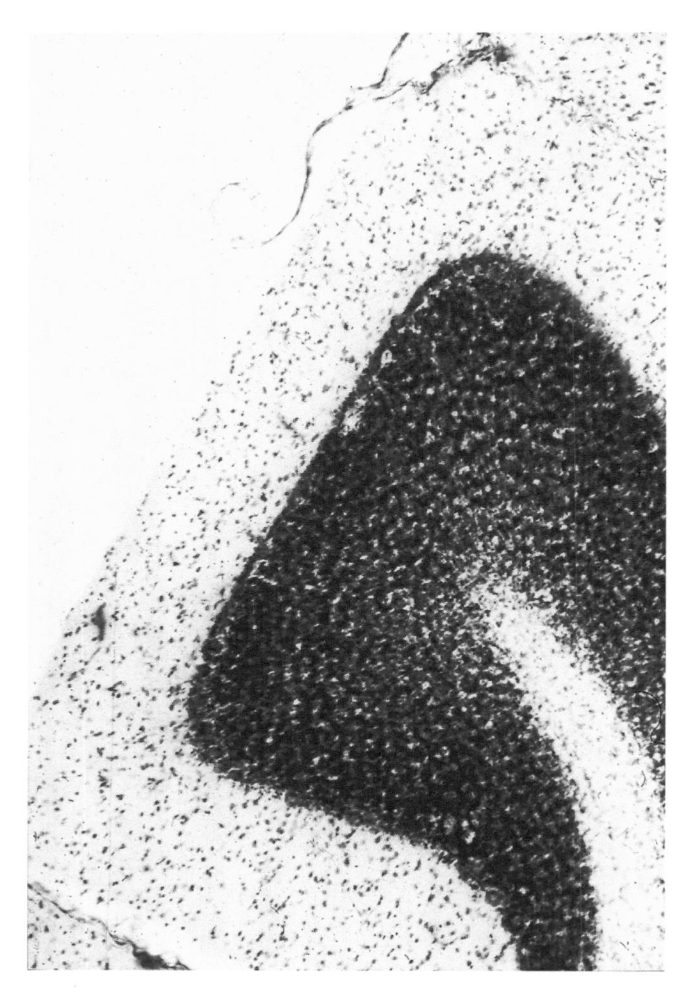

Figure 13.10
Cell-stained section of the cerebellum to show its three-layered structure. A very thin layer of individual Purkinje cells, stained purple, can just be seen. There is a dense array of granule cells below and a palely stained molecular zone above.

cerebral cortex. Corticopontine cells arise from a wide expanse of the cerebral cortex to connect to the pons. The pons is the source of the largest input to the cerebellar cortex.

Purkinje cells in the cerebellar cortex project principally to a set of cells located deep within the cerebellum, the so-called deep cerebellar nuclei. These in turn project to vestibular, reticular, and red nucleus and, via the ventral thalamus, back to the motor cortex.

Once again, our anatomical trail has led us back to the motor cortex. These are two major pathways, both originating in the cerebral cortex. One path leads via the basal ganglia back to the motor cortex; the other leads via pons and cerebellum also back to the motor cortex. Why this divergence of pathways? What different functions do these circuits have? It is likely that different sorts of actions are controlled predominantly by each pathway. For example, the basal ganglia may be especially related to the control of axial muscles of the face, the trunk, and the hips, whereas the cerebellum is more involved with control of distal muscles. The role of cerebellum in motor learning is discussed in chapter 14.

14

Learning and Memory

Memory is a broad concept. It can mean remembering a particular teacher you had in school, an especially happy day in your life, or the kind of car that your father drove many years ago. These are the so-called called *declarative memories;* they are what we usually think of when we hear the word memory. There are two kinds of declarative memory. One is episodic ("We went to Paris"), and the other is semantic ("Paris is the capital of France").

But there are other memories. We learn to drive a car or ski downhill without hitting a tree or banging into another skier. These are part of a large group of *nondeclarative memories.* You learned them, you can be aware of them, but they cannot be described in the same way as a declarative memory. There are further subdivisions. Your friend gives you a telephone number to ring to make a reservation in a restaurant or order a pizza. You hold the number in *working memory.* You can dial it right away, but unless you rehearse it or write it down, it will be lost. One important aspect of brain function is the constant filtering of short-term memories and the placing of important ones into more permanent storage.

Learning from Memory Loss

Patterns of loss in neurological patients can give us insights into how the brain stores the different classes of memories. One of the early clinical descriptions of memory loss was by Sergei Korsakoff (figure 14.1), a Russian physician born in 1854. Korsakoff studied medicine at the University of Moscow and was eventually appointed director of a psychiatric clinic at the University of Moscow. He described a characteristic form of memory loss in some of his longstanding alcoholic patients. One of the symptoms that he saw in chronic alcoholics was a profound disorder of memory. In 1889 he wrote:

Figure 14.1
Sergei Korsakoff (1853–1900) was a Russian psychiatrist who described a characteristic memory deficit in his long-term alcoholic patients. Courtesy U.S. National Library of Medicine.

Together with the confusion, nearly always a profound disorder of memory is observed, although the disorder of memory occurs in purer form. In such instances the disorder of memory manifests itself in an extraordinarily peculiar amnesia, in which the memory of recent events, those which just happened, is chiefly disturbed, whereas the remote past is remembered fairly well. . . . At first during conversations with such a patient, it is difficult to note the presence of psychic disorder; the patient gives the impression of a person in complete possession of his faculties; he reasons about everything perfectly well, draws correct deductions from given premises, makes witty remarks, plays chess or a game of cards, in a word comports himself as a mentally sound person. Only after a long conversation with the patient, one may note that at times he utterly confuses events and that he remembers absolutely nothing of what goes on around him: he does not remember whether he had his dinner, whether he was out of bed. . . . Patients of this type may read the same page over and again sometimes for hours, because they are absolutely unable to remember what they have read. In conversation they may repeat the same thing 20 times, remaining wholly unaware that they are repeating the same thing in absolutely stereotyped expressions. It often happens that the patient is unable to remember those persons whom he met only during the illness, for example his attending physician or nurse, so that each time he sees them, even though seeing them constantly, he swears that he sees them for the first time.[1]

Korsakoff's patients had impaired memory caused by longstanding alcoholism, often associated with thiamine deficiency. A similar memory loss can also be seen as an *acute* symptom of alcoholism, but in its initial stages such loss can get better with treatment. In longstanding cases it may be irreversible.

Other causes can lead to a similar devastating memory disorder. One of the most important cases was that of a man called HM, who had been studied for many years by Brenda Milner of the Montreal Neurological Institute and her colleagues. HM had suffered a brain injury in a bicycle accident as a child. In the years following the injury he developed an increasingly incapacitating series of epileptic seizures. In an attempt to relieve the epileptic symptoms, HM was operated on to remove a sizable portion of the hippocampus and adjacent structures of the temporal lobe on both sides.

HM lost the ability to form new memories almost entirely. When looking through a family photo album he would encounter a picture of an uncle he was particularly fond of. Each time when his mother told him that the uncle had died, HM wept yet again. HM had good recall for events that had occurred some years before the surgery, but he could not form new declarative memories. If a person left the room and returned some minutes later, that person was forgotten.

Although HM's brain lesion seemed to prevent him from laying down new memories, all memory acquisition was not lost. If HM was asked to do a simple task like drawing a mirror-maze, he was able to do it with some skill. The next day he drew it again and took less time to do it. After five days of testing HM had learned the task, but he denied ever having seen the apparatus before.

One form of nondeclarative memory was spared, and others were later discovered in further experiments. When the descriptions of HM's memory deficit were first published, it was often assumed that it was the hippocampus—and only the hippocampus lesion—that produced the severe impairment. But the lesion in HM's case extended beyond the hippocampus. In addition to the hippocampus it also damaged an adjacent area of the temporal lobe, including entorhinal, pararhinal, and parahippocampal cortex. This damage probably contributed to the severity of the memory deficit.

Place Cells and the Hippocampus

In rats the hippocampus contains a continuous record of the animal's position in space, and it is necessary for many types of spatial learning. John O'Keefe, a psychologist working at University College London, discovered that the rat hippocampus contains cells that are uniquely responsive to the animal's position. These so-called space cells form an obvious basis for learning a spatial task.

Localization in the Brain: The Cerebral Cortex and Memory

The cerebral cortex is a layer of nerve cells and fibers, supporting cells, and a rich blood supply. The cerebral cortex is one of the great constants of nature. Mammals that differ in size, like a mouse and an elephant, differ by only a factor of three in cortical thickness. In humans the cerebral cortex varies in different regions from 2 to 4 mm in thickness. It extends over a dense tangle of white matter. The cortex is deeply folded. If unrolled, it would be over 1 m in length and about one-sixth of a meter wide.

As we saw in previous chapters, the cerebral cortex has definite structural and functional subdivisions. The occipital lobe, at the back of the brain, contains a white stripe parallel to the cortical surface that is coextensive with the primary visual cortex. The visual cortex, when stimulated electrically, produces the sensation of light, and destruction of all or part of the visual cortex in humans or monkeys produces predictable deficits in vision.

A part of the cerebral cortex is motor in function. Long fiber tracts originate from this region of the brain and connect directly or via interneurons to motor neurons in the spinal cord. Electrical stimulation of this motor area elicits movement in humans and animals, whereas lesions impair movement. Another part of the cortex is the end station for a series of links that originate in the ear. Yet another cortical region receives its input from the skin senses.

But despite the evidence for cortical localization, a large percentage of the cerebral cortex did not seem to early investigators to be related directly to sensory or motor function. Electrical stimulation of these areas of cortex produced neither movement nor sensation. The failure to be associated with definite response on stimulation gave these areas one of their early names: "silent areas." More commonly, cortex that is neither clearly sensory nor motor was called "association areas," from an early belief that sensory messages were brought together or associated in these regions.

In the human brain there are two great regions of association cortex—a frontal area that lies in front of the motor areas of the frontal lobe and a posterior area that occupies much of the temporal and parietal lobes (see figure 2.10 for the location of the lobes.)

Although much remains to be learned about the functions of the association areas of the cortex, one of their functions is memory. These areas are involved, for example, in visual and tactile recognition of objects, short-term recall, and the understanding and production of language.

How does the cortex acquire and store learned information? As a first step toward answering this question, scientists asked *where* in the association areas of cortex learning was stored. Early attempts to answer questions about brain localization in learning were never very successful. But although these first attempts at localization of the memory trace often proved frustrating, the experiments helped set forth some of the problems with what seemed at first to have been a relatively straightforward question.

The pioneer in study of brain localization in learning was Karl Spencer Lashley (figure 14.2). Lashley was trained as a zoologist but identified himself as a psychologist, and he devoted most of his scientific

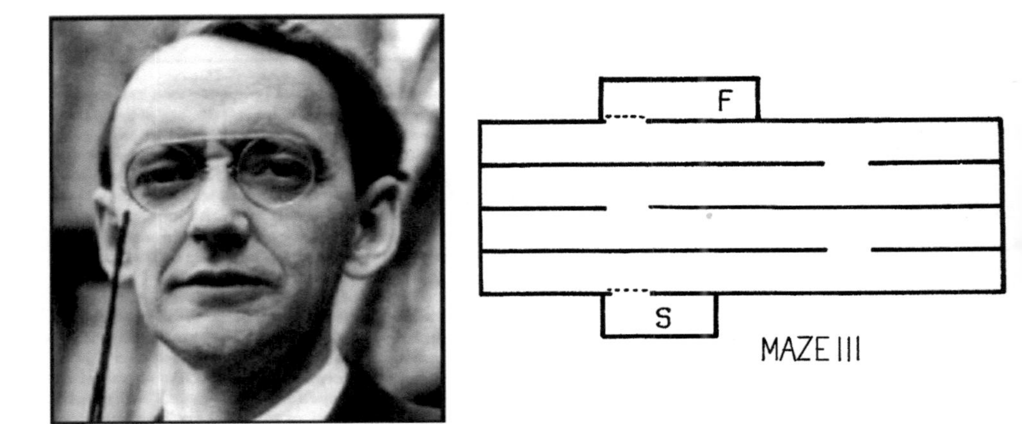

Figure 14.2
Karl Lashley (1890–1958) was a pioneering student of brain function. At right, a diagram of the maze used by Lashley. Courtesy University of Chicago Press.

career to the study of brain mechanisms in learning. Lashley began his studies at the Universities of Minnesota and Chicago, moving to Harvard in 1935.

While keeping his Harvard appointment as professor of psychology, Lashley also served as director of the Yerkes Laboratory for Primate Research near Jacksonville, Florida. Lashley taught and influenced a whole generation of physiological psychologists. Lashley's approach to the study of brain localization of memory seemed, at first, straightforward. He trained animals—usually rats—to master a specific task, and then he removed a part of their cerebral cortex. If memory for the task were localized in some specific region of cortex, he reasoned that only animals that had sustained a lesion of that area would show a loss of memory for the task. Alternatively, if a specific region of the brain were necessary for learning a given task, then a lesion of that region prior to training would severely impair successful acquisition. Lashley devised a series of simple mazes that rats could readily learn, like the one illustrated in figure 14.2.

Rats were placed in the start box, S, and allowed to run through the maze. When they reached the goal box, F, there was some food waiting for them. After they were allowed to eat a bit, they were once again picked up, replaced in the start box and allowed to find their way through the maze to the goal box again. Lashley measured two key aspects of the animals' performance: how long it took the rat to go from the start box to the goal box, and how many errors it made along the way—that is, how many times it stepped into a blind alley before reaching the goal.

The two measures are, of course, highly correlated, and either measure showed that the performance of normal animals improved with training. For example, initially animals would take an average of about 1,100 seconds from the time they were first placed in the start box until they reached the goal box. With successive training this time became progressively shorter. On retesting, the rats ran the maze in less than 100 seconds. Correlated with the increasing speed, the rats also made fewer and fewer errors until they could run through the entire maze without making a single error.

Lashley then set out to analyze how the brain stores the memory. He was searching for the hypothetical physical imprint known as the engram—a word that had been coined a few years earlier from the Greek for "that which is written in." He reasoned that if he trained a number of rats to run the maze, and if memory for the maze were stored at some definite locus, then only those animals in which the critical region was destroyed would show a deficit in maze performance when retested after the operation. By the same token, if an area were to be necessary for learning the maze, then only animals in which that area had been destroyed would be impaired in acquiring the maze habit.

Lashley's experiments did not yield a simplified pattern of results. He never found a single region of the cortex in which memory for the maze habit was strictly localized. Rather, it seemed as if the rats would show some deficit in acquisition or retention of the maze habit not in relation to *where* the lesion was made but in relation to *how much* of the cortex had been removed. Localization of the memory trace did not seem to be possible, and Lashley could merely plot how impaired the animals were. Figure 14.3 is from Lashley's classic monograph *Brain Mechanisms and Intelligence*.

Performance on the maze was impaired in proportion to the amount of brain removed, without any obvious relationship to any particular locus. Lashley hypothesized from these conclusions a general principle of brain function. He suggested that association cortex has no functional subdivision, but instead learning ability and intelligence in general are simply functions of the amount of cortex, not any specific region of cortex. Lashley called this principle *mass action*. Mass action seemed the only way in which Lashley could account for his experimental findings. It is probably wrong.

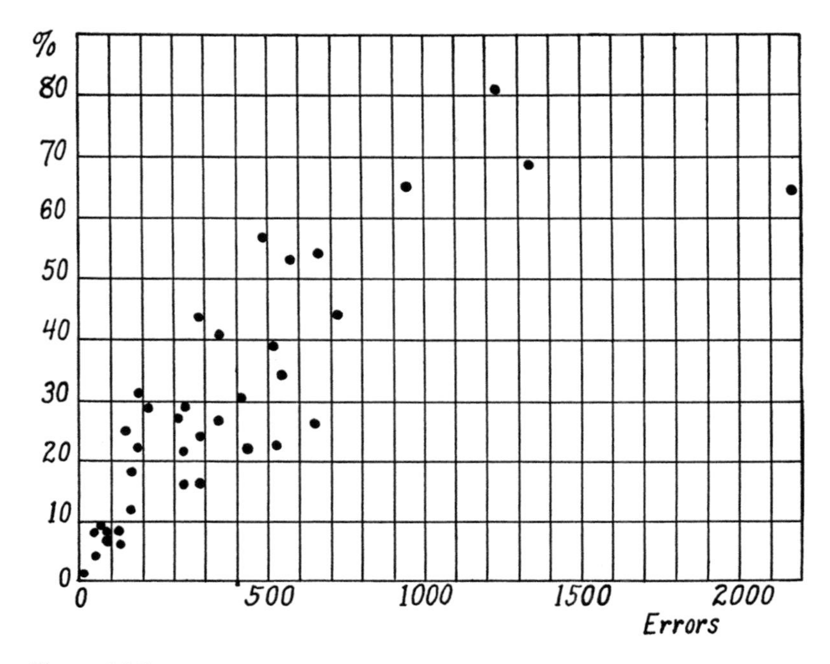

Figure 14.3
This figure shows that the rats that Lashley tested were impaired in learning the maze in relation to the amount of cerebral cortex lesion. Courtesy University of Chicago Press.

As Walter Hunter of Brown University pointed out in a critique that he wrote in 1930, Lashley's results could have come about for other reasons. For example, suppose that rats could learn the maze by several possible ways—visually, or by remembering a series of successive correct turns, or by feeling their way through the maze with their whiskers. If the maze was learned partly by vision, partly by whisker, and partly by learning successive turns, then Lashley's experiments could be explained without reference to mass action. Learning for each of these sensory channels could be precisely localized, but the multisensory nature of the maze test could obscure such localization. As we shall see, memory *can* be localized to some degree. Mass action is probably not a valid way to describe the way in which cortex works.

In addition to his experiments on association cortex and maze learning, Lashley studied the effects of brain lesions on other forms of learning. For example, he studied the ability of rats to learn and remember visual discrimination habits after lesions of the primary visual cortex.

Normal pattern vision was dependent on the integrity of the primary visual cortex. If all of primary visual cortex had been removed, rats were unable to learn a visual pattern. They could learn a visual pattern discrimination task if an area as small as one-sixtieth of the primary visual cortex remained intact. It seemed as if any portion of the visual cortex could serve for visual pattern discrimination learning independent of where it was in the visual cortex. Lashley called this ability of any region of visual cortex to serve for visual learning tasks the principle of *equipotentiality*.

Equipotentiality is sometimes confused with mass action, but they are not the same concept. Mass action refers to association cortex and a generalized ability to learn complex tasks. Equipotentiality refers to functions of specific sensory areas of the brain in which preservation of any small region may serve for acquisition of a task.

Localization Revisited: Interhemispheric Transfer and the Split Brain

Lashley's results presented more of a challenge than a solution for understanding how the brain stores memory. They never really satisfied him. He thought that perhaps a memory trace could be localized to one hemisphere or the other. It seems so self-evident that we do not think about it, but suppose you learn to recognize an object or a word with your left eye closed. Now open the left eye and close the right eye. How is it that there is such complete and immediate transfer between the eyes? The answer seems straightforward. Cells in the visual cortex receive input from both the left and the right eye.

But suppose that the sensory input could be restricted to a single side of the brain. Would the other side know the answer? Lashley suggested to a few of his students that they might try cutting the links between the two sides of the brain and testing for interocular transfer. One of his early colleagues, Clifford Morgan, summarized those studies. Morgan wrote:

> In one experiment with rats, the crossed fibres of the optic chiasm were cut and the uncrossed fibers left intact (Levine). In this way fibers from the left eye went only to the left side of the brain and those from the right eye went only to the right side of the brain. Animals prepared in this way and having one eye blindfolded were trained to make a discrimination. They were then tested with the other eye blindfolded. The result was perfect retention, i.e., perfect interocular transfer.
>
> But the trouble with this experiment is that it does not consider the connections from one visual cortex to the other via the corpus callosum. In the many studies

that have been carried out, the corpus callosum does not seem to be good for much, but it is possible that it figures in such functions as inter-ocular transfer. In another experiment, therefore, the same kind of training and testing was done in animals in which the posterior part of the corpus callosum had been sectioned (Lashley). In this case as in section of the optic chiasma there was inter-ocular transfer. The only trouble was that the animals in which the corpus callosum was sectioned were different from those in which the optic chiasma was partially sectioned. We end up then, without a crucial experiment, i.e., one that cuts all crossings of the visual system.[2]

Roger Sperry became acquainted with the problem of interocular transfer when he was a postdoctoral associate of Lashley's at the Yerkes laboratory from 1942 to 1946. Sperry later headed a laboratory at the University of Chicago in the 1950s where one of his associates, Ronald Myers, did the crucial experiment. Myers devised a simple leather mask that would allow a cat to see out of one eye at a time. If the optic chiasm is cut in the midline as illustrated in figure 14.4, each eye would now project only to the hemisphere on the same side.

The animals could still see well enough, and they readily learned to solve a visual discrimination task using either eye. Myers now trained cats on visual discrimination tasks with only one eye viewing the targets. When he tested whether they knew the correct choice using the untrained eye, there was high-level transfer between the eyes. Learning transferred between the eyes even though the chiasm had been cut. How is it that the untrained right eye knows the problem?

As Clifford Morgan's text had suggested, we must look for commissures—fiber tracts that cross between the two sides of the brain. The largest by far is a massive structure, the corpus callosum (Latin, the hard body). The callosum is big. There are many more fibers in the corpus callosum than all of the cranial nerves combined, and yet scientists had found no obvious symptoms that seemed to occur when it was cut. If a person or animal were unfortunate enough to have all of his or its cranial nerves cut, he or it would not smell, could not see, could not move his/its eyes, would have little sensation or movement throughout most of the head and face, could not hear, would have no sense of balance, and could not swallow or move his/its tongue, among other symptoms. And yet cutting a fiber tract far larger than all of these combined, the corpus callosum, seemed to cause no obvious symptoms.

Myers's experiments began to reveal the functions for this curious structure. When *both* the chiasm and the corpus callosum were cut, interocular transfer of learning was abolished. An animal could be trained on a discrimination task with its left eye open and its right eye closed

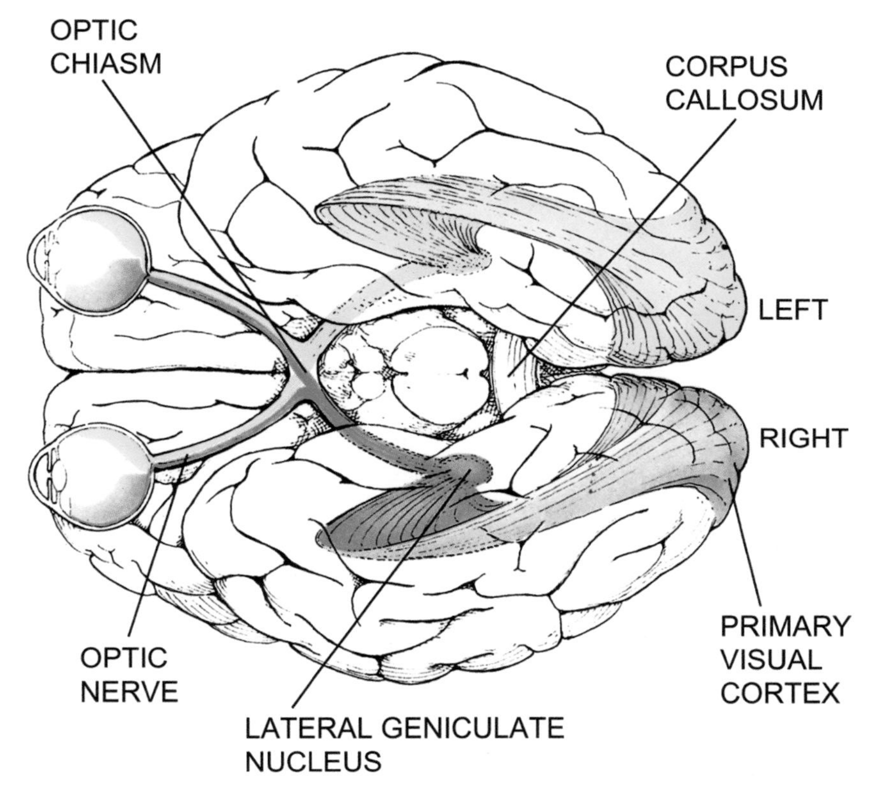

Figure 14.4
The pathway from eye to brain for visual fibers. Originally published in *Scientific American*.

and show no transfer when the mask was reversed so that the right eye viewed the problem. The animal could be trained to choose one pattern—say, a circle with the left eye open—and then choose the other pattern—say, a triangle with the right eye open. These two opposite solutions could be maintained, with the animal happily choosing the circle or triangle depending on which eye was open.

Showing that learning was localized in half of the brain seemed a modest enough gain. But these experiments reopened the question of localization. They demonstrated that the brain does work by definite rules—by orderly connections from one region to another. If only the corpus callosum and not the optic chiasm was cut, cats also showed high-level transfer of training between the two eyes. It became clear that there are at least two ways for the right eye to allow the left eye to know a problem that it has learned.

The corpus callosum does a similar job for other senses. The majority of the fibers carrying the sense of touch are crossed. The right hand projects primarily to the left hemisphere, and the left hand projects principally to the right. If a monkey is trained to feel two objects without seeing them, and to select the correct choice on the basis of touch alone, normal animals show strong and immediate transfer from the trained to the untrained hand. Animals in which the corpus callosum is cut show no evidence of transfer to the untrained hands. Just as in the case of vision, monkeys can learn opposite habits with the two hands.

In addition to the corpus callosum there are other possible pathways for interhemispheric transfer. Unlike the case in cats, if a monkey is trained to solve a visual task after its optic chiasm and corpus callosum are cut, there may still be high-level transfer between the eyes. In this case the anterior commissure can also serve as a pathway for interocular transfer. If the anterior commissure is cut along with optic chiasm and the corpus callosum, transfer is blocked.

The role of the anterior commissure in interocular transfer of form discrimination learning in monkeys can be understood in terms of its anatomical connections. In addition to linking some noncortical regions across the midline, the anterior commissure also has fibers that link the temporal lobe on the two sides. The temporal lobe has an established role in form vision, and the anterior commissure can serve to transfer knowledge of visual forms from one hemisphere to the other. Commissures in the midbrain can serve for interocular transfer of brightness discrimination.

The multiplicity of pathways that are capable of enabling transfer between the two sides of the brain can help to understand some of the difficulty of early brain localization experiments. If a task can be learned in several different ways—for example, by using one or another sense—then memories for the task might be stored in several places in the brain. In the same way, interocular transfer can be achieved either by way of the corpus callosum or the anterior commissure. Multiplicity of memory storage would tend to produce the sort of results that Lashley obtained in his studies of acquisition and retention of the maze habit after brain lesions.

Surgeons occasionally have sectioned the corpus callosum in man in an attempt to limit the spread of epileptic seizures through the brain. In general, humans, as well as other animals, show the same sort of independence of the two hemispheres after the corpus callosum has been severed. But studies of such corpus-callosum-sectioned people have revealed many new and fascinating principles of functional differences in the two hemispheres. We discuss the results of those studies in chapter 16.

Temporal Lobe and Vision

Beginning in the late nineteenth century, neurologists recognized a condition in which some patients with brain lesions suffered from difficulty in recognizing faces, a condition called *prosopagnosia* (from the Greek *prosop* for face and *agnosia* for failure to recognize). In a milder form it can even be present in otherwise normal humans. Heinrich Klüver and Paul Bucy, working in Chicago in the 1930s, made large lesions in the temporal lobes of monkeys. They found that, among other symptoms, their monkeys had difficulty in recognizing familiar objects after lesions of the temporal lobe.

K. L. Chow, working with Lashley at the Yerkes lab, and Mort Mishkin and Karl Pribram demonstrated that the deficit in visual learning and visual recognition was related to the damage to the inferior temporal cortex. Lesions of the inferior temporal lobe produce a lasting deficit in a monkey's ability to learn new visual problems. Visual tasks learned before the operation may be lost unless the animal had been overtrained for a long time before the surgery.

The role in vision is confirmed by recording from cells in the temporal lobe. Whereas visually responsive cells in the primary visual cortex respond to relatively simple targets, cells in the temporal lobe are often responsive to complex forms. Charles Gross and his colleagues at Princeton University in the 1970s found a class of cells in the temporal lobe that were active when a monkey was shown a picture of a face.

The cortex of the temporal lobe is vital for normal visual learning, and yet it does not receive a direct input from the lateral geniculate nucleus. The striate cortex connects to the temporal cortex by way of relays in the cortical areas between the occipital and temporal lobes. Damage to these prestriate relay areas lying between the striate cortex and the temporal lobe can produce deficits in visual discrimination learning that are as severe as those caused by destruction of temporal cortex.

Prefrontal Cortex and Immediate Memory

The temporal lobe is clearly involved in the acquisition and retention of visual memories. The other great extent of association cortex is in the prefrontal area of the frontal lobe. The prefrontal cortex is involved in a different sort of memory.

In the 1920s Walter Hunter of Brown University invented an ingenious behavioral task for studying short-term memory in experimental animals: *delayed response*. In a typical delayed response task a monkey is placed in a testing cage. In sight, but just beyond its reach, is a wooden tray with two plain blocks covering two shallow wells. The experimenter puts a bit of food in one of the wells while the monkey looks on. A block is placed over the well, and an opaque screen is lowered so that the animal cannot see either block during a delay interval. After a delay of usually between 5 and 15 s, the window is lifted and the tray slid forward so that the blocks are now in reach, and the animal is allowed to select one of the two blocks. If he is correct and remembers which of the two blocks covers the food, he is allowed to eat it. The tray containing the wells and blocks is withdrawn, and another trial is begun. Old World monkeys such as baboons and macaques can usually learn this task within a few days of training.

At the end of the nineteenth century there began to be systematic study of the effects of brain lesions on monkeys and other mammals. At first, postoperative observations were rather casual. How did the monkey react to threat? Could it reach for food? Was it able to walk as before? At the beginning of the twentieth century more rigorous behavioral studies began to be used. In a typical experiment an animal might be trained to perform a variety of skilled or semiskilled acts and tested for its ability to retain earlier training or learn new tasks after a brain lesion. The initial results of prefrontal lesions were puzzling. Although performance of some tasks seemed to be mildly affected, few reproducible effects could be uniquely attributed to removal of the frontal lobes.

The picture changed in the early 1930s. Carlyle Jacobsen (figure 14.5) was a psychologist working at Yale University with John Fulton. At first Jacobsen found only minimal disturbance caused by the ablation of the prefrontal cortex in his monkeys. But then he tried another test.

Jacobsen found that bilateral lesions of the prefrontal cortex abolished the monkeys' ability to successfully solve the delayed response task. Monkeys with prefrontal cortex lesions also show another form of memory deficit—*delayed alternation*. In delayed response the animal must remember which of two wells was baited and maintain the memory over an interval in which an opaque screen is placed between the monkey and the food wells. In the delayed alternation task the wells are baited alternately. If the monkey fails to select the correct well, a further delay

Figure 14.5
Carlyle Jacobsen, an American psychologist, discovered the memory deficit that follows lesions of prefrontal cortex in monkeys. Courtesy University of Iowa Archives, Papers of Carlyle F. Jacobsen.

is interposed, and the nonselected well is again baited. In the delayed response test the monkey must remember, "Where did he put the peanut?" In the delayed alternation task he must remember, "Where did I reach last?" Normal monkeys readily learn either task, and they can bridge a long temporal gap. Monkeys with dorsolateral frontal lesions cannot. Performance on both tasks is virtually at a chance level over the shortest of delays after bilateral prefrontal ablation.

Jacobsen began his experiments studying monkeys and extended his work to the study of chimpanzees, where he found emotional changes after frontal lesions. These are discussed in chapter 19. A summary of his work was published in 1936.

One function of the frontal lobes is the temporary storage of memory. Joaquin Fuster, working at UCLA, and Patricia Goldman-Rakic, working first at the National Institutes of Health in Maryland in the 1970s and later at Yale University, showed that neurons in the prefrontal cortex are directly involved in this form of short-term working memory. One class of neurons would fire continuously during the memory interval. Goldman-Rakic's work and life were cut short when she was struck by a car and died in 2003.

Short- and Long-Term Memory Storage

We live in a complex world of sensory impressions. Thousands of things are briefly noted, stored for a short time, and then forgotten. And yet some memories can persist from adulthood to old age. How do memories get into long-term storage? How do we go from sounds and sight to permanent memory? There must be at least two sorts of memory storage: some are transient, evanescent, and easily lost; the others are long-term and highly stable. The necessary transition from short- to long-term memories can be thought of as a sort of consolidation of those fleeting memories. The entire notion that memory is first briefly stored in one way and then permanently stored in another way is called the *consolidation hypothesis.*

One important piece of evidence for the consolidation hypothesis came from the study of people who had sustained sudden and severe head injuries. In most such cases patients may not only lose consciousness and memory for events after the accident but they also typically fail to remember things that have happened just prior to the accident. The bicyclist may remember his brakes failing on the top of the hill, but fail to recall his downward course and the truck that hit him at the bottom.

The loss of memory for events just prior to a head injury is known as *retrograde amnesia*. There always seems to be a definite period of time ranging from several seconds, hours, days, or even years prior to a major head injury for which memory is lost and cannot be recaptured.

Cerebellum and Learning: Pavlovian Conditioning

In Pavlov's early experiments, a neutral stimulus, for example, the sound of a metronome, would be followed by placing food in the animal's mouth. Repeated pairing of the metronome, called the *conditioned stimulus* (CS) followed by food, here called the *unconditioned stimulus,* (UCS) would eventually lead to the animal salivating to the sound of the CS alone.

In another example of Pavlovian conditioning a tone (CS) is followed by a puff of air to the eye (the UCS). After repeated pairing of tone with air puff, animals and people blink to the sound alone. In rabbits, in addition to blinking, the response includes closing the nictitating membrane—a third eyelid that acts as an additional protective device. Richard Thompson, a psychologist working first at Stanford University and then at the University of Southern California in the 1980s, showed that the conditioned response was abolished by large cerebellar lesions. Christopher Yeo, working at University College London in the 1980s, showed that the critical site was a definite region of the cerebellar cortex, the hemispheric component of lobule VI.

Cerebellum and Saccadic Adaptation

Saccades are the jerky movements of the eyes used to inspect a new item in the field of view. They are in constant use. We make as many saccadic eye movements in our lives as we have heartbeats. Saccades normally remain accurate from infancy to old age. It is clear from experiments that saccades are constantly being recalibrated.

The nature of that calibration can be revealed by playing a trick on the visual system. Suppose we sit in a chair and look at a fixation spot in front of us. We are asked to look at a different spot whenever it appears. People and monkeys are very skilled at this task. They respond rapidly and accurately with a saccade toward the new target. Suppose now, that after the eyes have started to move toward the new target, we shift its position; say from an initial 10 degrees to the right to 15 degrees. Humans and monkeys will show orderly compensation for the shift and

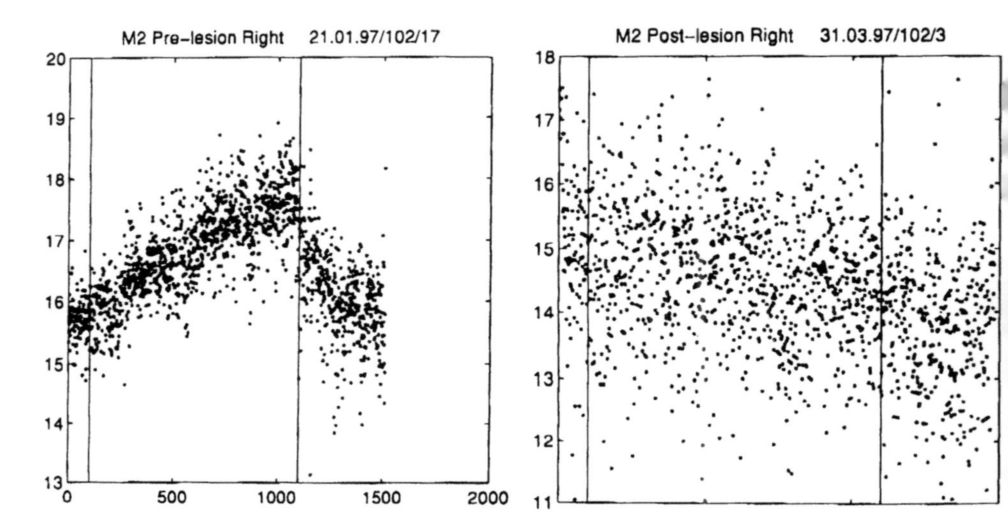

Figure 14.6
A monkey was trained to look at a target when it appeared. As soon as its eyes moved the target was shifted 5 degrees to the right. Each dot in the figure on the left represents the size of a series of eye movements. There is a gradual and steady adjustment to the displacement. The figure on the right shows the decrement in eye movement over time in a monkey with a lesion of the oculomotor cerebellum.

are unaware of the shift during the saccade. Initially the eyes will be directed to the target. When they reach the point where the target had originally been placed, they make a second 5-degree saccade to the new position. During a single testing session people and monkeys will increase the amplitude of the saccade by small increments in trial after trial. The cerebellum is necessary for the recalibration.

Figure 14.6 shows the results from one such experiment. A small lesion in the cerebellum abolishes the ability of monkeys to adapt to the displaced target.

Reflex Adjustment

Lengthening a muscle causes it to contract—the stretch reflex. Touching your finger to a hot stove causes your limb to withdraw—the flexion reflex. Both of these are examples of *closed-loop* reflexes in which the response acts on the stimulus that causes it.

Some reflexes are *open loop*. In the vestibuloocular reflex (VOR), for example, turning your head to the right causes a reflex movement of the eye in an equal and opposite direction to the left. The basic wiring of the VOR involves a three-neuron arc from the semicircular canals to the extraocular eye muscles. But the reflex does not feed back directly on the horizontal canal that originated it; thus it is an example of an open-loop reflex. Such open loop reflexes are in danger of getting too big or too small, and they must be continuously recalibrated. We can see the need to recalibrate if we wear spectacles. A new prescription will slightly change the size of the image on the retina and tend to make the VOR inaccurate. Experiments in which people or animals are tested with lenses that make the image larger or smaller show that the VOR is recalibrated to compensate and that such recalibration requires the cerebellum.

Learning as a Synaptic Process

So far we have considered the *locus* of memory: *Where* are memory traces stored in the brain? There is a more basic question: *How* is memory formed? Fundamental to memory storage is the synapse. The role of the synapse is basic to all learning. Cajal had emphasized that nerve processes touch one another, but they do not fuse. As we saw in chapter 3, Sherrington promoted the term *synapse* to describe the tiny gap between cells. Since the term *synapse* was first used it has become increasingly clear that the synapse is the critical site for simple and complex forms of learning.

But before the term *synapse* was coined, in 1893, the Italian psychiatrist and neurologist Eugenio Tanzi wrote:

> If now we think that the distances between the terminal arborisation of one neuron and the body of the next neuron constitute a resistance or . . . a kind of difficult passage that the nervous wave must overcome not without difficulty, it is evident that the conductivity of the nervous system will stand in inverse relation with the spaces between neurons. To the extent that exercise tends to shorten distances, it increases the conductivity of neurons in their functional capacity.[3]

Some years later the psychologist Donald Hebb suggested: "When an axon of cell A is near enough to excite cell B and repeatedly or persistently takes part in firing it, some growth process or metabolic change takes place in one or both cells such that A's efficiency, as one of the cells firing B, is increased."[4]

Learning involves modification at the synapse. Simple phenomena that resemble learning also involve modification at the synapse, and they may form the basis for more complex forms of learning. One of the simplest forms of behavioral modification is *habituation.* Your dog is sleeping by the fire. You give a brief whistle; he looks up at you. If nothing happens, he goes back to sleep. You whistle again. He looks, but for a shorter time. By the third or fourth whistle he carries on sleeping. Habituation is among the simplest forms of learning.

Aplysia, a marine mollusk, has about 20,000 neurons in all. Because they are large, individual neurons can often be identified for stimulation or recording. Suppose that you observed that every time you stimulated a specific neuron there was an observable twitch in one of the muscles. If the time between the stimulation and the muscle's response were very short, you would have evidence that the cell you had activated is a motor neuron.

Imagine that instead of stimulating an individual nerve cell, you record from it. Suppose that the neuron always fired a very short time after you brushed the skin on a definite region of the body surface. This would be evidence that the cell you were recording from is a *sensory* neuron.

American physiologist Eric Kandel and his collaborators, working first at New York University and later at Columbia beginning in the 1960s, studied the mechanisms of habituation of the gill withdrawal reflex in *Aplysia*. Touching a region of the skin causes the animal to withdraw the gill. If the same point is touched repeatedly, the animal habituates; it no longer responds to the tactile stimulus. As a first step Kandel and his team studied the region of skin whose stimulation triggers the reflex. They found that it was served by sensory neurons located in an abdominal ganglion of aplysia. By stimulation they identified the motor neurons that control the muscles involved in withdrawal of the gills. These motor neurons were also located in the abdominal ganglion of *Aplysia*. They found that there is a direct monosynaptic connection from the sensory neurons to the motor neurons controlling gill withdrawal. Branches of the sensory neurons in that patch of skin synapse directly on processes from the motor neurons that control gill withdrawal.

The machinery of the gill withdrawal could now be analyzed at a cellular level. They found that repeated stimulation of the sensory neuron at frequencies similar to those used in behavioral study of the reflex produces a gradual decrease in the amount of transmitter released by successive stimuli, with a decrease in the amplitude of the EPSP (excitatory post-synaptic potential) generated in the sensory neuron. Behavioral habituation is reflected at the cellular level by less release of transmitter and a consequent smaller EPSP.

Synaptic Changes Associated with Learning

If a pre- and a postsynaptic neuron are active simultaneously, it may result in *long-term potentiation* of the synaptic link. Activation of the presynaptic element will produce a stronger response in the postsynaptic element. The increase can last for hours or longer. In other situations pairing of a pre- and a postsynaptic element produces *long-term depression* at a synapse. These two phenomena come about by way of a variety of molecular mechanisms, and they are operative at many sites in the nervous system.

These phenomena posed a puzzle for scientific understanding. Electrophysiologists typically studied synaptic connections that work in less than a millisecond. Stimulation of the pyramidal tract causes a prompt release of transmitter substances onto motor neurons. For example, breathing is regulated by a powerful circuit that controls inspiration by alternately expanding and contracting the volume of the chest. But learning seems to involve much slower processes. It takes weeks before a young musician can learn to play a C-major scale on the cello, or her brother to play a short melody on the flute.

The work of Paul Greengard and his associates, working first at Yale and then at the Rockefeller University, has given insights into the way synapses can be modified by slower processes. These so-called *second messenger systems* are similar to hormone actions. They target pre- or postreceptor sites. But rather than acting directly, they work through slower biochemical pathways that increase or decrease the amount of transmitters released or the response of the postsynaptic elements to transmitter release. The work has also given important insights into why some drugs that might be effective may take several days before they start to work. Greengard proposed that we can think of the fast systems as the hardwiring of the nervous system. The slower second messenger systems are the brain's software.

15

Motivation

Claude Bernard (figure 15.1), the great nineteenth-century French physiologist, recognized the fact that most of the cells in the human and animal body are bathed in a fluid of remarkably constant character. This "milieu interne," or internal environment as he was to call it, is made up of the liquid portion of blood and the fluid that surrounds cells. "It is as though the organism had enclosed itself in a kind of hothouse where the perpetual changes in external conditions cannot reach it." The exact constituents of this internal environment were not fully known in Bernard's time. We now know that, in addition to water and oxygen, there are precise proportions of salts, sugar, and organic molecules dissolved in those fluids.

Homeostasis is the word that Bernard coined to describe the tendency of the organism to maintain physiological functions at some constant level. Typically functions are controlled homeostatically both by physiological and by behavioral adjustments. Ultimately we rely on behavior: we must eat and drink to meet energy requirements and maintain fluid balance. *Motivation* is the term for prompting behavior that serves homeostasis. Motivation is closely related to emotion, a topic that we discuss in chapter 18 of this volume.

Autonomic and Behavioral Regulation: The Hypothalamus

The regulation of functions such as fluid and food intake has been studied intensely over the years, and much has been learned about how they work. Here are some examples of the questions that have been asked. How do mammals and birds maintain their body temperature so precisely? How is it that animals maintain such remarkably constant weight in spite of great variation in the quality and amount of available food and the energy demands placed on them? How does the body

Figure 15.1
Claude Bernard (1838–1878) was a French physiologist who studied autonomic regulation. His study of the effects of the poison curare showed that it did not block motor nerves but abolished reflex function. Courtesy Wellcome Library, London.

maintain the precise concentration of blood and body fluids in the face of continuous loss of water to the environment?

Many of these functions are done quietly by the autonomic nervous system (see chapter 2 of this volume) without requiring any obvious behavioral intervention. But there are limits. Temperature can be controlled automatically unless the outside air becomes too cold. A small loss of blood volume can be compensated by balancing the fluid inside and outside cells.

But ultimately all of these adjustments must enlist behavior. We must wear clothing or seek a warmer place; we must drink to compensate for loss of fluid from sweat and breathing. The hypothalamus is the master switchboard for all of these functions. The hypothalamus is the head ganglion of the autonomic nervous system (figure 15.2).

Functions such as fluid balance and temperature regulation are typically controlled at two levels—the physiological and the behavioral. Automatic physiological regulatory systems deal with moment-to-moment adjustments. The kidney, for example, maintains the correct proportions and relative concentrations of water and salts in the blood. But over time we lose fluids and salts in sweat and in urine, and such losses must be made up behaviorally. We must eat enough food to balance the energy that we expend. Behavior is necessary for preserving the species. Like eating and drinking, sexual behavior is controlled by hormonal factors and the brain, essentially by the hypothalamus. About the size of a human thumbnail, the hypothalamus regulates eating, drinking, maintaining a constant temperature, and sexual function.

Thirst and Drinking

The control of fluid intake can serve as a model system for understanding regulatory mechanisms in general. Most of the vertebrate body is made up of water. Substances such as salts, sugars, and proteins are dissolved in that water. The concentrations of such solutes are usually kept within narrow and precise limits, even though we are continuously losing fluid in urine, sweat, and expired air. These losses must ultimately be replaced by drinking. How does fluid loss lead to the sensation of thirst and produce drinking? In order to answer that question, we must first understand how water and the solutes in it are distributed within the body.

Each cell of the body is surrounded by a thin membrane that is freely permeable to water but not equally permeable to all of the solutes dissolved within that water. Physiologists divide body water into an

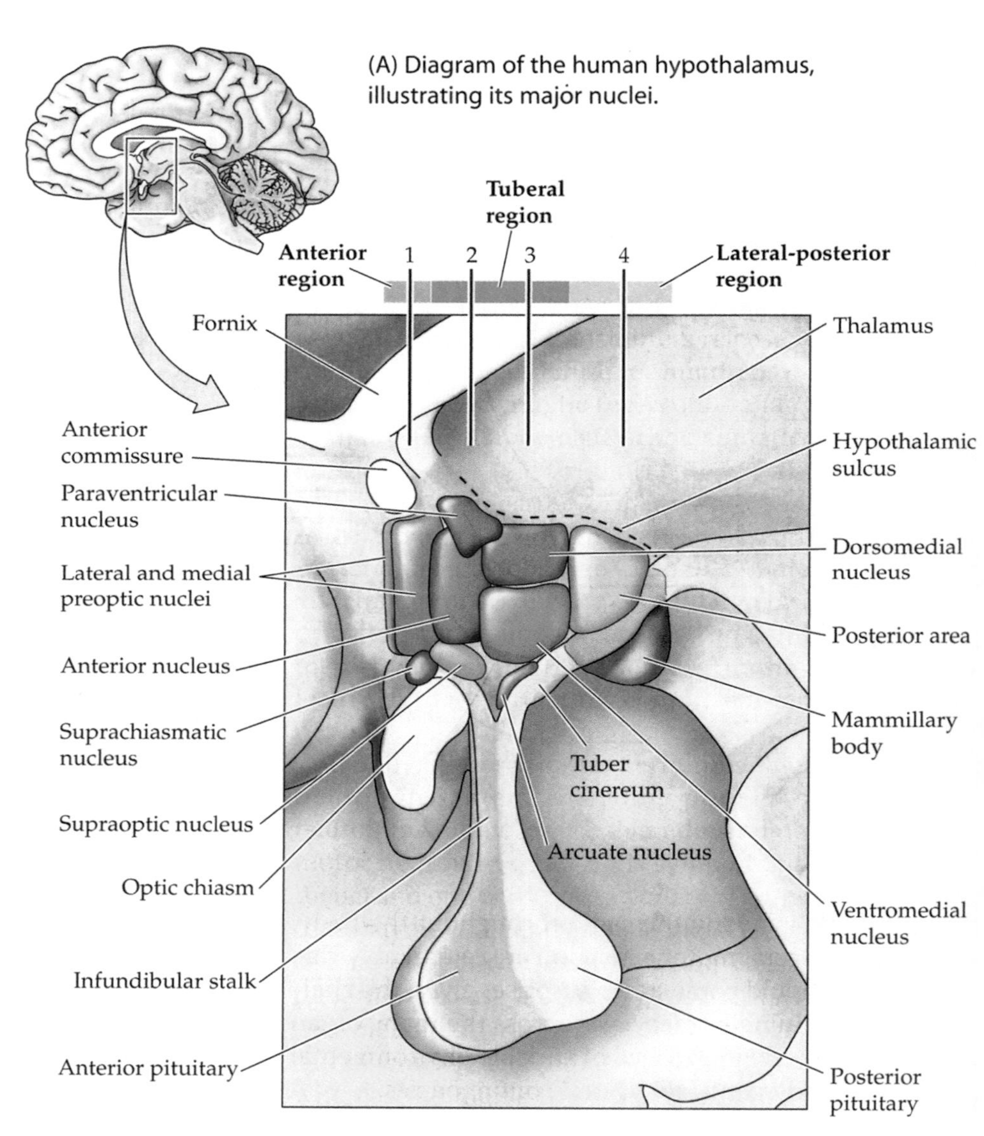

Figure 15.2
Location and subdivisions of the hypothalamus.

intracellular compartment, which comprises the water contained within all the cells of the body, and an *extracellular* compartment made up of the fluid portion of blood and the *interstitial fluid* that surrounds and bathes these cells.

Now suppose the contents of the blood or interstitial fluid were to change. For example, water is lost as water vapor is expelled in the expired air or directly in sweat. Water loss poses two challenges. A certain volume of blood is necessary to maintain normal blood pressure and, hence, normal nutrition of all of the cells in the body. The lost fluid must be recovered. There is a second challenge posed by fluid loss. Loss of water would raise the concentration of solutes in the extracellular fluid and make it *hyperosmotic*, thus drawing water from the cells of the body. Loss of fluid volume through *hypovolemia* and *cellular dehydration* are both potent stimuli to thirst, which allows losses to be made up by drinking.

Thirst and the Brain

All day I've faced a barren waste

Without the taste of water, cool water

Old Dan and I with throats burnt dry

And souls that cry for water

Cool, clear, water

—Song by Bob Nolan, 1941

What are the mechanisms for thirst? Why does one drink when water is lost from the body? One of the early explanations for thirst related it to the sensation arising from a dry mouth. If we are deprived of water our mouth feels dry. Perhaps this simple consequence of dehydration could account for thirst. But the dry mouth theory of thirst fails as a complete explanation on several counts. Even if the sensory input arising from the lips, tongue, and mouth is blocked by cutting the afferent nerves' supply from those areas, thirst and drinking are not strongly affected. Removal of the salivary glands in an experimental animal, although it produces constant dryness of the mouth, does not change greatly the net amount of water the animal consumes. Also, merely moistening the mouth or throat without ingesting water into the stomach has no lasting effect on thirst. A dry mouth is not a necessary condition for thirst, nor is a wet mouth sufficient for satiety.

Thirst cannot be simply accounted for by the presence of a dry mouth. There must be some central mechanism, some structure or structures in the brain that sense the need for water and drive us to drink it. A clue was provided by a case in Germany. In 1881 H. Nothnagel, a physician based in Jena, saw a patient who suffered a head injury that caused him serious thirst. He wrote:

> A thirty-five-year-old stonemason, whose home is in Lichtenhain near Jena, was admitted to the clinic on the 25th of July 1881 at 11:30 at night. Father had died (unknown cause). Mother alive, four living sisters. One sister had died from scarlet fever, another from tuberculosis. The patient had a "lung inflammation" at twenty-five and had recurrence of the same illness since twice; most recently two years ago. Since then, he has been basically healthy.
>
> On the 25th of July between 7:30 and 8:00 the patient was riding an unsaddled horse which was for sale. The animal became wild and threw him off three times. Without intending to, he held on to the horse's tail. The horse struck out and hit him in the left half of the stomach with its hind foot. Because of that he fell down backwards hitting his head on the hard ground, and hitting his right ear at the same time on a piece of fallen wood. He did not lose consciousness, but shortly thereafter he had an attack of vomiting. He had a feeling of dampness in the head. Afterwards he could not stand up on his own. He had to be lifted several times because of severe pain in his body, his neck, and the back of his head.
>
> Soon after the accident—according to the patient himself and the people nearby as far as is known, since the people did not have access to the exact time among them, the most common time reported was "one half hour after the event" the patient felt a strong thirst, so that in the next 3 hours until he was admitted to the clinic around 11:30 he drank a great deal, water and beer one after the other, in all at least three liters. He urinated around 11 o'clock for the first time, and the only time until he was admitted, thus two and a half hours after the onset of the thirst. Thus, while travelling the short distance from Lichtenhein to Jena he had to get down from the wagon to urinate.[1]

Nothnagel argued that the delayed onset of urine makes it clear that the primary cause of drinking was probably related to the brain injury the patient had sustained. The paper does not specify the exact site of the possible brain lesion, but it is clear that thirst can be produced directly as a consequence of brain injury.

This unfortunate patient had an irresistible urge to drink following his accident. People or animals with a lesion of the hypothalamus or the pituitary gland to which it is attached also suffer a similar thirst. After damage to a slightly different region of the hypothalamus, they fail to drink. Lesions can interrupt the normal balance between thirst and satiety centers for water balance. What are the signals from the body that normally affect those brain structures?

Osmotic Thirst

A change in the osmotic pressure of the blood and interstitial fluid is one link in a chain that leads to thirst and drinking. Suppose that an organism loses water in water vapor in expired air. If the volume of solutes dissolved in the blood and the interstitial fluid remained the same, a net increase in the osmotic pressure of blood and interstitial fluids would result. Water would be drawn out of cells, thereby shrinking them. A crucial step in the chain producing thirst might be cellular dehydration.

Specialized cells located in the hypothalamus are especially sensitive to the minute change in volume associated with loss of water. If so, damage to these cells might impair the animal's ability to regulate body fluids. Indeed, destruction of cells in the lateral hypothalamus produces an immediate failure of rats to drink, a condition called adipsia. It seems very likely that some of the cells in the hypothalamus that are destroyed and that render the animal adipsic are those that are specialized to detect cellular dehydration and initiate drinking. But even if the lateral hypothalamus is destroyed, animals can recover and begin to regulate body fluids again by drinking. There must be another mechanism that can produce thirst.

Hypovolemic Thirst: Renin and Angiotensin

In the last part of the nineteenth century, it was discovered that the kidney produced a substance that can cause a constriction of peripheral blood vessels and a consequent increase in blood pressure. This substance released by the kidney was named *renin*. Renin is an enzyme that releases a hormone in the blood called *angiotensin I*. Angiotensin I in turn leads to the production of another hormone, *angiotensin II,* which acts to constrict peripheral blood vessels, thereby raising the blood pressure.

In order to maintain normal blood pressure, blood volume must be kept relatively constant. This enzyme pathway maintains normal blood pressures even in the face of lowered blood volume, from bleeding for example. Angiotensin II is a powerful constrictor of peripheral blood vessels, but in addition to its effect on blood pressure, angiotensin II makes animals thirsty. Whether injected intravenously or directly into the brain, angiotensin II produces immediate drinking even in a satiated animal. Thus, a loss of blood volume activates an enzyme pathway that produces both a compensatory increase of blood pressure and also the stimulus that is needed to restore blood volume by drinking.

Thus, in regulating drinking, peripheral factors detect the need for fluids, and central factors signal thirst or satiety. The ultimate switch is in the hypothalamus. An imbalance between the centers for thirst and satiety led to the incessant thirst of the patient described above. Different hypothalamic lesions can produce the opposite effect and render a person or animal unwilling to drink.

Hunger and Body Weight

Animals drink to recover water that is lost. They also must eat to balance expenditure of energy and growth. If they eat too little to balance energy and growth, they lose weight; if they eat too much, they gain weight. How is hunger regulated? Two aspects of this question can be separated. One deals with the long-term aspect of food intake. How is food intake regulated over weeks or months? The other question relates to short-term controls. What makes us hungry, and what makes us feel satiated during any single meal?

The Guiness Book of Records lists the fattest people whose weights have been carefully documented. One man whose picture was taken when he weighed a mere 670 pounds went on to gain 400 more pounds and achieve the record for being the heaviest person ever at 1,069 pounds. People can live with such extreme overweight, but they may die at a younger than normal age, often from the complications of their obesity.

Overweight and obesity are among the most common health problems. The percentage of overweight people varies in different parts of the world, but no country is free of them. According to one American study, 32 percent of the men and 46 percent of the women interviewed described themselves as being too heavy, and most of those at some time of their lives actively attempted to reduce their weight by diet. In almost all cases the great bulk of the excess weight is due to a larger than normal amount of fat tissue.

The human digestive system is remarkably efficient at extracting available energy from the food we eat. Ideally, there is a neat balance between the energy consumed in the form of food and the energy expended in the form of physical activity. If we consume more energy than we use, we store that excess as fat. The storage of fat is a relentless process. The more the energy consumed exceeds the energy used, the more fat we store. One expert estimated that an extra pat of butter on our bread each morning, if continued, would increase our weight by 10 pounds in one year and 100 pounds in ten years. One obvious way to correct overweight or obesity is diet—restricting the total amount of food or eating food with fewer calories. The hope is that the energy taken in from foods will balance that expended or, in a weight-loss diet, to eat even less, thus drawing on the reserves of energy from the available fats and thereby losing weight.

Because some people are extremely obese, surgical measures are sometimes taken. For example, one operation simply limits the volume of the stomach, whereas another bypasses much of the small intestine, thus limiting the amount of nutrient absorbed by the gastrointestinal tract.

One other extreme measure has been to treat very obese people with total fasting. That is, they are given enough fluids and necessary vitamins and minerals but little or no food. Under medical control, such fasting can be effective. But most patients, over time, regain the lost weight. This surprising obstinacy of obesity appears to reflect an underlying physiological process. People and animals seem to have a kind of built-in set point, which sets the level of their weight. When disrupted from the set point they restore their original weight over time.

If people of normal weight starve for any length of time, they lose weight, but when they are allowed to eat freely, they rapidly gain to precisely the same weight at which they started. The converse is also true. People who volunteered to eat a greatly excessive amount of food—by heroic efforts they ate as much as 8,000 calories a day—predictably gained weight. When such people are put back on free choice, they quickly lose back to almost their original weight. The same sort of thing is seen in experimental animals over an even wider range of conditions. Animals deprived of food become very thin. When allowed food again they gain back to the level of free-feeding control animals within a short period of time. Similarly, animals that are force-fed an excess amount of food gain weight, but they quickly lose the weight by diminishing their food intake when they are no longer force-fed.

Although there is a great deal of variability among people, each individual tends to stay very close to the same weight for most of his or her adult life. There appears to be a set point for weight that is partially under control of the brain.

The Mechanisms of Hunger and Satiety

An early suggestion about the sensation of hunger was that hunger pangs are simply contractions of the stomach. The physiologist Walter Cannon, working at Harvard in the early twentieth century, attempted to demonstrate the significance of stomach contractions when he had volunteer subjects swallow a recording balloon that was filled with air while it was in the stomach. Stomach contractions, as measured by the balloon, seemed to correlate with the subjects' reports of their hunger sensations. But the method of swallowing a balloon produces an atypical pattern of stomach motility. With less invasive methods it became clear that people cannot reliably detect contractions. Moreover, people whose stomachs have been surgically removed still report feeling hungry. Stomach contractions cannot be the sole stimulus for hunger.

S. W. Ranson and his co-workers at Northwestern University during the 1930s found that very small brain lesions in the hypothalamus can produce dramatic changes in feeding. Lesions restricted to the ventromedial nucleus of the hypothalamus of experimental animals led to a period of intense overeating, or hyperphagia, with a consequence that the animal soon became obese, going up to twice its normal body weight. This report of hypothalamic hyperphagia was paralleled by the discovery that bilateral lesions more laterally placed in the hypothalamus had the reverse effect. Animals were rendered unwilling or unable to eat (aphagia) and might not drink as well (adipsia).

From such observations of hypothalamic hyperphagia following ventromedial hypothalamic damage and aphagia caused by damage to the lateral hypothalamus, there arose a plausible scheme of the basic organization of appetite control. Feeding was thought to be initiated and controlled by a set of neurons in the lateral hypothalamus. This lateral hypothalamic feeding center was thought to be under inhibitory control from the ventromedial hypothalamic nucleus. Thus, direct destruction of the lateral hypothalamic feeding center would be expected to impair normal feeding in proportion to the severity of the damage, and destruction of the ventromedial nucleus would be expected to remove an inhibitory "brake" on the feeding center and thereby produce an animal that overeats.

Although this theoretical interpretation of food intake has some experimental support, as originally stated, it is oversimplified. Like the thermostat in a home, these nuclei contain cells that somehow sense the body weight and motivate feeding behavior when the weight becomes too low, or stop feeding if the weight is too high.

Animals prepared so that the contents of the stomach are immediately drained may nonetheless continue to eat for hours. Thus, the presence of food in the mouth or the act of eating is not enough to signal satiety. Some signal from the stomach or gut must be normally present after a meal. Leptin and ghrelin are two hormones secreted by the stomach. Ghrelin is associated with hunger. Leptin is secreted from the stomach when it has food. In obese people the circulating level of leptin is increased. One possible cause of obesity is related to leptin resistance.

One dramatic example of adaptive mechanisms in the control of feeding is the phenomenon of learned taste aversion. The California psychologist John Garcia, working in the 1950s, discovered that if rats are given a saccharin-flavored solution, they will normally drink it. If rats are given such a solution for the first time, and later made ill, they will no longer drink saccharin-flavored solution. Rats learn to associate a distinctive new flavor with the subsequent illness even though many hours may elapse from the first taste of food to when they become ill. This mechanism is the basis of "bait-shyness" also known as the "Garcia effect." When faced with new foods wise rats will nibble a bit. If they become sick, they then avoid that food. Most people, when asked, report the same effect for a particular food.

Temperature Regulation

Most animals function well only within a specific range of body temperatures. Extremes of heat or of cold can be lethal. In many animals, the *poikilotherms* or cold-blooded animals, there are no automatic regulatory mechanisms to keep internal temperature constant. Thus, all temperature regulation by such animals—fish, amphibians, and lizards—must be behavioral. The animal must move to a warmer or cooler place to avoid extremes of temperature.

Birds and mammals are homeotherms. They maintain their body temperature within relatively narrow limits by physiological adjustments. But these automatic temperature regulatory mechanisms themselves work only over a definite range, and so the constancy of our internal temperature must be maintained by behavioral as well as by physiological means. We may provide a warm spot on a cold night by building a fire or by adjusting the thermostat that controls a furnace.

Automatic Temperature Regulation

The bodies of humans and other animals continuously exchange heat with their environment. Heat is lost from the skin to the surrounding air on a cold day. At all air temperatures heat is produced by the body as a byproduct of metabolism, and a great deal of such heat is produced by muscles during heavy activity.

Temperature is continuously automatically regulated by changes in the diameter of peripheral blood vessels. On a cold day blood vessels in the skin are constricted. Indeed, heat loss is further controlled by conserving the heat in the blood that goes to superficial tissues. Arteries that distribute blood to the body surface are close to veins. Returning venous blood is directly heated by warmer arterial blood so that the arterial blood is cooled before it reaches the periphery. The blood's heat is not wasted. This "countercurrent" mechanism for heat preservation is especially well developed in mammals with a large surface area:volume ratio such as bats, whose wings have a huge surface area when compared to total body volume.

Just as peripheral blood vessels constrict in cold temperatures, so they expand in warmer air, facilitating heat loss. Humans can cool their skin by sweating, thereby increasing evaporative heat loss from the body surface in warm temperatures.

Behavioral Temperature Regulation

Humans heat or ventilate their homes. Animals will actively seek an appropriate temperature for maximal comfort. In warm weather a dog finds the shady side of a building for her nap. In cold weather she prefers to lie near the fire. Rats, when placed in a cold chamber, will work actively to heat it by pressing a lever to maintain air temperature at a comfortable level. Harry Carlisle, a psychologist working at the University of California at Santa Barbara, showed that rats will work to cool a hot chamber by pressing a lever that will give them a blast of cold air. They will even press a lever to give themselves a cold shower when their cage is too warm (figure 15.3).

The Hypothalamus and Temperature Regulation

The hypothalamus, critical for maintenance of thirst and appetite, is also necessary for controlling internal temperature. Rats with the entire forebrain removed, including cortex and basal ganglia, still have automatic temperature-regulating mechanisms intact. But if the hypothalamus is destroyed, or if the brainstem is transected just below the hypothalamus, such temperature regulation is lost irrevocably.

How does the hypothalamus sense temperature? How is it aware of the temperature of the body? There are two major ways in which information about warmth is communicated to the hypothalamus. Peripheral receptors in the skin and elsewhere feed information to the hypothalamus by neural pathways that ascend in the spinal cord. The hypothalamus

Figure 15.3
Showering rat diagram. Courtesy Hannah Glickstein Synan.

itself also has temperature sensitive cells whose firing rate is determined by the temperature of the blood and that are capable of triggering both automatic and behavioral temperature-regulating mechanisms.

Cells such as these are receptors for heat that can monitor the temperature of the blood circulating through the hypothalamus and initiate appropriate physiological and behavioral responses.

If a small heating probe is inserted in the hypothalamus, and if the temperature at the tip of the probe is raised just slightly, this direct thermal stimulus to the brain can initiate automatic physiological temperature-regulating responses. Heating of the hypothalamus initiates automatic physiological temperature regulation, and it can also serve to initiate behavioral responses. If the hypothalamus is directly heated, rats will press a lever for a blast of cool air even in a cool environment.

Sexual Structure and Function

For an individual to survive it must eat, drink, and maintain its temperature within an acceptable range. For the species to survive its members must reproduce their own kind. In most mammals ova are produced by females, and these are fertilized internally by males. In the late nineteenth century, biologists were aware of *chromosomes*—the odd-looking strands inside a cell's nucleus. Most chromosomes appeared to come in pairs. At first it seemed that there was an extra chromosome, called the "X" chromosome. By 1905 it had become clear that the X chromosome is typically paired with a smaller chromosome, now called the Y chromosome. A gene that is normally located in the Y chromosome promotes the development of testes in the embryo.

Geoffrey Raisman and Pauline Field, working in Oxford in the 1960s, showed that there is a subtle structural difference in the hypothalamus of male and female animals.

The length of the estrous cycle varies in different female mammals. The cycle in rats is four days; in humans it is twenty-eight days. In most mammals there is a direct effect on behavior by the hormones produced by the ovaries in the female. At the same time the hormones are causing ovulation, they are making the female more receptive to mating. Indeed, direct injection of estrogen into the appropriate area of the hypothalamus of female cats and rats produce a marked change in behavior. In both species of animal females become more receptive to mating.

Male hormone levels are not cyclical. The testes produce a constant amount of androgen in most males. Men castrated prior to puberty never develop normal secondary sex characteristics. Their voices remain high (the "castrati" of Italian music), and they seldom engage in any sexual behavior.

During prenatal development, the embryo normally develops female sexual characteristics. Androgen in the embryo and newborn produces the characteristic male sexual development. Indeed a single injection of testosterone—the principal male sex hormone—in newborn female rats affected all of their later sexual development. Female rats so treated never showed normal ovulation or sexual cycling when they reached adulthood. Genetically male rats castrated at birth will show normal female sexual receptivity when given female sex hormone in adulthood. It is tempting to generalize to the human condition. I believe, for example, that many cases of cross-gender identity are related to the early intrauterine environment, but the evidence is speculative.

16

Language and the Brain

The Oxford English Dictionary defines language as: "The system of spoken or written communication used by a particular country, people, community . . . typically consisting of words used within a regular grammatical and syntactic structure."

Language: Universally and Uniquely Human

Language as defined above is present in all humans. People can produce as well as recognize words and sentences. Many animals can respond to spoken commands. Dogs are particularly clever and responsive animals. The American psychologist John Pilley recently taught Chaser, a six-year-old border collie, to fetch one object among a large number of specific objects, and she seems to understand hundreds of words. Chaser even appears to be able to infer that a new word that she never heard before must be associated with a new object that she had not seen before. Chaser seems to be about half-way to having acquired language.

Chimpanzees have been the principal subjects in attempts to teach animals language. For example, Laura and Winthrop Kellogg moved from Indiana to Florida in the 1930s, where they raised a baby chimpanzee, Gua. They compared Gua's early development with that of their son Donald. Gua developed motor coordination well before Donald. She learned to respond to directions such as "Where is your nose?" but she never gave evidence of acquiring speech. Twenty years later, Keith and Cathy Hayes, like the Kelloggs, were also based at the Yerkes Primate Research Laboratory in Florida. They raised Vicki, making a concerted effort to try to teach her to speak. Vicki managed to say a word or two—for example, "cup" if she was thirsty.

Vicky and all chimps lack the vocal apparatus of a normal human. Allen and Beatrix Gardner, working first at the University of Nevada in the 1960s, raised Washoe as if she were their own deaf child, using American Sign Language to communicate with her. According to the Gardners, Washoe learned over three hundred signs and used them appropriately. But others have questioned their conclusions. Psychologist David Premack, a professor at the University of Pennsylvania, also tested language ability in a chimpanzee. Premack listed a number of criteria for faculties that underlie the evolution of language. Although chimps seem to have acquired some of these mental and physical skills, he concludes convincingly that they do not have language in the human sense: ". . . chimpanzees can represent what they perceive, whereas humans can represent what they imagine."[1]

Language Localization

By the beginning of the nineteenth century it had become clear that lesions of the human or animal cerebral cortex can cause deficits in sensation, movement, and thought. There was little known about how these functions might be regulated and controlled by different parts of the brain. Functional localization in the brain was asserted by some and denied by others. Franz Josef Gall and Johann Caspar Spurzheim, physicians working in Austria in the late eighteenth and early nineteenth centuries, developed an elaborate scheme of localization whereby sensation, feelings, and talents were thought to be localized in different parts of the brain. Although Gall was an excellent anatomist, his speculations and those of his followers went far beyond available evidence. Gall proposed that the functions of different parts of the brain would be reflected in subtle differences in the shape of the overlying skull. Changes in skull structure, in turn, might affect the position of the eyes. In discussing people with protruding eyes Gall wrote,

> Persons who have eyes like this not only possess an excellent memory for words but they have a particular disposition for the study of languages and in general judgment of all that pertains to literature. Functions of librarians or conservator; they gather the scattered riches of all the centuries; they compile learned volumes; they scrutinize antiquities and, no matter how little they may have of the other faculties, they are the admiration of all the world on account of their vast erudition.

On language localization, Gall wrote,

> I regard the organ of memory of words that part of the brain which rests on the posterior half of the orbital roof (the central portion of the posterior third of the orbital surface of the cerebral hemisphere, immediately anterior to the tip

of the temporal lobe). In the engravings we have not given special numbers to the part of the faculty of speech.[2]

Gall and Spurzheim called their "science" *cranioscopy*, later called *phrenology* by his followers. Phrenology became popular among lay people as well as physicians in the early nineteenth century.

As mentioned in chapter 8 of this volume, The French physiologist Pierre Flourens (figure 16.1) disagreed with Gall's pronouncements on the principle of localization in the cerebral cortex. Flourens made lesions in the brains of experimental animals and studied the resultant effects on behavior. He correctly identified the cerebellum as playing a critical

Figure 16.1
Pierre Flourens (1794–1867) made lesions in the brain of experimental animals and looked for evidence of localized functions. Although he affirmed that the cerebral cortex is necessary for normal vision, hearing, movement, and thought, he was unable to find localization for any of those functions. "Unity is the great principle."

role in the coordination of movement. Although he found that the cerebral cortex was directly involved in sensation and movement, despite searching for evidence of localization, he found none. He wrote: "The cerebral lobes are the exclusive site of sensations perceptions and volitions. . . . All these sensations, perceptions, and volitions concurrently occupy the same area in these organs. Therefore, the ability to feel, perceive, and to desire constitute only one essentially single faculty."[3]

Despite Flourens' failure to find evidence, there were hints of evidence of localization for speech and language in the writings of some French physicians in the early nineteenth century. But definitive proof was lacking.

The French physician Jean-Baptiste Bouillaud wrote the following in 1825:

General Conclusions

1. In man the brain plays an essential role in the mechanism of a large number of movements; it directs all those which are subject to the control of intelligence and volition.
2. There are in the brain several special organs, each of which directs specific muscular movements.
3. The movements of the organs of speech in particular are controlled by a special distinct and independent cerebral center.
4. This cerebral center occupies the anterior lobes.
5. The loss of speech depends now upon the loss of the memory of words, now upon the loss of muscular movements by which speech is composed, or, what comes perhaps to the same thing; now upon a lesion of the gray matter and now upon that of the white matter of the anterior lobes.
6. The loss of speech does not involve loss of the movements of the tongue considered as an organ of prehension, mastication, or swallowing of food, or the loss of taste; which presupposes that the tongue has three sources of distinct action in the nerve center, a hypothesis, or rather a fact, which agrees admirably with the presence of a triple nervous organ in the tissue of the tongue.[4]

Bouillaud's son-in-law, Simon Alexandre Ernest Aubertin, qualified in medicine in Paris in 1852. Nine years later, Aubertin described a patient seen by himself and his father-in-law. Speaking in 1861 at the Anthropological Society in Paris, Aubertin said,

For a long time during my service with M. Bouillaud I studied a patient, named Bache, who has lost speech but understood everything said to him and replied with signs in a very intelligent manner to all questions put to him. This man, who spent several years at Bicêtre, is now at the Hospital for Incurables. I saw him again recently and his disease has progressed; slight paralysis has appeared but his intelligence is still unimpaired and speech is wholly abolished. Without doubt this man will soon die. Based on the symptoms that he presents, we have diagnosed softening of the anterior lobes. If, at autopsy, these lobes are found to be intact, I shall renounce the ideas that I have expounded to you; however, I can only argue according to the facts which exist in science today. Now, to the best of my knowledge, no one has ever seen a lesion limited to the middle and posterior lobes that has destroyed the faculty of speech. . . I shall content myself with the examination of one particular function of the anterior lobes of the brain, for it is sufficient to demonstrate a single localization in order to establish the principle of localizations, and I ask no more for this.[5]

The critical evidence came that same year, in April 1861, and was observed and reported by Pierre Paul Broca (figure 16.2). Trained in

Figure 16.2
Pierre Paul Broca (1824–1880) was a French physician who described the post-mortem location of a lesion in the left hemisphere associated with a severe loss of the ability to speak. Courtesy Wellcome Library, London.

medicine in Paris, Broca was on the faculty of The University of Paris when a patient, M. Le Borgne, was transferred to his ward. M. Le Borgne suffered from a profound defect in speech. He was called "Tan" by the other patients and the nursing staff. Although he appeared to understand all that was said to him, he was unable to say anything other than "Tan" in reply.

M. Le Borgne aka "Tan" died in April 17, 1861, and the following day the lesion in his brain was described by Broca to a meeting of the Anthropological Society in Paris. Broca wrote of this case,

> On the eleventh of April, 1861, to the general infirmary of the Bicêtre, to the service of surgery, was brought a man 51-years-old called Leborgne who had diffused gangrenous cellulitis of the whole right inferior extremity, from the foot to the buttocks. To the questions which one addressed to him on the next day as to the origin of his disease he responded only by the monosyllable "tan," repeated twice in succession and accompanied by a gesture of the left hand. I tried to find out more about the antecedents of this man, who had been at Bicêtre for 21 years. I asked his attendants, his comrades on the ward, and those of his relatives who used to see him, and here is the result of this inquiry.
>
> Since his youth he was subject to epileptic attacks, but he could become a last-maker at which he worked until he was 31 years old. At that time he lost the ability to speak, and that is why he was admitted at the Hospice of Bicêtre. One could not find out whether this loss of speech came on slowly or fast, nor whether some other symptom accompanied the beginning of this affliction.
>
> When he arrived at Bicêtre he could not speak for 2 or 3 months. He was then quite healthy and intelligent and differed from a normal person only by the loss of articulate language. He came and went in the Hospice where he was known under the name of "Tan." He understood all that was said to him. His hearing was actually very good. Whatever question one addressed to him, he always answered, "tan, tan," accompanied by varied gestures, by which he succeeded in expressing most of his ideas. If one did not understand his gestures, he usually got irate, and added to his vocabulary a gross swear word. . . ."

Broca went on to describe the brain at autopsy,

> . . . On the lateral side of the left hemisphere . . . lies . . . a large and profound depression of the cerebral substance . . . .When this fluid was evacuated by a puncture, the pia mater became profoundly depressed, and there resulted a long cavity about as large as a hen's egg, which corresponded to the Sylvian fissure, hence separated the frontal and the temporal lobes. It reached backwards as far as the sulcus of Rolando which separates, as one knows, the frontal from the parietal convolutions. The lesion is therefore completely in front of the sulcus, the parietal lobe is intact. . . .
>
> In our patient, the original seat of the lesion was in the second or third frontal convolution, more likely in the latter. It is therefore possible that the faculty of

articulate speech is in one or the other of these two convolutions. However, it is difficult to know it at present, because the former observations do not describe the state of each convolution, and one cannot even forecast it because the principle of localization by convolution is not yet firmly established.

Broca concluded,

This abolition of speech in individuals, who are neither paralyzed nor idiots, is a sufficiently important symptom so that it seems useful to designate it by a special name. I have given it the name Aphemia, for what is missing in these patients is only the faculty to articulate words: they hear and understand all that is said to them, they have all their intelligence and they emit easily vocal sounds.[6]

Broca's designation "aphemia" was later changed to "aphasia" by Armand Trousseau, a contemporary of Broca's in Paris.

Carl Wernicke and Wernicke's Aphasia

Broca's patient could not speak, but he seemed to understand everything that was said to him. There had always been parallel cases in which patients lost the ability to understand language, but often these were dismissed as simply reflecting a general loss of intellect. The German psychiatrist Carl Wernicke in a classic monograph linked such cases to damage to a particular locus in the temporal lobe. Wernicke, who was born in 1848 in Tarnowitz in Upper Silesia and studied medicine at the University of Breslau (now Wroclaw), argued that there is more than a single kind of aphasia. He identified a different brain locus for the other kind of deficit. He wrote:

Case I

Suzanne Adam. Nee Sommer, a labourer's widow age 59 years, suddenly became ill from an unknown cause on March 1, 1874 with symptoms of vertigo and headache. However, there was no loss of consciousness of the type which is often accompanied by confused speech. Correct speech usage was infrequent, and her answers to questions were completely confused. Although she was able to express her complaints regarding headache and dizziness quite accurately, the meaningless word "begraben" (buried) was interjected into whatever she said.

The patient had gone to bed after lunch, which was her custom, and on the next day was admitted to an inner ward of the Allerheiligen Hospital. There her condition was diagnosed as dementia, and since no physical illness was apparent, she was placed on a psychiatric ward.

On March 7, 1874, the following status was recorded. A frail, middle-aged woman, demonstrating a senile cataract on the right and a coloboma of the iris in the left. The patient's facial expression shows intelligence and cooperation. No

Figure 16.3
Carl Wernicke (1848–1904). Courtesy U.S. National Library of Medicine.

disturbance is noted in gait. Hand grasp is bilaterally weak, with greater weakness demonstrated on the left. Some generalized decrease in sensation, tested by pinprick, is observed with minimal response demonstrated in fingers toes and face. Marked tortuosity of the veins, which are hard as ropes to palpation, including the superficial temporal arteries is evident. Heart and lungs are essentially normal. Hearing, tested by holding a watch to the ear, is well preserved bilaterally. Glaucomatose excavation of the pap. optica is demonstrated on the left.

The patient could comprehend absolutely nothing that was said to her. However, caution was necessary to avoid giving clues by gesture. When called, she responded to unfamiliar names as well as her own by turning round and answering "yes."

> Superficially she gave the impression of dementia, not only because her answers to questions were inappropriate, but in that the sentences themselves were incorrectly produced, containing meaningless and garbled words. And yet the overall meaning of a sentence, which could be grasped in a general way, was always reasonable. There was no trace of flight-of-ideas. Her behavior was calm and appropriate, while dementia, on the contrary, is generally accompanied by severe psychic deterioration.

Wernicke concluded,

> In agreement with the preceding cases which consider speech development from the standpoint of conscious movement, a priori reasoning would consider that restricting of the speech center to a single area, namely Broca's gyrus, would be highly improbable. A consideration of the anatomic structure as described above, the support of numerous necropsy findings and finally the variability in the clinical picture of aphasias all strongly lead us to the following interpretation of the data. The entire region of the first primordial convolution, the gyrus surrounding the Sylvian fissure in association with the insular cortex serves as a speech center. The first frontal gyrus (Leuret) which is motor in function acts as a center of motor imagery. The first temporal gyrus which is secondary in nature, may be regarded as the center of acoustic images: the fibrae propriae converging onto the insular cortex from the mediating reflex arc.[7]

Broca, Wernicke, and Conduction Aphasia

Wernicke thus suggested a division of functions. Wernicke's temporal lobe area, near the sensory cortex, would be a center for speech recognition. Broca's frontal lobe area, adjacent to motor areas of the cerebral cortex, would be a center for the production of speech. Patients with damage to Broca's area have difficulty in producing speech, but they understand it. Because of its characteristic symptoms it is called an expressive aphasia. Patients with damage to Wernicke's area are said to have *receptive aphasia*. They are often capable of fluent speech, but what they say may make little sense, and they fail to understand what is said to them.

Wernicke suggested that there must be an anatomical linkage between the area that he identified in the temporal lobe and Broca's area in the frontal lobe. Patients with a milder form of aphasia, conduction aphasia, have difficulty in repeating sentences or words the examiner says to them, indicating the absence of a link between the two areas. The anatomical basis of the condition has been ascribed to an interruption of a broad bundle of nerve fibers called the *arcuate fasciculus*, which connects the temporal to the frontal lobe. One interpretation is that the fiber cuts prevent linkage between the brain area responsible for recognizing an object with the brain area responsible for naming it.

The Corpus Callosum, Language, and Hemispheric Specialization

The corpus callosum is the massive system of nerve fibers that links the two sides of the brain. As we saw in chapter 14 of this volume, the callosum serves the interhemispheric transfer of memory. In the 1930s the neurosurgeon Walter Dandy, working at Johns Hopkins University, had a patient who presented with a tumor of the pineal gland. Dandy cut through the callosum in order to reach and remove the tumor. He saw no obvious symptoms caused by the surgery. He wrote,

> The corpus callosum is split longitudinally, from its posterior extremity to a point anteriorly where the third or lateral ventricle comes into view; this incision is bloodless. Usually this incision takes most, and sometimes all of the structure to its downward bend. No symptoms follow its division. This simple experiment at once disposes of the extravagant claims to function of the corpus callosum.[8]

It doesn't.

A year later, two other neuroscientists at Johns Hopkins, J. H. Trescher and F. R. Ford, studied a patient of Dandy's who had been operated on to drain a cyst of the third ventricle. The surgical approach in this case required cutting the corpus callosum. Although the patient appeared to be in many respects entirely normal, careful study revealed an important clue about the functions of the callosum—and of brain mechanisms in language. When presented with letters or words in the left visual field, the patient was unable to name them. There were no problems in the right visual field. The same inability to read in the left visual field in two callosum-sectioned patients was reported eleven years later by the Italian neurosurgeon Paolo Maspes, working at the University of Torino.

Neurologists and neurosurgeons are sometimes confronted with a patient with intractable epilepsy. Seizures persist despite drug treatments. In some of those patients it seems as if a healthy hemisphere is being subjected to continuous abnormal brain activity by the damaged side. In 1940 the surgeon W. P. Van Wagenen, working at the University of Rochester in upstate New York, described the results of his attempts to relieve the condition by cutting the corpus callosum. Despite extensive follow-up by the neurologist Andrew J. Akelaitis at the University of Rochester, there seemed to be no obvious symptoms that could be ascribed to the surgery.

Another important neuroscientist in the study of the corpus callosum was Joseph Bogen. He had been a postdoctoral student with Anton van Harreveld at the California Institute of Technology, where he became familiar with the work on interhemispheric transfer of Roger Sperry and his collaborators. As a resident in neurosurgery, Bogen had reviewed the earlier studies by Van Wagenen and his colleagues and reopened the question of whether complete section of the corpus callosum and the smaller anterior commissure might be beneficial to patients with intractable seizures. In 1962, along with the chief of neurosurgery at the White Memorial Hospital, Philip Vogel, he performed their first commissure sections.

By good luck, Bogen and Sperry were joined by Michael Gazzaniga, who was a graduate student at Cal Tech. It was Gazzaniga who initiated much of the testing and drew many of the conclusions from careful study of the early patients. His observations not only confirmed the inability of callosum-sectioned patients to read or name objects presented in the left visual field—that is, to the right hemisphere—but their work also opened a new view of cerebral lateralization. In a 1977 summary of the work of Gazzaniga and others, Sperry wrote,

> Meantime, the non-vocal mute "minor" right hemisphere can show by the use of manual signs, nonverbal pointing, or tactual retrieval that it perceives and comprehends correctly the same stimulus input for which the speaking hemisphere disclaims any awareness. Further, the mute hemisphere displays a corresponding lack of ability to respond to stimulus input restricted to the vocal hemisphere. The foregoing applies in general to all sensory modalities thus far tested that can be lateralized, including visual, somesthetic, olfactory and auditory.[9]

Despite this disability of the right hemisphere for speech and language, Gazzaniga later found that there can be a residual capacity for single-word utterances controlled by the right hemisphere.

In the great majority of humans the critical areas for speech are localized in the left hemisphere. As Michael Gazzaniga, now at the University of California at Santa Barbara, and his colleagues showed, patients in whom the great mass of fibers that links the two hemispheres—the corpus callosum—is cut may be unable to name an object presented to them in the left visual field, although they may be able to use it appropriately. Unable to name a pair of scissors or a spoon, they can show how it is used.

As Marjorie Lorch, a neuroscientist and historian at Birkbeck College in London showed recently, the discovery of an association between a specific brain lesion and language deficit had an important influence on the law relating to mental capacity. Previously patients who had lost the power of speech were relegated to a class of idiocy. Importantly, Tan and other patients, although they lost the ability to speak, were not demented.

Speech and Language Development

Newborn infants are born with the ability to learn to speak and understand. There is a typical sequence whereby infants begin by babbling. By the age of a year infants show evidence of understanding simple commands such as "Show me your nose." By the age of two years the typical child's vocabulary consists of fifty or more words. In normal children this capacity to speak and understand reflects great storage capacity in the left hemispheres.

Philosophers have speculated about whether a child who never heard speech could develop normal language. Alleged cases of feral children raised by wolves or gazelles were sometimes cited as evidence. One case of the alleged wolf-raised children is probably simply a fraud. In the 1920s Amala and Kamala, two girls from India, were allegedly raised by wolves. Although they made it into the introductory sociology and psychology texts, the evidence suggests that their history was almost certainly misrepresented.

There are cases of children mistreated by parents or caretakers—for example, a girl locked in a room by herself without human contact. In all such cases language is reported to be absent, and if language is acquired at all when the child is subsequently adopted, it is with enormous difficulty. None of these stories of such severe deprivation can be authenticated. The abused children are typically mentally handicapped to start with, and the severity of the earlier deprivation is often exaggerated.

In most adults the control of language is centered in the left hemisphere. There is however, firm evidence that the right hemisphere can acquire normal language in brain-injured children. The evidence comes from cases of children who have sustained massive lesions of the left hemisphere early in life. Sometimes a family will be confronted with a stark choice, in which a young child's left hemisphere is so damaged by scarring or tumor, that it is advisable to remove it entirely. Surprisingly, if this is done early enough, the child may develop normal language, presumably mediated by the right hemisphere.

Eric Lenneberg, a psychologist working at Harvard in the 1960s, discussed documented cases showing that, if the damage occurs before the age of eight years, there can be complete or near complete recovery of language. By the age of twelve to fourteen damage always results in a residual language deficit. The evidence shows that, up until the age of twelve, the brain retains some plasticity for language.

Study of language and the brain has underscored a principle that began to emerge slowly in the nineteenth century. Control of functions such as vision, hearing, movement, and language are precisely localized in the cerebral cortex. Of these functions, language is typically controlled by the left hemisphere, and it acquires that localization early in life.

Some Conclusions

From the above we can draw some general conclusions about language:

- Language is universal among all humans, and it is a uniquely human ability.
- The abilities to speak and to understand language depend on localized regions in the cerebral cortex. Damage to those areas causes difficulty in understanding and/or speaking language.
- In the great majority of people language is represented in the left hemisphere. If links from visual or somatosensory cortex to the language areas are interrupted, people cannot report verbally what they see or feel.
- There is a critical period in early childhood during which language is usually acquired.

17

Neurological Disease

Like other organs of the body, the brain and spinal cord are subject to damage and disease. This chapter is about diseases of the nervous system. The chapter attempts to convey the nature of several of the major neurological diseases, where possible relying on direct quotes from classical description of those diseases. The history reveals a typical sequence from first recognizing a disease entity and relating it to obvious gross structural damage to analysis of defects in transmitter release and uptake, and more recently of membrane channels or abnormal proteins.

How Are Diseases Named?

Dorland's medical dictionary defines *sign* as "any objective evidence of a disease," *symptom* as "any functional evidence of a disease or of a patient's condition . . . ," and *syndrome* as "a complex of symptoms which occur together." Diseases are sometimes named for their sign or symptoms. Most were named before modern understanding of their causes, often by physicians who were classically educated, and so the names of the disease might reflect the symptoms as expressed in Greek or Latin.

In the nineteenth century there was a great increase in the study of pathology, with the aim of establishing postmortem *why* a patient died. Disease names might then reflect structural changes that are associated with the disease. In assigning a name there has always been a tendency to recognize a particularly clear description of a collection of the symptoms that constitutes a disease. Such diseases may be named for the physician who gave an early description. On some rare occasions a disease might be named for a patient who suffered from it.

Amyotrophic lateral sclerosis means "abnormal muscle tone associated with a hardening of the lateral white columns of the cord." In France the disease was usually called *Charcot disease* after the nineteenth-century neurologist who described its symptoms and the appearance of the brain and spinal cord postmortem. Because it particularly attacks the motor system, it is also called *motor neuron disease*. Lou Gehrig, one of the most famous baseball players in the United States, began to lose his great ability as a fielder and hitter and died from the disease in 1939. Thus, the disease is often called *Lou Gehrig's disease*.

Epilepsy

Epilepsy, from the Greek meaning seizure, was among the very first neurological diseases to be recognized as a brain disorder. Epilepsy manifests itself as a brief period of unconsciousness, often accompanied by vigorous movement. The Greek physician Hippocrates lived on the island of Cos, surrounded by a group of colleagues and students, many of whom wrote on medical questions. The body of Hippocratic writings dates from a period around 400 BC. At the time many physicians believed that the disease of epilepsy was caused by direct influence of the gods. Hippocrates disagreed. He wrote:

> I do not believe that the "Sacred Disease" is any more divine or sacred than any other disease but, on the contrary, has specific characteristics and a definite cause. Nevertheless, because it is completely different from other diseases, it has been regarded as a divine visitation by those who, being only human, view it with ignorance and astonishment. . . . So far from this being the case, the brain is the seat of this disease, as it is of other very violent diseases.

The text goes on to discuss the manifestations and the causes of the disease.

> Patients who suffer from this disease have a premonitory indication of an attack. In such circumstances they avoid company, going home if they are near enough, or to the loneliest spot they can find if they are not, so that few people as possible will see them fall, and they at once wrap their heads up in their coats. . . .
>
> Small children, from inexperience and being unaccustomed to the disease, at first fall down wherever they happen to be. Later, after a number of attacks, they run to their mothers or to someone whom they know well when they feel one coming on. This is the fear and fright at what they feel, for they have not yet learnt to be ashamed. Adults neither die from an attack of this disease, nor does it leave them with palsy.[1]

As in much of the Hippocratic writings, his clinical descriptions often remain valid today. The writer identifies the fact that many patients experience an aura, a premonition that an attack is imminent. The description of the protective act of wrapping their head with their coat is also of interest. If they know that they are going to fall, they might hope to prevent damage to their head.

By the nineteenth century the manifestations of epilepsy were well recognized and described. The British neurologist W. R. Gowers wrote in his textbook of neurology:

> Epileptic attacks are commonly divided into two classes, "major" or severe, or "minor" or slight. These two forms, although clearly distinguished in their general characters, are not separated by any sharp demarcation. In the major attacks (grand mal) there is loss of consciousness and severe muscular spasms. In the minor attacks (petit mal) there is usually brief loss of consciousness, often without any muscular spasm, sometimes with a slight spasm, and very rarely there is a slight spasm or some sudden sensation without loss of consciousness. In severe attacks the patient, if standing, falls to the ground (hence the old English name "falling sickness"). In slight attacks he may or may not fall. In very severe fits, muscular spasm and loss of consciousness are simultaneous in onset, but in less severe fits the muscular spasm may commence before consciousness is lost; the patient is aware of the onset. Still more frequently the spasm and loss of consciousness are preceded by some sensation. The sensation, or commencing spasm, which informs the patient of the oncoming attack, constitutes the "warning" or "aura" of the fit.[2]

In addition to the distinction between grand-mal and petit-mal seizures, a further class of epileptic fits was described by the English neurologist John Hughlings Jackson in the last half of the nineteenth century. In one form of epilepsy the muscular spasms may be limited to one side of the body. Moreover, the contractions might involve successive, neighboring groups of muscles, now called "Jacksonian epilepsy." Jackson wrote:

> Then in unilateral convulsions the "aura" nearly always begins in the hand, sometimes, however, in the side of the face and rarely in the leg. So the speculation is that although each movement is everywhere represented, there are points where particular movements are specially represented.

And later he wrote:

> I think the mode of beginning makes a great difference as to the march of the fit. When the fit begins in the face, the convulsion in involving the arm may go down the limb. . . . When the fit begins in the leg, the convulsion marches up: when the leg is affected after the arm the convulsion marches down the leg.[3]

Jackson's astute observations were made about the same time as Fritsch and Hitzig's studies of the effects of direct electrical stimulation of the cerebral cortex (see chapter 13 of this volume). Although Jackson at first believed the source of the disorder to be localized in the basal

Figure 17.1
British neurologist John Hughlings-Jackson (1835–1911) described a characteristic form of epileptic seizure in which a series of adjacent body parts becomes successively involved in the fit. Courtesy Wellcome Library, London.

ganglia, he later identified the cerebral cortex. Jackson's observations also were an important element in recognizing the somatotopic organization of a motor map in the brain. The fact that the fits in Jacksonian epilepsy follow a course in which they affect successively neighboring areas seems to confirm that the control of nearby body parts should be by adjacent structures in the brain.

Multiple Sclerosis

Multiple sclerosis—sclerosis refers to the hardening of central white matter tracts—is a disease of the central nervous system that attacks the myelin coating of axons. The disease may strike at any fiber system of the brain or spinal cord. The disease has probably been present in the population for hundreds of years, but it was first identified in the medical literature by the French neurologist Jean-Martin Charcot in the late nineteenth century. Typically first appearing in young adults, the disease is characterized by periods of remission and progresses slowly. Damage to the myelin coating may impair or abolish the functions of a given fiber tract and eventually lead to the death of the fiber; thus, the symptoms are variable, depending on where in the brain or spinal cord the disease is active. The periods of remission may be associated with some restoration of myelin in the affected tracts.

Two famous cases in the literature are among those that predate Charcot's late nineteenth-century identification. The earliest report of a disease that was almost certainly multiple sclerosis afflicted Lidwina of Schiedam in The Netherlands. As a young adolescent, Lidwina fell while skating. Thereafter she developed headaches and tooth pains, with a slow increase of difficulty in walking and control of her facial muscles. Most authors agree that the progression of her symptoms indicate that she was suffering from multiple sclerosis. Lidwina took her troubles calmly, feeling that the suffering brought her closer to Jesus. After her death she was canonized as Saint Lidwina by the Roman Catholic Church.

A much better-documented case was that of Augustus D'Este (figure 17.2). Augustus was born in 1794, a grandson of King George III of England. His father was Prince Augustus Frederick, the sixth son of George III. His mother was Lady Augusta Murray, the daughter of the earl of Dunmore. The parents met while both were traveling in Italy, and they married in that country. The marriage outraged the King. Despite a second wedding ceremony in London, the King instructed the Court of Arches to annul the marriage, and so young Augustus was technically illegitimate, although he traveled in aristocratic circles. At the age of fourteen he was sent first to Harrow, and a year later to Winchester School. At the age of eighteen he joined the army and reached the level of colonel in his regiment of fusiliers. Augustus kept a diary in which he recorded the progression of symptoms caused by his disease.

Augustus wrote:

In the month of December 1822 I travelled from Ramsgate to the Highlands of Scotland for the purpose of passing some days with a Relation for whom I had the affection of a son. On my arrival I found him dead. I attended his funeral: there being many persons present I struggled violently not to weep. I was however unable to prevent myself from so doing: Shortly after the funeral I was obliged

A Boy
Said to be Sir Augustus Frederick d'Esté (1794–1848)
by Richard Cosway: signed and dated 1799

Miniature in the Victoria and Albert Museum

THE CASE OF
AUGUSTUS D'ESTÉ

BY

DOUGLAS FIRTH
M.A., M.D. (Cantab.), F.R.C.P. (Lond.)
Fellow of Trinity Hall, Cambridge

CAMBRIDGE
At the University Press
1948

Figure 17.2
August D'Este was from an aristocratic family who kept notes describing the symptoms of his progressive disease, almost certainly multiple sclerosis. Courtesy Cambridge University Press.

to have my letters read to me, and their answers written for me, as my eyes were so attacked that when fixed upon minute objects indistinctness was the consequence: Until I attempted to read, or to cut my pen, I was not aware of my Eyes being in the least attacked. Soon after I went to Ireland, and without anything having been done to my Eyes, they completely recovered their strength and distinctness of vision.[4]

The subsequent history was one of ever increasing disability. Augustus consulted a number of physicians and surgeons, who prescribed various drugs, series of baths, and electric stimulation. He died at the age of fifty-four (figure 17.3).

The progression and character of his symptoms make it almost certain that Augustus suffered from multiple sclerosis. Among the most frequent early signs of the disease are visual symptoms, called "optic neuritis." In

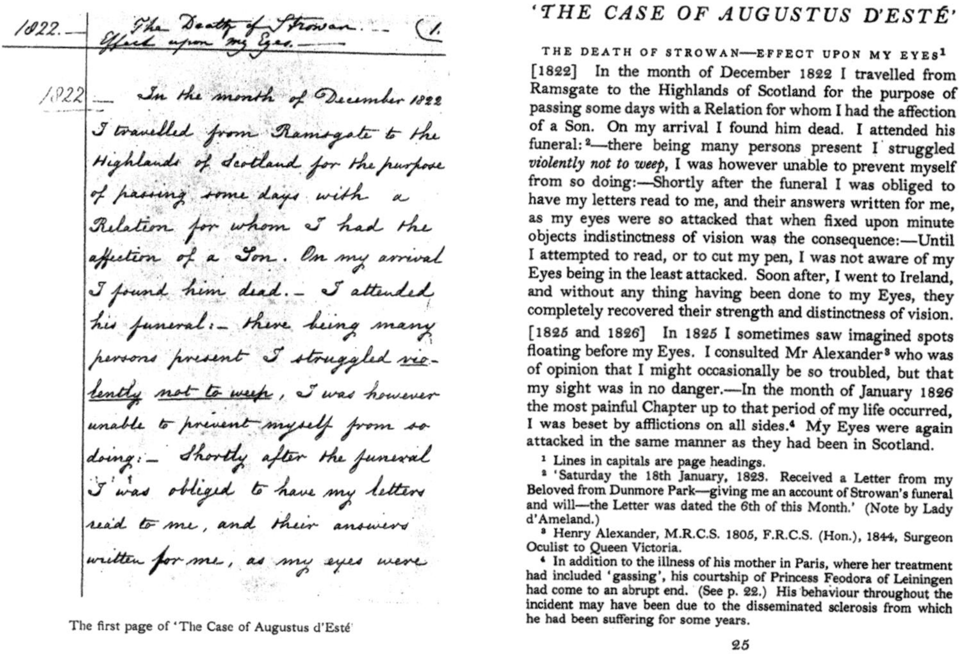

1822. — The Death of Strowan. — Effect upon my Eyes. — (1.

1822 — In the month of December 1822 I travelled from Ramsgate to the Highlands of Scotland for the purpose of passing some days with a Relation for whom I had the affection of a Son. On my arrival I found him dead. — I attended his funeral: — there being many persons present I struggled violently not to weep, I was however unable to prevent myself from so doing: — Shortly after the funeral I was obliged to have my letters read to me, and their answers written for me, as my eyes were

The first page of 'The Case of Augustus d'Esté'

'THE CASE OF AUGUSTUS D'ESTÉ'

THE DEATH OF STROWAN—EFFECT UPON MY EYES[1]

[1822] In the month of December 1822 I travelled from Ramsgate to the Highlands of Scotland for the purpose of passing some days with a Relation for whom I had the affection of a Son. On my arrival I found him dead. I attended his funeral:[2]—there being many persons present I struggled *violently not to weep*, I was however unable to prevent myself from so doing:—Shortly after the funeral I was obliged to have my letters read to me, and their answers written for me, as my eyes were so attacked that when fixed upon minute objects indistinctness of vision was the consequence:—Until I attempted to read, or to cut my pen, I was not aware of my Eyes being in the least attacked. Soon after, I went to Ireland, and without any thing having been done to my Eyes, they completely recovered their strength and distinctness of vision.

[1825 and 1826] In 1825 I sometimes saw imagined spots floating before my Eyes. I consulted Mr Alexander[3] who was of opinion that I might occasionally be so troubled, but that my sight was in no danger.—In the month of January 1826 the most painful Chapter up to that period of my life occurred, I was beset by afflictions on all sides.[4] My Eyes were again attacked in the same manner as they had been in Scotland.

[1] Lines in capitals are page headings.

[2] 'Saturday the 18th January, 1823. Received a Letter from my Beloved from Dunmore Park—giving me an account of Strowan's funeral and will—the Letter was dated the 6th of this Month.' (Note by Lady d'Ameland.)

[3] Henry Alexander, M.R.C.S. 1805, F.R.C.S. (Hon.), 1844, Surgeon Oculist to Queen Victoria.

[4] In addition to the illness of his mother in Paris, where her treatment had included 'gassing', his courtship of Princess Feodora of Leiningen had come to an abrupt end. (See p. 22.) His behaviour throughout the incident may have been due to the disseminated sclerosis from which he had been suffering for some years.

25

Figure 17.3
Letter and text from *The Case of Augustus D'Este*. Courtesy Cambridge University Press.

addition, the transient nature of Augustus's visual impairment is typical of the disease, in which symptoms can appear, followed by remission.

Parkinson Disease

The English physician James Parkinson was born in Hoxton, near London, in 1755, and trained in medicine at the London Hospital. Over the years he observed and recorded the symptoms of the disease that bears his name. In 1817 he wrote his famous essay summarizing the symptoms and course of the disease.

History

So slight and nearly imperceptible are the first inroads of this malady, and so extremely slow its progress, that it rarely happens, that the patient can form any recollection of the precise period of its commencement. The first symptoms perceived are, a slight weakness with a proneness to trembling in some particular

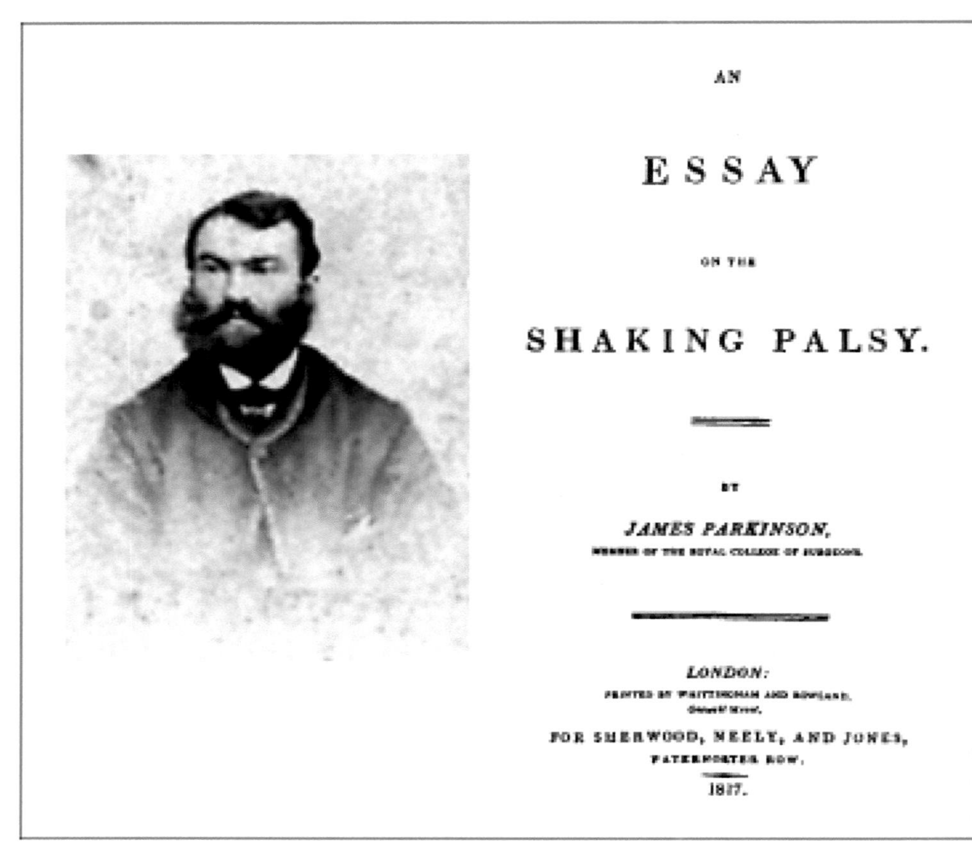

AN

ESSAY

ON THE

SHAKING PALSY.

BY

JAMES PARKINSON,

LONDON:

FOR SHERWOOD, NEELY, AND JONES,

PATERNOSTER ROW.

1817.

Figure 17.4
Portrait of James Parkinson (1755–1824). Parkinson described the symptoms and the course of the disease that bears his name.

part: sometimes in the head, but most commonly in one of the hands and arms. These symptoms gradually increase in the part first affected: and at an uncertain period, but seldom in less than twelve months or more, the morbid influence is felt in some other part. Thus, assuming one of the hands and arms to be first attacked, the other, at this period becomes similarly affected. After a few more months the patient is found to be less strict than usual in preserving an upright posture: this being most observable whist walking, but sometimes whilst sitting or standing. Sometime after the appearance of this symptom, and during its slow increase, one of the legs is discovered slightly to tremble, and is also found to suffer fatigue sooner than the leg of the other side: and in a few months this limb becomes agitated by similar tremblings, and suffers a similar loss of power.

Parkinson goes on to describe the slow increase in the severity of the symptoms:

At this period the patient experiences much inconvenience, which unhappily is found daily to increase. The submission of the limbs to the direction of the will can hardly ever be obtained in the performance of the most ordinary offices of life. The fingers cannot be disposed of in the proposed directions, and applied with certainty to any proposed point. As time and the disease proceed, difficulties increase: writing can now be hardly at all accomplished: and reading, from the tremulous motion, is accomplished with some difficulty. Whilst at meals the fork not being duly directed frequently fails to raise the morsel from the plate: which when seized, is with much difficulty conveyed to the mouth.

As the debility increases and the influence of the will over the muscles fades away, the tremulous agitation becomes more vehement. It now seldom leaves him for a moment: but even when exhausted nature seizes a small portion of sleep, the motion becomes so violent as not only to shake the bed-hangings, but even the floor and sashes of the room. The chin is now almost immoveably bent down upon the sternum. The slops with which he is attempted to be fed, with the saliva, are continually trickling from the mouth. The urine and faeces are passed involuntarily; and at the last, constant sleepiness, with slight delirium, and other marks of extreme exhaustion, announce the wished-for release.[5]

Substantia Nigra and the Neurochemistry of Certain Diseases

The neuropathology of Parkinson disease remained unclear until the early twentieth century. In 1919 the Russian neuropathologist Konstantin Tretiakoff, in his thesis for the doctorate in medicine in Paris, described characteristic changes in the *substantia nigra*, a structure in the midbrain (figure 17.5). In normal life one subdivision of the substantia nigra, the *pars compacta,* is characterized by cells that contain a dark melanin granule.

These granules give the structure its name as the black substance. In postmortem studies it was recognized that these cells were dead or dying. Traditionally, physicians tended to think about neurological disease in terms of the place in the brain affected by a disease. Damage to the optic nerve, the lateral geniculate nucleus, or the visual cortex produces complete or partial blindness. But brain lesions may also be chemical. A disease may affect all of the neurons that share a common transmitter or those that have receptor sites sensitive to a given transmitter. Suppose an enzyme that is crucial for production of a certain transmitter in the brain were lacking or present in insufficient amount. Not enough of the transmitter would be produced, and hence, cells that use that transmitter would not function normally.

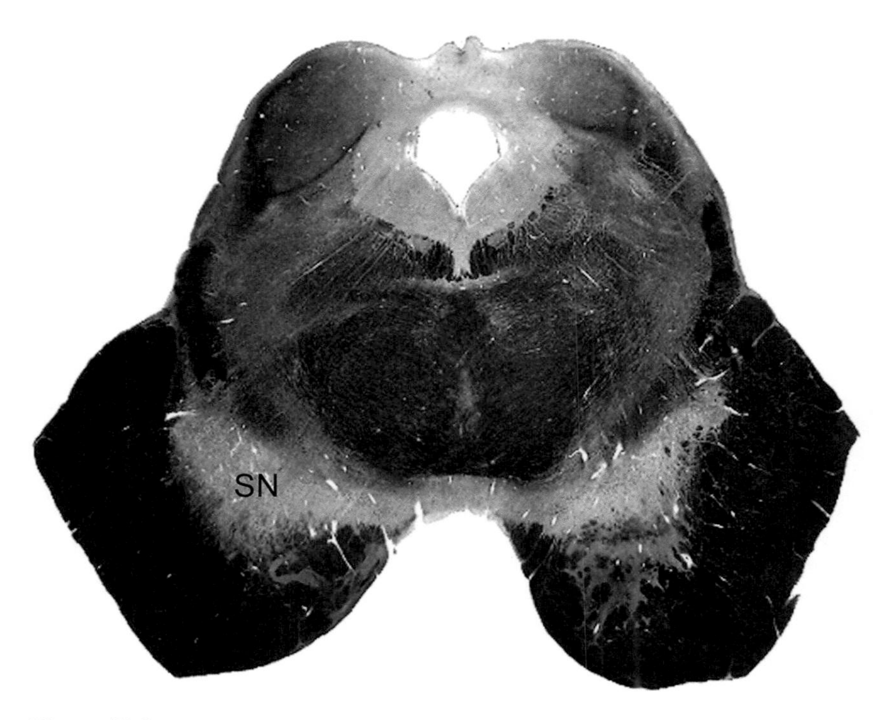

Figure 17.5
Cross section through the human midbrain. The cerebral peduncles are seen at the base of the brain. Just above the peduncles is the substantia nigra. Courtesy University of Washington.

Parkinson disease is a classic example of a transmitter-based disease. Patients are impaired in performing a wide variety of simple movements. They may have a resting tremor, and their facial muscles are often immobile, resulting in a characteristic mask-like expression. Although at autopsy cells in the substantia nigra were found to be abnormal and fewer in number, these lesions did not seem severe enough to account for such devastating symptoms.

Crucial additional evidence came in the 1950s when Arvid Carlsson, a Swedish pharmacologist, noted that rats that had been given high doses of the antipsychotic drug chlorpromazine developed motor deficits that resembled closely the symptoms of patients who suffered from Parkinson disease. This was an important clue to the nature of the disease because one of the effects of chlorpromazine is to block selectively receptors in the brain for dopamine, one of the known transmitter substances. Perhaps Parkinson disease resulted from an insufficient quantity of dopamine.

This interpretation of Parkinson disease received support when the Austrian biochemist Oleh Hornykiewicz, working in Vienna, determined the dopamine content of the brains of Parkinson patients who had died. Sharply reduced dopamine levels were found in the brains of these patients at autopsy. These discoveries suggested that patients with Parkinson disease might be helped if normal dopamine levels could be restored. It would be useless, however, to administer dopamine directly since it does not pass freely from blood vessels into the brain.

George Cotzias, a Greek-born neurologist working at the Brookhaven National Laboratory on Long Island, began testing the effect of administering L-dopa, an immediate precursor to dopamine that is converted within the brain into dopamine. L-dopa worked. It relieved the motor symptoms in Cotzias's patients and has now become the cornerstone of modern therapy for the treatment of Parkinson disease.

Management of Parkinson disease with L-dopa depended for its discovery on a number of prior discoveries. The dramatic success of the combination of basic research and clinical application serves as a model and an inspiration for further research into the mechanisms and causes of other diseases that may be based on deficits in the production, storage, or release of neurotransmitters.

Huntington Disease

In 1872 George Huntington read a paper at a medical society in Middleport, Ohio. He described a mother and daughter with a hereditary disease occurring in adult life and consisting of chorea and mental deterioration. Huntington's grandfather was a physician who had first observed the family in 1797 when he began medical practice in East Hampton, Long Island. Postmortem, the main observations were serious loss of cells in the caudate nucleus and putamen.

Huntington read his essay before the Meigs and Mason Academy of Medicine at Middleport, Ohio, on February 15, 1872. It begins with a characterization of the usual forms of chorea and then goes on to describe a special type of chorea:

> And now I wish to draw your attention more particularly to a form of the disease which exists, so far as I know, almost exclusively on the east end of Long Island. It is peculiar in itself and seems to obey certain fixed laws. In the first place, let me remark that chorea, as it is commonly known to the profession, and a description of which I have already given, is of exceedingly rare occurrence there. I do not remember a single instance occurring in my father's

practice, and I have often heard him say that it was a rare disease and seldom met with by him.

The hereditary chorea, as I shall call it, is confined to certain and fortunately a few families, and has been transmitted to them, an heirloom from generations away back in the dim past. It is spoken of by those in whose veins the seeds of the disease are known to exist, with a kind of horror, and not at all alluded to except through dire necessity, when it is mentioned as "that disorder." It is attended generally by all the symptoms of common chorea, only in an aggravated degree, hardly ever manifesting itself until adult or middle life, and then coming on gradually but surely, increasing by degrees, and often occupying years in its development, until the hapless sufferer is but a quivering wreck of his former self.

It is as common and is indeed, I believe, more common among men than women, while I am not aware that season or complexion has any influence in the matter. There are three marked peculiarities in this disease: 1. Its hereditary nature. 2. A tendency to insanity and suicide. 3. Its manifesting itself as a grave disease only in adult life.

1. Of its hereditary nature. When either or both the parents have shown manifestations of the disease, and more especially when these manifestations have been of a serious nature, one or more of the offspring almost invariably suffer from the disease, if they live to adult age. But if by any chance these children go through life without it, the thread is broken and the grandchildren and great-grandchildren of the original shakers may rest assured that they are free from the disease. This you will perceive differs from the general laws of so-called hereditary diseases, as for instance in phthisis, or syphilis, when one generation may enjoy entire immunity from their dread ravages, and yet in another you find them cropping out in all their hideousness. Unstable and whimsical as the disease may be in other respects, in this it is firm, it never skips a generation to again manifest itself in another; once having yielded its claims, it never regains them. In all the families, or nearly all in which the choreic taint exists, the nervous temperament greatly preponderates, and in my grandfather's and father's experience, which conjointly cover a period of 78 years, nervous excitement in a marked degree almost invariably attends upon every disease these people may suffer from, although they may not when in health be over nervous.

2. The tendency to insanity, and sometimes that form of insanity which leads to suicide, is marked. I know of several instances of suicide of people suffering from this form of chorea, or who belonged to families in which the disease existed. As the disease progresses the mind becomes more or less impaired, in many amounting to insanity, while in others mind and body both gradually fail until death relieves them of their sufferings. At present I know of two married men, whose wives are living, and who are constantly making love to some young lady, not seeming to be aware that there is any impropriety in it. They are suffering from chorea to such an extent that they can hardly walk, and would be thought, by a stranger, to be intoxicated. They are men of about 50 years of age, but never

> let an opportunity to flirt with a girl go past unimproved. The effect is ridiculous in the extreme.
>
> 3. Its third peculiarity is its coming on, at least as a grave disease, only in adult life. I do not know of a single case that has shown any marked signs of chorea before the age of thirty or forty years, while those who pass the fortieth year without symptoms of the disease, are seldom attacked. It begins as an ordinary chorea might begin, by the irregular and spasmodic action of certain muscles, as of the face, arms, etc. These movements gradually increase, when muscles hitherto unaffected take on the spasmodic action, until every muscle in the body becomes affected (excepting the involuntary ones), and the poor patient presents a spectacle which is anything but pleasing to witness. I have never known a recovery or even an amelioration of symptoms in this form of chorea; when once it begins it clings to the bitter end. No treatment seems to be of any avail, and indeed nowadays its end is so well-known to the sufferer and his friends, that medical advice is seldom sought. It seems at least to be one of the incurables.

Huntington identified a characteristic form of a hereditary disease that still bears his name. Unlike most such diseases, which are recessive in character, Huntington disease is based on a dominant gene. "But if by any chance these children go through life without it, the thread is broken and the grandchildren and great-grandchildren of the original shakers may rest assured that they are free from the disease."[6] Huntington wrote these words at a time when Mendel's work on genetic transmission was unrecognized. And yet his description of the familial transmission is clearly that of a dominant gene. Note that the symptoms appear later in life, after the typical child-bearing age. There would have been selection against so profound a disorder if it were to appear earlier in life.

In the past few years a particular gene has been identified that is involved in making the protein *huntingtin,* which plays a role in the normal functioning of some neurons.

Alzheimer Disease

Alois Alzheimer was a German psychiatrist born in 1864. He completed his studies in medicine at the University of Tübingen. In 1901 he examined a fifty-one-year-old woman who had been admitted to the psychiatric hospital in Frankfurt. Her illness consisted of memory defects, hallucinations, and delusions. Within five years it led to profound dementia and death. She died in 1906, by which time Alzheimer had moved to the University of Munich, where he studied her brain. Alzheimer wrote:

Clinically the patient presented such an unusual picture that the case could not be categorized under any of the known diseases. Anatomically the findings were different from all other known disease processes.

A woman, aged 51 years old showed jealousy towards her husband as the first noticeable sign of the disease. Soon a rapidly increasing loss of memory was noticed. She could not find her way around in her own apartment. She carried objects back and forth and hid them. At times she would think that someone wanted to kill her and she would begin to shriek loudly.

In the institution her behavior bore the stamp of utter perplexity. Periodically she was totally delirious, dragged her bedding around, called her husband and her daughters, and seemed to have auditory hallucinations. During her subsequent course, the phenomena that were interpreted as focal symptoms were at times less noticeable. But they were only slight. The generalized dementia progressed, however.

After 4½ years of the disease death occurred. At the end the patient was completely stuporous; she lay in her bed with her legs drawn up under her.[7]

In Alzheimer disease, the postmortem brain seems atrophied, with the brain shrunken to a weight of less than 1,000 g (figure 17.6). Importantly, one particular target is the hippocampus, which doubtless accounts for some of the memory deficit. But the disease is not selective, and in

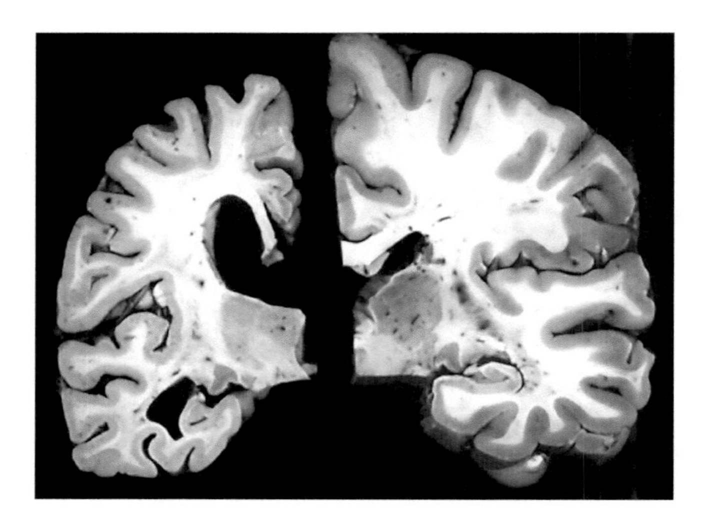

Figure 17.6
Brain slice of a seventy-year-old brain with Alzheimer disease, left, and a normal seventy-year-old brain, right. Courtesy Nigel Cairns, Department of Neurology, Washington University Medical School.

cases where the patient survives for many years there are few healthy cells left in the brain.

Amyotrophic Lateral Sclerosis: "Lou Gehrig's Disease"

As described earlier, Lou Gehrig was one of the great baseball players of his day. He began his major league career as a first-baseman with the New York Yankees in 1923. He amassed a record as a great fielder and is remembered for his prowess as a long-ball hitter. He was a member of the Yankee team until 1939, but in his last year his playing became increasingly impaired by amyotrophic lateral sclerosis.

The disease was identified by the French neurologist Jean-Martin Charcot (figure 17.7). In 1865, in a presentation to the Paris Hospitals Medical Society, Charcot described a case of a young woman who developed progressive weakness. In describing the pathology of the disease he wrote:

> On careful examination of the surface of the spinal cord, on both sides in the lateral areas there are two brownish gray streak marks produced by sclerotic changes. These grayish bands begin outside the line of insertion of the posterior roots and their anterior approaches, but do not include the entrance area of the anterior roots.[8]

The disease manifests itself in wasting and weakness of muscles; it is a specific assassin, seeking out and killing *motor neurons* and motor-related spinal tracts.

Myasthenia Gravis

Myasthenia gravis, literally "serious muscle weakness," was first recognized as a distinct clinical entity by Thomas Willis, the seventeenth-century English physician. Willis is sometimes called the father of clinical neuroscience. He first clearly described the cerebral arteries. (Willis's portrait is seen in chapter 9 of this volume, figure 9.1.) Willis wrote that those who suffer from the disease

> ". . . are distempered with Members very much loosened from their due vigour and strength, and with a languishing of their Limbs; so that though they are well in their stomach, and have a good and laudable pulse and urine, yet they are as if they were enervated, and cannot stand upright, and dare scarce enter upon local motions, or if they do, cannot perform them long: yea, some without any notable sickness, are for a long time fixed in their Bed, as if they were everyday about to dye; whilst they lie undisturbed, talk with their Friends, and are

Figure 17.7
A French neurologist, Jean-Martin Charcot (1825–1893) described the symptoms and the neuropathology of motor neuron disease. Courtesy Wellcome Library, London.

chearful, but they will not, nor dare not move or walk; yea, they shun all motion as a most horrid thing wherefore the sick are scarce brought by any perswasion to try whether they can go or not. Nevertheless, those labouring with a want of spirits, who will exercise local motions as well as they can, in the morning are able to walk firmly, to fling about their Arms hither and thither, or to take up any heavy thing, before noon the stock of Spirits being spent, which had flowed into the Muscles, they are scarce able to move Hand or Foot."[9]

The most important symptom of the disease is muscle weakness. Myasthenia gravis is an autoimmune disease. In this case the immune system attacks a normal transmitter system. Acetylcholine serves as the transmitter that is released at the terminals of motor axons onto skeletal muscles. An immune attack on the receptors for the transmitter in the muscle leads to the severe weakness.

Channelopathies

Nerve and muscle cells are separated from the environment around them by a cell membrane. Cells regulate their size and their activity by communicating with their surroundings via channels in the membrane. There are a large number of specialized nerve cells making and responding to different combinations of transmitters. These differences are typically based on their set of channels that regulate the substances that can go through them.

As we saw in chapter 4 of this volume, the pioneering work of Alan Hodgkin and Andrew Huxley on the squid axon showed that two types of membrane channels are involved in the action potential—the sequence of transiently increased sodium and potassium permeability that constitutes the response. The fact that the poison *tetrodotoxin* blocked the sodium phase of the action potential made it seem possible ultimately to construct a picture of the nature of the sodium channels.

In the years since, there was continued progress in recording from many cell types. In the 1980s a new technique appeared that made possible the study of *individual* channels and their response to activation. Erwin Neher and Bert Sakmann, working at the Max Planck Institute in Göttingen in Germany, developed a *patch clamp* technique with which experimenters could observe the opening and closing of single channels in the membrane.

In parallel with this electrophysiological advance, molecular biologists learned how to determine the molecular structure of the channels. It is now recognized that both genetic and acquired diseases may be expressions of a modification of a single type of channel. These discoveries are of profound importance for understanding—and hopefully for leading to the ultimate cure—of many of the devastating neurological diseases, some of which are discussed in this chapter.

Scrapie, Kuru, and Creutzfeld-Jakob Disease

As if genes, viruses, and bacteria didn't give us enough troubles, enter prions.

For hundreds of years shepherds and farmers have recognized a severe disease of sheep characterized initially by skin lesions and odd behavior, leading to eventual death of the animal. The condition was called "scrapie" because of a tendency of the sheep to rub against surfaces, as if to quiet an itchy place on their skin. In 1920 Hans-Gerhardt Creutzfeldt, a German physician, described the case of a twenty-three-year-old woman who developed a variety of severe mental and neurological symptoms and died within six weeks of first presenting with the symptoms. Shortly thereafter, a series of cases with similar symptoms and postmortem findings were presented by another German doctor, Alfons Maria Jakob. Rare but devastating, *Creutzfeld-Jakob disease* (CJD) involves a miserable set of neurological symptoms and a death sentence for its sufferer.

In 1950s a disease called "kuru" by the Fore tribe of Papua, New Guinea, was described by Australian medical officers. Kuru led to increasingly severe deficits in movements, inappropriate laughter, increasing weakness, and death. More recently a similar sequence of severe neurological symptoms and rapid course of the disease began to infect British cattle. In "mad cow disease" or *bovine spongiform encephelopathy,* the brains of the affected animals became spongy and involved massive cell loss.

CJD, kuru, and mad cow disease all share similar symptoms and similar postmortem findings of cell loss with a characteristic spongy texture of the brain. CJD can arise spontaneously; kuru and mad cow disease can be transmitted by eating infected tissue—especially brain and spinal cord.

Daniel Carlton Gajdusek, an American physician and scientist, lived for several years with the people of New Guinea. He was one of the first to identify kuru as being similar to CJD. Anthropologists had noted that the Fore tribe had a history of cannibalism. Men ate the best bits; brains were left for women and children, which accounted for the much higher incidence of the disease in women. Gajdusek reasoned that it was this practice that had perpetuated the disease. He further demonstrated that kuru could be transmitted to other species by injecting into a chimpanzee a small amount of the brain of a woman who had died from kuru. After a long incubation period, the injected chimp fell ill with the disease.

Scrapie, CJD, kuru, and mad cow disease do not behave in a typical fashion for an infectious disease. In most cases the time from exposure to developing the disease can be very long. Gajdusek believed that it was a slow virus that carried the disease, since at the time, bacteria and viruses seemed to be the only mechanism capable of transmitting an infectious disease.

A few years later Stanley Prusiner, working at the University of California in San Francisco, demonstrated that scrapie could be transmitted from sheep to a hamster. Resistant to heat and ultraviolet radiation, the infective agent did not behave like any known bacteria or viruses. Moreover, there was no hint of DNA or RNA in the infective agent. Prusiner suggested that the infective agents in these diseases are neither viruses nor bacteria. He proposed that a naturally occurring protein—which he called a prion—can undergo a change in its folded structure to become infectious. The damaged protein could inflict similar changes to other prions, in a malicious cascade. Despite initial resistance to accepting the prion, it is now considered to be the most likely cause of infection in these and several other neurological diseases. Alzheimer disease, Prusiner recently argued, may be based on a similar process.

On Cures and the Future

The diseases described represent a spectrum of things that go wrong with the brain. Although there are at present no cures for any of these diseases, increasing understanding of their molecular basis holds out the promise of designing drugs and strategies for preventing, managing, and ultimately curing many of these conditions.

18

Personality and Emotion

We are our brains. If we are by nature happy or sad, shy or outgoing, is that fact likely to be reflected in some aspect of brain structure or function?

Questions like that were asked, and procedures to address them were introduced, in the latter part of the nineteenth century and the first half of the twentieth. The sorts of evidence used to address questions about the functions of different parts of the brain came from measurements of head dimensions in living subjects, postmortem study of human brains and skulls, animal experiments, and accidental or war-wound lesions.

Despite attempts to show differences in the gross size or fissural patterns of the human brain, no strong relationship has ever been established in a reliable way between the size or fissural pattern and any aspect of human talent. Nor, for that matter, has a relationship ever been established between size or fissural pattern and the various human races. Nevertheless, there is evidence of how profoundly our personality is changed by brain injury.

Frontal Lobes

The human cerebral cortex has four major subdivisions, one of which is the frontal lobe. The frontal lobe itself contains both motor and nonmotor areas. Motor areas of the frontal lobe give rise to long descending tracts extending into the spinal cord. In front of the motor areas is the prefrontal cortex, a region whose functions are more difficult to characterize. Study of the functions of the prefrontal area is best understood in the wider context of brain function.

What has been learned about the functions of various divisions of the human cerebral cortex? What consequences have followed from these claims?

The history of cortical localization follows a pattern that can be understood by studies of cortical processing of vision. In 1948 Stephan Polyak, an anatomist working at the University of Chicago, produced a massive and scholarly book, The Vertebrate Visual System. The 1,300-page book described in detail the way in which visual information is projected from the eye by way of the lateral geniculate body to the striate cortex, where "vision" takes place. It seemed as if the question of cortical representation of vision was over. It is as if a picture is projected from the eye to the primary visual cortex. But Polyak never asked the next question, "Who looks at the picture?" It is now clear that visual information has to be interpreted by other parts of the brain. Further analysis is carried out by a great number of visual areas beyond the primary visual cortex—thirty-one such areas by one count. Each region is responsible for analysis aspects of the visual scene. A similar complexity is true for the motor areas. A single "motor cortex" is now known to be surrounded by a number of areas that are involved in the regulation of movement: some with a direct output via descending tracts to the spinal cord; others indirectly via connections to the motor or premotor cortex. There probably are at least as many subdivisions of the prefrontal cortex.

How have people interpreted the functions of different regions of the brain? In ancient and Renaissance times it was the ventricles with their cerebrospinal fluid that seemed to be the essential part of the brain (figure 18.1). Early ventricular localization speculation did not allocate a specific region for control of aggression. The anterior ventricles were thought to be the locus of fantasy and imagination.

In the nineteenth century identification of the criminal brain at autopsy was thought to be an easy task. Cesare Lombroso (figure 18.2), an Italian anatomist, believed that the criminal brain was more asymmetric than that of nonviolent people. He was also convinced that, lacking direct evidence of that excessive asymmetry, the easy road to identify a criminal is simply to look at him. Tattoos would indicate criminality for Lombroso, or just a careful look at the face (figure 18.3).

The phrenologists Franz Joseph Gall and Johann Spurzheim found zones on the head that they associated with a tendency to commit violent acts. The frontal lobes, however, were thought to be the locus of memory—a guess, but not a bad one (figure 18.4).

Monkeys have a greater development of frontal lobe than cats; chimpanzees greater than monkeys; people greater than chimpanzees. If there are differences among species, are there also differences among people? Postmortem study of the brains of gifted and talented people has never

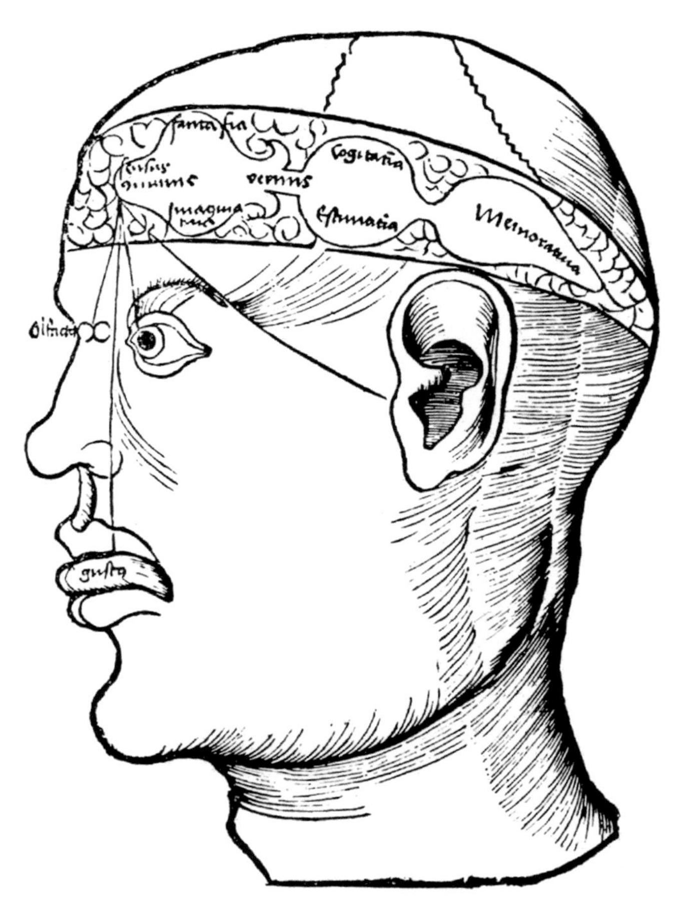

Figure 18.1
Medieval view of the brain. The ventricles were thought to be the essential part of the brain, each with a presumed distinct set of functions. Courtesy Wellcome Library, London.

Figure 18.2
Cesare Lombroso (1836–1909) was an eminent Italian psychiatrist with views about the relationship among facial features, brain symmetry, and criminality. Courtesy Wellcome Library, London.

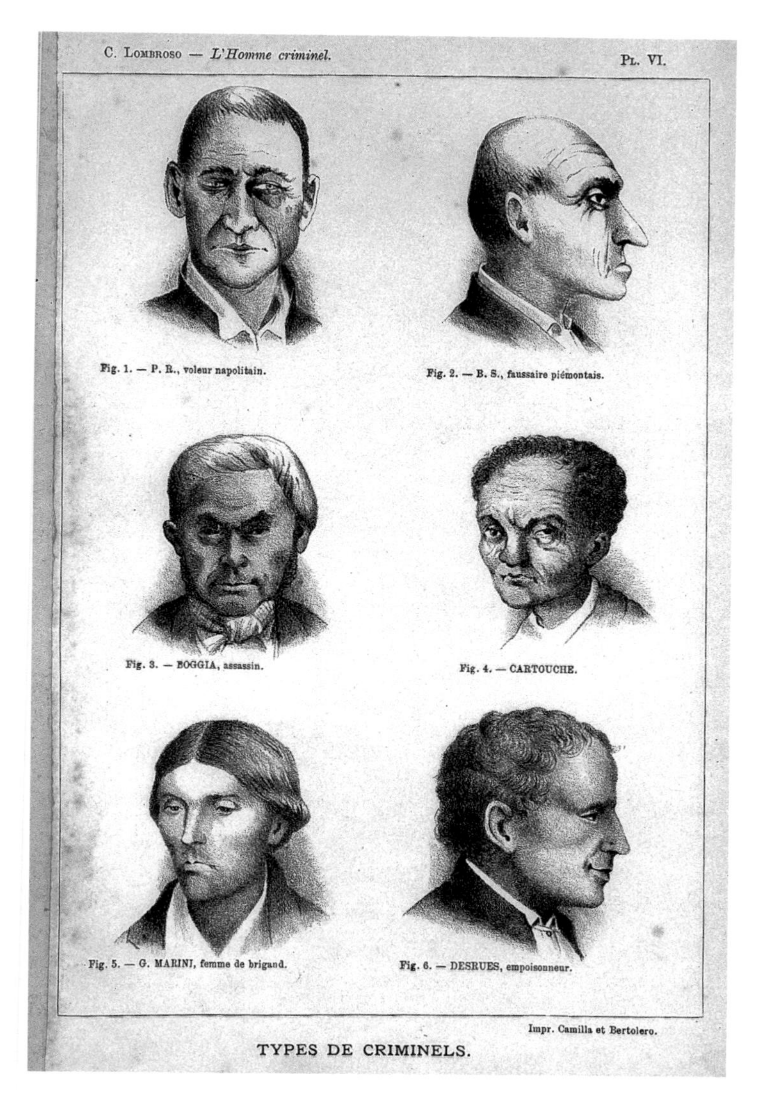

Figure 18.3
Lombroso's identification of criminal faces with alleged personality traits. Lombroso first published his work in 1876 in *L'uomo deliquente* (The Criminal Man). The illustration here is from a later French edition, *L'Homme Criminel.* Courtesy Wellcome Library, London.

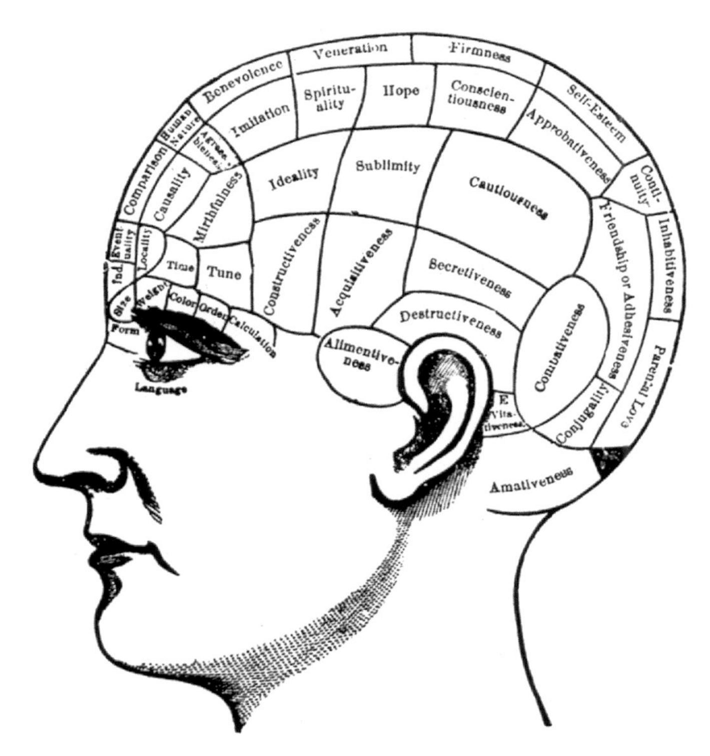

Figure 18.4
A nineteenth-century phrenology diagram. The phrenologists took the idea of brain localization to extremes, with little evidence for their assertions. Courtesy Wellcome Library, London.

yielded a consistent story. Accidents that involve brain injuries, on the other hand, have helped us to understand localization in the cerebral cortex.

Frontal Lobes and Personality

One of the most important events leading to our views of the function of the frontal lobes happened in the small town of Cavendish, Vermont, in 1850. Phineas Gage was a church-going, competent young man, foreman of a crew laying track for a new railroad in Central Vermont. Just over a mile from the center of the village of Cavendish stood a small rock outcropping that lay in the path of the intended railroad. The way of dealing with such obstructions was to remove them by

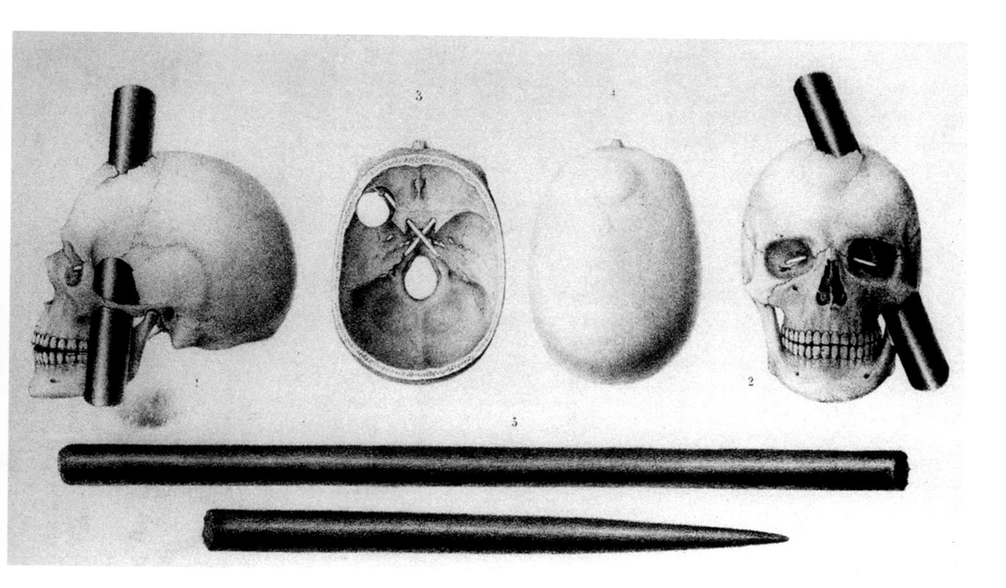

Figure 18.5
Phineas Gage's skull and crowbar.

blasting. A hole was drilled into the rock and a charge of gunpowder was placed into the hole along with a fuse. Sand and cotton were placed over the gunpowder, and the entire assembly was tamped down with a long bar of iron. The hole was dug and the gunpowder put in place, but not the cotton or sand. Phineas, unaware of this, placed his tamping iron in the hole to push down on what he assumed would be the sand. It wasn't there. His iron rod struck the rock and caused a spark that ignited the gun powder. The rod, being in a sealed chamber, was discharged skyward like a rifle bullet. The rod entered Phineas's head just under the left eye and emerged from the top of the skull on the right side (figure 18.5).

Few could believe Dr. John Harlow's report to the Massachusetts Medical Society (figure 18.6). No one, they thought, could survive such a devastating accident. In order to confirm his story Harlow did two things. He arranged for his patient to be examined by the professor of surgery at Harvard, Henry Bigelow. In later reports Harlow added testimony from various worthy citizens of Cavendish, including the minister of the Congregational Church and the justice of the peace (figure 18.7). He also included testimony from "an Irishman." Nonetheless, many still believed that Harlow's story was a "Yankee invention."

THE

BOSTON MEDICAL AND SURGICAL JOURNAL.

VOL. XXXIX. WEDNESDAY, DECEMBER 13, 1848. No. 20.

PASSAGE OF AN IRON ROD THROUGH THE HEAD.

To the Editor of the Boston Medical and Surgical Journal.

DEAR SIR,—Having been interested in the reading of the cases of "Injuries of the Head," reported in your Journal by Professor Shipman,
of Cortlandville, N. Y., I am induced to offer you the notes of a very
severe, singular, and, so far as the result is taken into account, hitherto
unparalleled case, of that class of injuries, which has recently fallen under my own care. The accident happened in this town, upon the line
of the Rutland and Burlington Rail Road, on the 13th of Sept. last, at
4½ o'clock, P. M. The subject of it is Phineas P. Gage, a foreman,
engaged in building the road, 25 years of age, of middle stature, vigorous physical organization, temperate habits, and possessed of considerable
energy of character.

It appears from his own account, and that of the by-standers, that
he was engaged in charging a hole, preparatory to blasting. He had
turned in the powder, and was in the act of tamping it slightly before
pouring on the sand. He had struck the powder, and while about to
strike it again, turned his head to look after his men (who were working
within a few feet of him), when the tamping iron came in contact with
the rock, and the powder exploded, driving the iron against the left
side of the face, immediately anterior to the angle of the inferior
maxillary bone. Taking a direction upward and backward toward the
median line, it penetrated the integuments, the masseter and temporal
muscles, passed under the zygomatic arch, and (probably) fracturing the
temporal portion of the sphenoid bone, and the floor of the orbit of the
left eye, entered the cranium, passing through the anterior left lobe of the
cerebrum, and made its exit in the median line, at the junction of the
coronal and sagittal sutures, lacerating the longitudinal sinus, fracturing
the parietal and frontal bones extensively, breaking up considerable
portions of brain, and protruding the globe of the left eye from its socket,
by nearly one half its diameter. The tamping iron is round, and rendered comparatively smooth by use. It is pointed at the end which entered first, and is three feet, seven inches in length, one and one quarter
inch in diameter, and weighs 13¼ pounds. I am informed that the
patient was thrown upon his back, and gave a few convulsive motions
of the extremities, but spoke in a few minutes. His men (with whom

20

Figure 18.6
Phineas Gage's injury: the original 1848 report of Dr. Harlow in the *Boston Medical and Surgical Journal.*

and brains," several rods behind where Mr. Gage stood, and that they washed it in the brook, and returned it with the other tools; which representation was fully corroborated by the greasy feel and look of the iron, and the *fragments* of *brain* which I SAW upon the rock where it fell.

(Signed) JOSEPH ADAMS,
Justice of the Peace.

CAVENDISH, *Dec.* 14, 1849.

The Rev. Joseph Freeman, whose letter follows, informed himself of the circumstances soon after the accident.

CAVENDISH, *Dec.* 5, 1849.

DEAR SIR—I was at home on the day Mr. Gage was hurt; and seeing an Irishman ride rapidly up to your door, I stepped over to ascertain the cause, and then went immediately to meet those who I was informed were bringing him to our village.

I found him in a cart, sitting up without aid, with his back against the fore-board. When we reached his quarters, he rose to his feet without aid, and walked quick, though with an unsteady step, to the hind end of the cart, when two of his men came forward and aided him out, and walked with him, supporting him to the house.

I then asked his men how he came to be hurt? The reply was, "The blast went off when he was tamping it, and the tamping-iron passed through his head." I said, "That is impossible."

Soon after this, I went to the place where the accident happened. I found upon the rocks, where I supposed he had fallen, a small quantity of brains. There being no person at this place, I **passed on to a blacksmith's shop a few** rods beyond, in and about which a **number of Irishmen were collected. As I came up to them, they pointed me to the iron, which has since attracted so much attention,** standing outside the shop-door. They said they found it **covered with brains and dirt,** and had **washed it in the brook.** The *appearance* **of the iron corresponded** with this story. It had a greasy appearance, and was **so to the touch.**

After hearing their statement, as there was no assignable motive for misrepresentation, and finding the appearance of the iron to agree with it, I was compelled to believe, though the result of your examination of the wound was not then known to me.

I think of nothing further relating to this affair which cannot be more minutely stated by others.

Very respectfully, yours,
(Signed) JOSEPH FREEMAN.

Dr. J. M. HARLOW.

Dr. WILLIAMS first saw the patient, and makes the following statement in relation to the circumstances :—

NORTHFIELD, *Vermont, Dec.* 4, 1849.

Dr. BIGELOW: Dear Sir—Dr. Harlow having requested me to transmit to you a description of the appearance of Mr. Gage at the time I first saw him after the accident, which happened to him in September, 1848, I now hasten to do so with pleasure.

Dr. Harlow being absent at the time of the accident, I was sent for, and was

Figure 18.7
Testimony of Joseph Adam and the Reverend Freeman, from Dr. Bigelow's 1850 article in the *American Journal of the Medical Sciences*. These worthy residents of the local town in Vermont were asked to confirm that they had witnessed the results of the accident to Phineas Gage.

Despite his physicians' treatments—poking their fingers along the track of the iron bar, insertion of a metallic probe, blood letting, silver nitrate application, and purging—Phineas lived for twelve years after the accident. But his character and personality were radically altered. He no longer went to church, he swore, and he continuously changed his plans. He made a living in various ways, one of which was lecturing about his case while showing off his iron rod. Phineas Gage died in San Francisco, almost certainly in status epilepticus. His family arranged for his skull to be sent back East, where it was displayed for years (along with the iron rod) in the Warren Anatomical Museum at Harvard.

The challenge of understanding the functions of the prefrontal areas was now in place. Phineas could see, he could speak (within minutes of his accident), he could walk. It was his personality that was profoundly altered by the injury. Friends said he was "no longer Gage." The course of the iron rod through Phineas's skull probably maximally damaged cortex and connections in the medial region of the frontal cortex and probably included the anterior cingulate area. This case was cited whenever the functions of the prefrontal areas were described.

At the end of the nineteenth century—some four decades after Phineas's death–there began to be systematic study of the effects of brain lesions on monkeys and other mammals. At first, postoperative observations were rather casual. How did the monkey react to threat? Could it reach for food? Was it able to walk as before? In the first part of the twentieth century, more rigorous behavioral study began. In a typical experiment an animal might be trained to perform a variety of skilled or semiskilled acts, and then, after a brain lesion was made, the animal was tested for its ability to retain earlier training or to learn new tasks. Initially, the results of prefrontal lesions only added to the puzzle. Although performance of some tasks might be mildly affected, there seemed to be few or no reproducible effects that could be uniquely attributed to removal of the prefrontal cortex.

The picture changed somewhat in the early 1930s. Carlyle Jacobsen was a psychologist working with the physiologist John Fulton at Yale University. At first Jacobsen found only minimal disturbance caused by the ablation of the prefrontal cortex in his monkeys. But then he tried another test. As we noted in chapter 12, Jacobsen found that bilateral lesions of the prefrontal cortex forever abolishes the monkeys' ability to successfully solve the delayed response task.

Jacobsen then extended his work to the study of chimpanzees. He trained two female chimps, Becky and Lucy, in a variety of behavioral tasks, and made lesions in a similar region of prefrontal cortex. Although mildly impaired in some of these tasks, Jacobsen's chimps, like his monkeys, were also impaired at solving the delayed response task. Memory deficits are more subtle in humans with frontal lesions, but they are there. One function of the frontal lobes is the temporary storage of memory.

Animal experiments including those of Jacobsen in the 1930s gave rise to the idea that certain psychiatric conditions might be ameliorated or even cured by removing the prefrontal lobes or disconnecting them from the rest of the brain. In chapter 19 of this volume we discuss the history of prefrontal lobotomy, how it came to be used as a psychotherapeutic tool, and how its practice fell into decline.

Emotion and the Brain

In the second half of the nineteenth century, there was increasing interest in localization in the cerebral cortex. As we have seen, Pierre Paul Broca had shown that damage to a restricted region of the frontal cortex caused a permanent aphasia in the patient M. Le Borgne, known as "Tan" (see chapter 16 of this volume). Gustav Fritsch and Eduard Hitzig had shown that electrical stimulation of a region of the frontal cortex produces movement on the opposite side of the body (see chapter 13 of this volume).

New evidence arose in the late nineteenth century about the role of the temporal lobe in emotion. Brown and E. A. Schäfer, working at University College London, studied the effects of large cortical lesions on monkeys. In two cases (their cases VI and XII), the lesions removed the temporal lobe bilaterally. Brown and Schäfer wrote:

> These severe operations were recovered from with marvelous rapidity, the animal appeared perfectly well even so early as a day after the establishment of the second lesion. A remarkable change is, however, manifested in the disposition of the Monkey. Prior to the operations he was very wild and even fierce, assaulting any person who teased or tried to handle him. Now he voluntarily approaches all persons indifferently, allows himself to be handled, or even teased or slapped, without making any attempt at retaliation or endeavouring to escape. His memory and intelligence seem deficient. He gives evidence of hearing, seeing and of the possession of his senses generally, but it is clear that he no longer clearly understands the meaning of the sounds, sights, and other impressions that reach him. Every object with which he comes in contact, even those with which he was previously most familiar, appears strange and is investigated with curiosity. This is not the case only with inanimate objects, but also with persons and with his fellow Monkeys.[1]

An almost identical result was reported some years later by the psychologist Heinrich Klüver and the surgeon Paul Bucy working at the University of Chicago. Two of the symptoms—failure to recognize objects by sight and a striking lessening of fear and aggression—were very similar to those described by Brown and Schäfer. Klüver and Bucy's monkeys also tended to place objects in their mouths ("oral tendencies") and were hypersexual. Although Klüver considered the effects of the temporal lobe lesion a unitary effect, it is now clear that components of the Klüver-Bucy syndrome are related to the particular structures that were damaged. Destruction of the amygdala, which was damaged both in the Brown and Schäfer monkeys and those of Klüver and Bucy, probably accounts for the tameness following the surgery. The lesion of the temporal neocortex would have produced the observed deficit in visual recognition.

The Papez Circuit, Limbic System, and Emotion

In 1937 James Papez (rhymes with grapes) was an anatomist working at Cornell University. Papez reviewed the available evidence about the brain and emotion. He noted that several authors had demonstrated an important role for the hypothalamus in emotion. Starting with the hypothalamus Papez proposed a circuit of interconnected structures through the brain that might be involved in the organization of emotional responses. Papez identified the most caudal portion of the hypothalamus—the mammillary bodies—as an important link in his circuit. From the mammillary bodies there arises a prominent tract linking them to the anterior nuclei of the thalamus. The anterior nuclei of the thalamus in turn project to the cingulate cortex. Cingulate cortex projects to the hippocampus, and hippocampus projects to the mammillary bodies. Papez recognized that each of these links might be reciprocal. Papez summarized his proposal:

> An attempt has been made to point out certain anatomic structures and correlated physiologic symptoms which, taken as a whole, deal with various phases of emotional dynamics, consciousness, and related functions. It is proposed that the hypothalamus, the anterior thalamic nuclei, the gyrus cinguli, the hippocampus and their interconnections constitute a harmonious mechanism which may elaborate the functions of central emotion as well as participate in emotional expression. This is an attempt to allocate specific organic units to a larger organization dealing with a complex regulatory process. The evidence presented is mostly concordant and suggestive of such a mechanism as a unit within the larger architectural mosaic of the brain.[2]

Subsequent studies have led to modifications of some aspects of Papez's proposal. For example, the amygdala, which was not included in Papez's original formulation, is a vital element in emotional control. The hippocampus, which Papez did include, is far more related to spatial memory. Nevertheless, Papez's proposal is important because it suggests that a *group* of interconnected brain structures, rather than a single nucleus, might be critical. The connections that Papez cited are there, and the proposal was an early attempt to understand complex functions in terms of a set of interconnected brain nuclei.

A slightly different grouping of nuclei and their interconnections is the *limbic system*—limbic from *limbus,* Latin for "an edge" (figure 18.8). In addition to including many of the structures in Papez's circuit, the limbic system importantly includes the amygdala, parahippocampal gyrus, and orbitofrontal cortex.

Amygdala and Fear

Although Papez did not include the amygdala (figure 18.9) in his circuit, its role in fear is unequivocal. If a neutral tone is followed by a mild foot shock, rats develop an obvious fear response; they crouch and freeze in anticipation of the shock. If the animal is subjected to a bilateral lesion of the amygdala, the freezing response to the tone is abolished. In this and other examples it seems as if the experience of fear is greatly attenuated. In the temporal lobe lesions of the monkeys studied by Brown and Schäfer, and those studied by Klüver and Bucy, the amygdala was included. Those animals also behaved as if they had lost fear.

Animals cannot tell you that they are not afraid, but a person can. In very recent times a woman known in the literature as SM[3] has a rare genetically based condition that involves a virtual absence of the amygdala, with the rest of the brain essentially normal. SM is normal on tests of IQ, memory, language, or perception. She experiences almost the full gamut of human emotion—at times sad and lonely, angry and upset, happy and cheerful, but never afraid. When directly tested on her response to fear-producing stimuli, she showed none. Although she said she did not like snakes, she cheerfully handled one. SM lives in an area with a high incidence of crime, danger, and drugs, and she herself has often been the victim of crime–perhaps because she fails to experience fear of the threats around her.

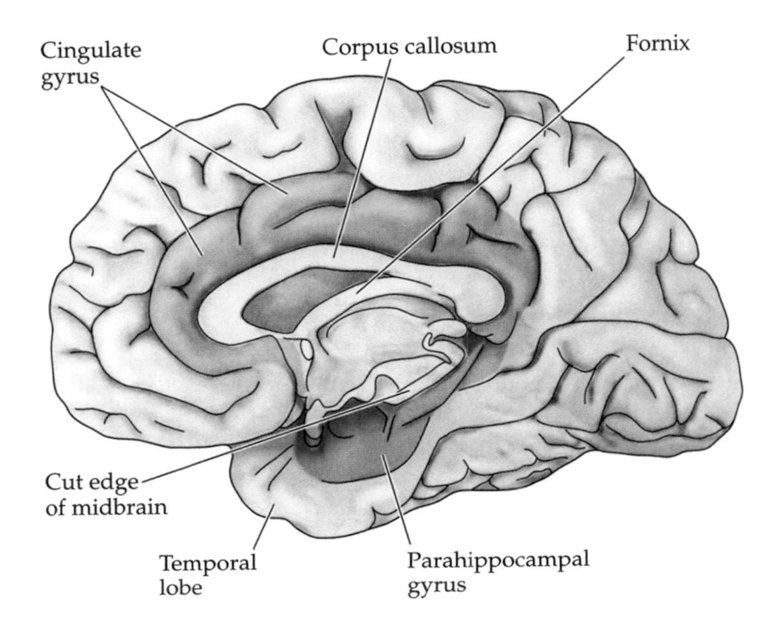

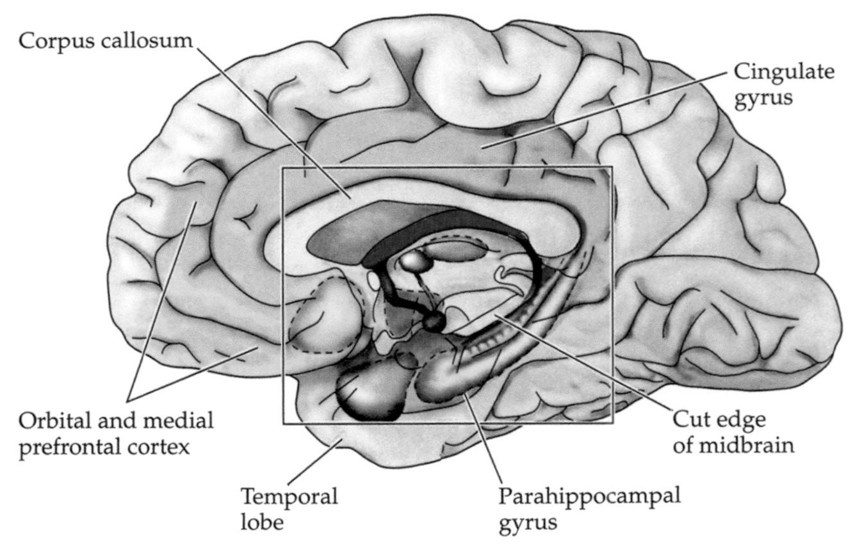

Figure 18.8
The limbic system. Courtesy Dale Purves and Sinauer Associates.

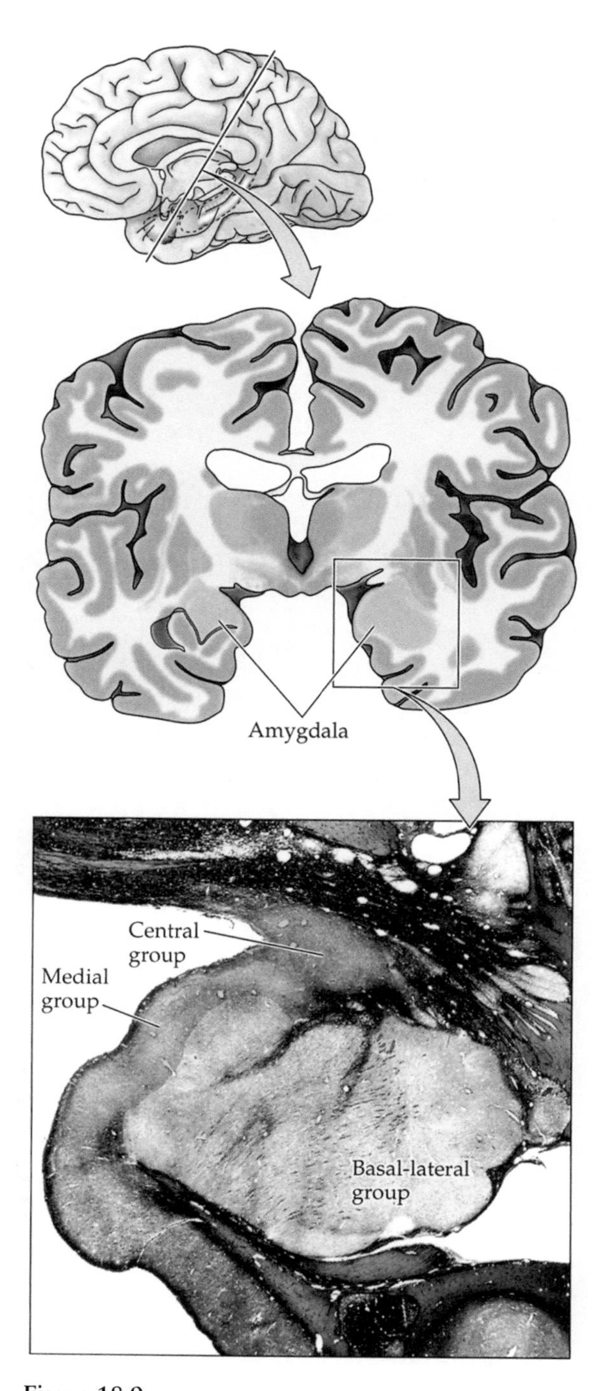

Figure 18.9
The amygdala is an essential brain component mediating response to fear. Courtesy Dale Purves and Sinauer Associates.

19

Mental Illness and the History of Surgical and Drug Treatment

Up to 5 percent of the people in the world are at some time subject to a severe mental illness. The presentations are broadly classed as disorders of cognition, such as schizophrenia, and disorders of mood, such as bipolar disorder or unipolar depression. Schizophrenia, originally called dementia praecox, and bipolar disorder (manic depression) typically have an onset in late adolescence or early adulthood. These two constitute the great majority of all mental illness.

There are many subdivisions of these broad categories, and there are other, milder forms of mental illness. The Diagnostic and Statistical Manual of Mental Disorders of the American Psychiatric Association (called the "Gray Book" in the trade) runs to nearly a thousand pages and lists the names of hundreds of contributors. There are countless diagnostic subdivisions, but schizophrenia and mood disorders dominate.

Mental illness has been recognized from the earliest recorded human history. Hippocrates described symptoms of madness and identified the brain as its source. The physician and scientist Thomas Willis was one of the great contributors to our understanding of the structure of the brain and its blood supply (figure 19.1). Willis began his career as professor at Oxford, where his anatomical studies of the brain and its circulation formed the basis for our modern understanding of blood circulation in the brain.

Willis later moved to London, where he had a successful medical practice. In his book *Two discourses concerning the soul of brutes* (1683), Willis described many of the symptoms of mental illness, distinguishing between melancholy and madness. In his chapter titled "Of Melancholy," Willis wrote:

> Although the universal Distemper of Melancholy contains manifold Delirious Symptoms, yet they chiefly consist in these three: 1. That the distemper'd are almost continuously busied in thinking, that their Phantasie is scarce ever idle or

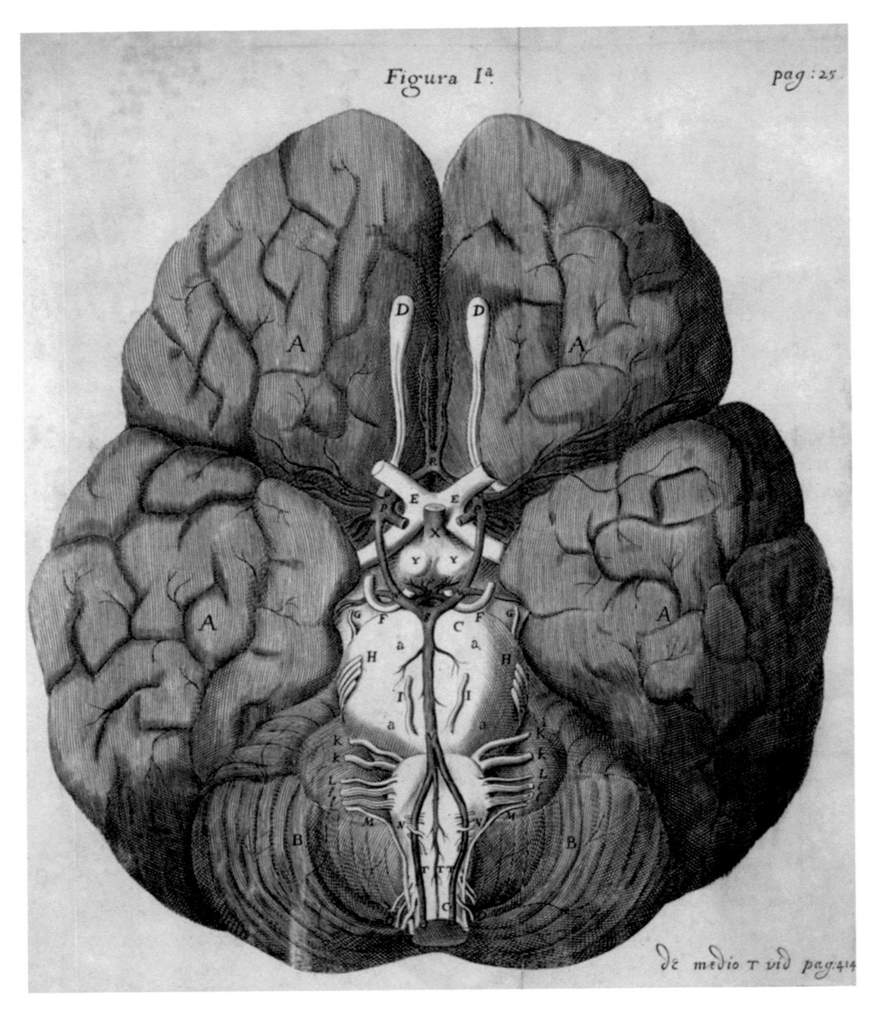

Figure 19.1
Thomas Willis (1621–1675) described the surface features of the human brain and its vasculature, as pictured in this engraving. Courtesy Wellcome Library, London.

quiet. 2. In their thinking they oftentimes comprehend in their mind fewer things that before they were wont, that oftentimes they roll about in their mind day and night the same thing, never thinking of other things that are sometimes of far greater moment. 3. The Ideas of objects or conceptions appear often deformed, and like hobgoblins, but are still represented in a larger kind or form; so that all small things seem to them great and difficult.

On the subject "Of Madness," Willis wrote:

It is observed in Madness that these three things are almost common to all: viz. First, That their Phantasies or Imaginations are perpetually busied with a storm of impetuous thoughts, so that night and day they are muttering to themselves various thing[s], or declare them by crying out, or by bauling out aloud. Secondly, that their notions or conceptions are either incongruous, or represented to them under a false or erroneous image. Thirdly, to their Delirium is most often joyous Audaciousness and Fury. Contrary to Melancholicks, who are always infected with Fear and Sadness.[1]

The symptoms described by Willis might be partially regrouped by a modern psychiatrist, but they underline two fundamental aspects of mental illnesses: disorders of thought and disorders of mood. Willis's reference to delirium raises a further class of mental disorder associated with infection or other toxic effects on the brain. These are very common causes of psychiatric presentations in countries with high rates of untreated infections.

The correct identification of psychiatric disorders as we now understand them awaited the adoption of asylum care, where the chronic long-term course of mental illness could be established. In the nineteenth and early twentieth centuries the German psychiatrist Emil Kraepelin (figure 19.2) was director of the psychiatric clinic at the University of Heidelberg. Kraepelin observed the patients under his care over a long period of their illness. Like Willis, he distinguished two different forms of psychiatric illness. One seemed to affect young people and appeared to be nonreversible. Kraepelin called this disorder *dementia praecox* because of its typical appearance in young people and its effects on their mental function.

Of a patient with dementia praecox, Kraepelin wrote in 1894:

Gentlemen you have before you today a strongly built well-nourished man, aged 21 who entered the hospital a few weeks ago. He sits quietly looking in front of him, and does not raise his eyes when he is spoken to, but evidently understands all our questions very well, for answers quite relevantly, though slowly and often only after repeated questioning. From his brief remarks, made in a low voice, we gather that he thinks he is ill, without getting any more precise information about the nature of the illness and its symptoms. The patient attributes his malady to

Figure 19.2
Emil Kraepelin (1856–1926) was a German psychiatrist, in charge of a major mental hospital. Kraepelin followed psychiatric patients over many years and described the categories and the time course of mental illness. Courtesy Wellcome Library, London.

the masturbation he has practiced since he was 10 years old. He thinks that he has thus incurred the guilt of sin against the sixth commandment, has very much reduced in his power of working, has made himself feel languid and miserable, and has become a hypochondriac. Thus as the result of reading certain books, he imagined that he had a rupture and suffered from wasting of the spinal cord, neither of which was the case. He would not associate with his comrades any longer because he thought they saw the results of his vice and made fun of him. The patient makes all these statements in an indifferent tone, without looking up or troubling about his surroundings. His expression betrays no emotion: he only laughs for a moment now and then. . . .

At first sight perhaps the patient reminds you of the states of depression which we have learned to recognize in former lectures. But on closer examination you will easily understand that in spite of certain isolated points of resemblance we have to deal with a disease having features of quite another kind. The patient makes his statements slowly and in monosyllables, not because his wish to answer meets with overpowering hindrances but because he feels no desire to speak at all. He certainly hears and understands what is said to him very well, but he does not take the trouble to attend to it. He pays no heed, and answers whatever occurs to them without thinking. No visible effort of the will is to be noticed. All his movements are languid and expressionless, but are made without hindrance or trouble. There is no sign of emotional dejection, such as the one would expect from the nature of his talk, and the patient remains quite dull throughout, experiencing neither fear nor hope nor desires. He is not at all deeply affected by what goes on before him, although he understands it without actual difficulty. It is all the same to him who appears or disappears where he is, or who talks to him, and takes care of him, and he does not even once ask their names.

This peculiar and fundamental want of any strong feeling of the impressions of life, with unimpaired ability to understand and to remember, is really the diagnostic symptom of the disease we have before us. . . .

He occasionally composes a letter to the doctor, expressing all kinds of distorted and half-formed ideas, with a peculiar and silly play on words, in very fair style but with little connection. . . .

These scraps of writing as well as his statement that he is pondering over the world, of putting himself together a moral philosophy, leaves no doubt that besides the emotional barrenness there is also a high degree of weakness of judgment and flightiness, although the pure memory is suffered little if at all. We have a mental and emotional infirmity to deal with, which reminds us only outwardly of the states of depression previously described. This infirmity is the incurable outcome of a very common history of the disease, to which we will provisionally give the name of dementia praecox.[2]

A few years after Kraepelin's characterization of the disorder, essentially the same mental illness was renamed *schizophrenia* by the Swiss psychiatrist Eugen Bleuler. Bleuler argued that schizophrenia was not a unitary disease and that the many symptoms of the disorder required independent description. Moreover, he questioned the claim that the disorder was unrecoverable, or that, as the name implies, inevitably led to dementia. The term schizophrenia as Bleuler used it referred to a splitting of the patient's experience between irreconcilable thoughts or impulses and the ordinary demands of a normal life, not the "split personality" of some popular writing (figure 19.3).

Manic Depression

In addition to schizophrenia, Kraepelin also described manic depression:

Manic-depressive insanity includes on the one hand the whole domain of so-called periodic and circular insanity, on the other hand simple mania, the greater part of the morbid states termed melancholia and also a not inconsiderable number of cases of amentia. Lastly, we include here certain slight and slightest colourings of mood, some of them periodic, some of them continuously morbid, which on the one hand are to be regarded as the rudiment of more severe disorders, on the other hand pass over without sharp boundary into the domain of personal predisposition. In the course of the years I have become more and more convinced that all the above-mentioned states only represent manifestations of a single morbid process. . . . A further common bond which embraces all the morbid types brought together here and makes the keeping of them apart practically almost meaningless, is their uniform prognosis. There are indeed slight and severe attacks which may be of long or short duration, but they alternate irregularly in the same case. This difference is therefore of no use for the delimitation of different diseases. A grouping according to the frequency of the attacks might much rather be considered, which naturally would be extremely welcome to the physician. It appears, however, that here also we have not to do with fundamental differences, since in spite of certain general rules it has not been possible to separate out definite types from this point of view. On the contrary the universal experience is striking, that the attacks of manic depressive insanity within the delimitation attempted here, never lead to profound dementia, not even when they continue throughout life almost without interruption.

. . . In members of the same family we frequently enough find side by side pronounced periodic or circular cases, occasionally isolated states of ill temper or confusion, lastly very slight, regular fluctuations of mood or permanent conspicuous colouration of disposition.

. . . Accordingly we distinguish first of all manic states with the essential morbid "symptoms of flight of ideas, exalted mood, and pressure of activity, and melancholia or depressive states with sad or anxious moodiness and also sluggishness of thought."[3]

LEHRBUCH
DER PSYCHIATRIE

VON

DR. E. BLEULER
O. PROFESSOR DER PSYCHIATRIE AN DER UNIVERSITÄT ZÜRICH

ZWEITE, ERWEITERTE AUFLAGE

MIT 51 TEXTABBILDUNGEN

BERLIN
VERLAG VON JULIUS SPRINGER
1918

Figure 19.3
Egon Bleuler (1857–1939) was head of a large psychiatric institution where he had occasion to observe the onset and course of psychiatric diseases. He coined the term “schizophrenia” to mean a splitting off from reality, not the “split personality” of popular use of the term. Courtesy Wellcome Library, London.

It seems likely that these psychiatric conditions must ultimately be understood in terms of altered brain function. The early psychiatrists were often also trained as anatomists, and they sought to understand mental symptoms as related to structural changes in the brain. Somewhat later, when synaptic transmission was better understood, the research focus shifted to anomalies in the action of neurotransmitters or the formation of synaptic connections themselves.

The Frontal Lobes and Frontal Lobotomy

As discussed in chapter 18 of this volume, animal experiments in the 1930s began to clarify some of the functions of the prefrontal cortex. Carlyle Jacobsen discovered the effect of lesions of prefrontal cortex on short-term memory. Jacobsen showed that monkeys in which the prefrontal cortex had been ablated were unable to solve delayed response tasks.

His observations also gave rise to the idea that certain psychiatric conditions might be ameliorated or even cured by removing the prefrontal lobes or disconnecting them from the rest of the brain. In addition to demonstrating memory impairment, Jacobsen made another important observation on his chimps. He noted that one of them, Becky, who previously had tantrums and bad behavior when she failed at any task, seemed more placid after a bilateral frontal lesion, "as if she had placed her troubles in the hand of the Lord."

Jacobsen reported his results at the 1935 meeting of the International Society of Neurologists. Egas Moniz (Antonio Caetano de Abreu Freire) was in the audience. Although Moniz later claimed to have been planning for several years to try to relieve psychiatric conditions by surgical lesions of the brain, it is clear from the testimony of people who knew him at the time that it was Jacobsen's description of the change in his chimp's behavior that triggered the plan to begin the radical practice of frontal lobotomy.

Egas Moniz (figure 19.4, left) was among the most eminent citizens of Portugal—at one time Portuguese ambassador to Spain and later foreign minister. He had pioneered the study of intra-arterial injections of radiopaque substances into the cerebral vasculature for imaging the blood vessels for diagnosing space-occupying lesions and aneurysms. Moniz was, however, plagued with gout throughout his life and thus was incapable himself of performing surgery. He chose for his colleague a young neurosurgeon, de Almeida Lima. Together they developed an operation they called frontal leucotomy (figure 19.5).

EGAS MONIZ

TENTATIVES OPÉRATOIRES
DANS LE TRAITEMENT
DE CERTAINES PSYCHOSES

MASSON & Cie, ÉDITEURS

Figure 19.4
Egas Moniz (1874–1955), a Portuguese neurologist, initiated the surgical treatment of mental illness. His monograph, on the right, was published in 1936. Portrait courtesy of the U.S. National Library of Medicine.

Within a few months of their operations, Moniz wrote up his first cases in the monograph shown on the right in figure 19.4, claiming great improvement in many of his patients. The operation seemed to him to be particularly effective in relieving the symptoms of agitated depression and anxiety. But there was, at best, only casual postoperative follow-up. Schizophrenic patients were, for the most part, not improved.

Most of the follow-ups were done just a few weeks after the surgery. The book is, however, filled with a number of before-and-after pictures designed to demonstrate the effectiveness of the procedure (figures 19.6 and 19.7).

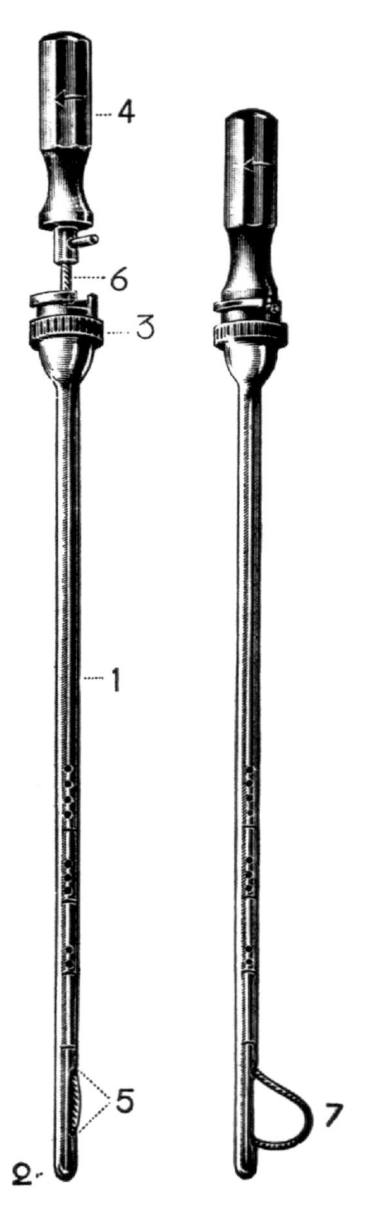

Figure 19.5
The leucotome used for Egas Moniz's surgery.

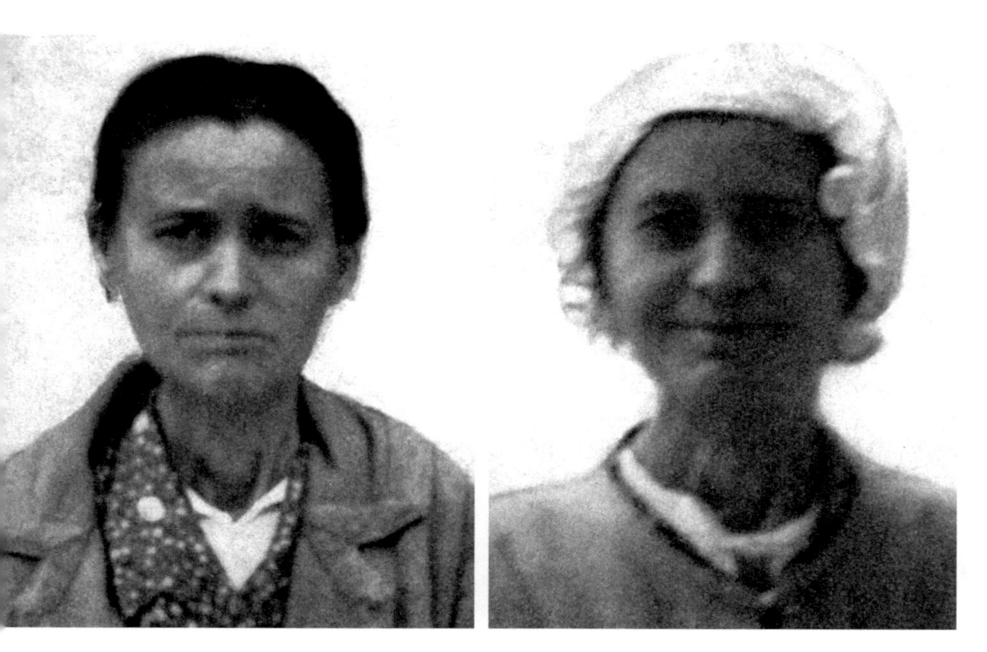

Figure 19.6
From Moniz's monograph. A forty-seven-year-old woman with "anxious melancholia" before and after prefrontal leucotomy.

Figure 19.7
From Moniz's monograph. A fifty-one-year-old man before and after prefrontal leucotomy.

RÉSULTATS

Dans chacun de ces groupes, on a obtenu les résultats suivants:

Mélancolie involutive	1
Guérison clinique	1
Syndrome de Cottard	1
Amélioration	1
Mélancolie anxieuse	5
Guérison clinique	4
Améliorations	1
Neurose d'angoisse	3
Guérison clinique	1
Améliorations	2
Manie	3
Guérison clinique	1
Améliorations	1
Sans résultat	1
Schizophrénie et paraphrénie	7
Améliorations	2
Sans résultat	5

Figure 19.8
Egas Moniz's table of results of the surgery.

Judging by the happy smiles, the procedure may have improved the patients' psychological state (figure 19.8). It seemed to be the answer to treatment of a class of hitherto untreatable diseases. The procedure soon took root, and variants of the operation were tried in centers in Europe, Latin America, and the United States.

One of the most enthusiastic champions of the procedure was the American psychiatrist Walter Freeman (1895–1972). Freeman, with a younger colleague James Watts, devised a surgical approach to lobotomy by inserting a probe to a measured depth through holes that had been drilled in the skull and then rotating it through an arc designed to destroy cells as well as fibers connected to the prefrontal cortex. In a later variant of the procedure Freeman, this time on his own, used an ice pick, entering the brain through the thin, easily breakable bones of the eye socket; anesthesia was induced by a prior electroconvulsive shock.

Transorbital lobotomy could be performed by any medical practitioner—even a psychiatrist—in his office suite, without elaborate sterile procedures. Freeman demonstrated his transorbital lobotomy in dozens of hospitals and clinics in the United States and abroad. Freeman was one of the people who proposed Egas Moniz for the Nobel Prize which he shared in 1949.

The operation of prefrontal lobotomy continued to spread widely in the United States and abroad. It was not until psychoactive drugs became available in the 1950s that there was a decline in the number of operations performed for the relief of psychiatric conditions. Although the frontal lobes remain a subject of active study, lobotomy is seldom performed.

One established function of the frontal lobe and its thalamic circuitry must be in the storage of a form of short-term memory. There is little evidence that surgical destruction or disconnection can reliably improve psychiatric illness. The history shows that a mere chance remark, such as Jacobson's observation about the chimpanzee Becky's improvement in disposition after a bilateral frontal lesion, can trigger an unforeseen attempt at therapeutic intervention. When a disease is widespread and untreatable, the standards of proof go down. Claims for therapeutic benefit of one or another procedure may be accepted with inadequate evidence. Frontal lobotomy as a treatment for psychosis took hold with minimal evidence for its efficacy. Desperation and poor methodology compounded the problem. In recent years a more rigorous methodology has been developed and gradually applied, the Randomized Controlled Trial (RCT), which is the standard by which treatments are evaluated.

Shock Therapies

In addition to prefrontal lobotomy other treatments originally based on dubious grounds were developed in the early twentieth century. Ladislas Meduna was a neurologist and neuropathologist based in Budapest in the 1920s and 1930s and later in Chicago. Meduna studied the brains postmortem of a number of schizophrenic and epileptic patients. He found a difference in the number and density of glial cells, with fewer in the schizophrenics' brains than in those of epileptics. Believing that the difference might reflect a difference in the energy level of the two classes of patient, he went on to try a number of drugs that induced seizures. He settled on inducing seizures by injecting pentylene tetrazol (metrazol, cardiazol). He reported success in half of his patients, nearly all of whom were suffering from a catatonic form of schizophrenia. The drug had many unwelcome side effects, the most striking of which was the induction of extreme anxiety.

Drug-induced seizures were soon replaced by electrically induced seizures, a procedure developed in Rome in the 1930s by the psychiatrist Lucio Bini and the neurologist Ugo Cerletti. Electroconvulsive therapy (ECT) is still in use as a psychiatric treatment for severe depression. Controlled trials show that ECT is probably the most effective treatment for severe depression. But at the same time, its action inevitably produces disorientation in the short term and subtler autobiographical memory disturbance in the long term—a consequence that remains poorly understood.

The Drug Therapies and the Mechanism of Schizophrenia

In the early 1950s in the course of searching for a calming agent for his patients, the French neurosurgeon Henri Laborit tried a number of existing antihistaminic drugs. One drug, chlorpromazine, seemed particularly effective as a calming agent. The success of chlorpromazine in calming suggested that it might also be of use in treating psychiatric illness, and the drug was soon tested on psychiatric patients. When prescribed in high doses, chlorpromazine was found to be effective in relieving symptoms in schizophrenia and mania.

Chlorpromazine and similar drugs were christened *neuroleptics* from Greek stems meaning to "grasp neurons." Neuroleptics not only calmed the patients but also appeared to reduce the frequency and severity of psychotic symptoms; hence, they were also called *antipsychotics*. But the improvement in these patients soon came to be contaminated by an unwanted side effect of the treatment, *tardive dyskinesia*. The symptoms are a variety of uncontrolled tic-like movements typically of the tongue or lips. The symptoms are strongly associated with high doses of antipsychotic treatment and may begin to occur only a few months after the treatment begins. Doctors and their patients thus face the dilemma of stopping the drug, which may then allow psychiatric symptoms to reappear.

The side effects of these drugs also provided a clue as to their mechanism of action, and this, in turn, led to a new view of the underlying basis of mental illness. The drugs blocked the response of brain structures to the neurotransmitter substance dopamine. Indeed the efficacy of drugs in controlling schizophrenic symptoms is closely related to their ability to block the response to dopamine, suggesting that the most florid symptoms of the illness may be associated with a poor regulation of dopamine transmission.

Drugs, Anxiety, and Depression

The discovery of drugs to treat anxiety and depression came about by chance observation. One of the drugs used in the treatment of tuberculosis, iprozianid, in addition to combatting tuberculosis, seemed to elevate mood. The drug soon came into use for treating depression.

Ancient remedies also sometimes proved useful. For many centuries people in Africa and India relied on the snakeroot plant for a variety of ailments, including hysteria. Swiss chemists experimenting with the plant in treating high blood pressure discovered that the plant also seemed to have a calming effect. They produced a major tranquilizer, reserpine. Although reserpine had some success in the treatment of schizophrenic symptoms, the drug led to depression in some patients, probably by *depleting* the availability of the neurotransmitters dopamine and serotonin at the synapse. The antidepressant imipramine, on the other hand, was shown eventually to block the reuptake of serotonin and noradrenaline thus *increasing* the availability of these transmitters at the synapse.

An alternate class of antidepressive drugs is related to the concentration of serotonin. The concentration of serotonin is normally closely regulated at the synapse by the balance between release and reuptake of the transmitter. Drugs that influence this reuptake mechanism selectively for serotonin are called the SSRIs—selective serotonin reuptake inhibitors—and they maintain the concentration of serotonin at the synapse high by selectively inhibiting its uptake. The prototype of this drug class in the United States (fluoxetine or brand name Prozac) achieved iconic status due to overenthusiastic marketing by its manufacturers and a few of its prescribers and takers.

On Treatment: Talking Therapies

Mental symptoms seem to demand understanding and relearning of abhorrent behaviors and associations. Accordingly, a psychological or social approach to treatment has long seemed appropriate and ethically desirable. Unfortunately, proponents have struggled to find a coherent methodology that works and has a scientific basis. Instead, over the years psychological treatments, starting with Freud, have taken a variety of sometimes strange paths. They have in common a tendency to reflect the preoccupations of the therapists rather than to meet the need to treat severely ill patients.

Purely social interventions were important in the deconstruction of asylum-based care, which resulted in far too many vulnerable people being warehoused in large impersonal institutions. The desirable elements of this change in social provision were at times undermined by illogical attempts to suggest that institutions caused mental illness in the first place. Once discharged from a long-stay psychiatric hospital, of which there remain a great many throughout the less developed world, patients face major problems coping with the demands of living, even in an accepting community.

Systematic psychological approaches that are based on a coherent theory and can be shown to work in suitably designed clinical trials are a very recent innovation. However, cognitive behavior therapy, behavioral activation, psychoeducation, and cognitive remediation are all finding a place in the management of severe mental illness. Modern neuroscience is helping to recombine psychology and biology in addressing the major challenge that treatment of severe mental illness presents.

Conclusions

Mental illness is a major cause of misery and economic loss. Psychotherapy can be of help in relieving depression. There are also a number of drugs that are effective in the control of some of the symptoms of psychiatric disease. In most cases these drugs were discovered by chance—from observing their effects when they were used to treat other conditions. In the past few years there has been a great increase in our understanding of the nature of transmitter action, the functions of receptors in the brain, and how networks of cells work together to underpin behavior. From this knowledge will come new and more effective ways to help people who suffer from these devastating conditions.

20

Consciousness and the Techniques for Study of the Human Brain

Consciousness for neuroscientists used to be the elephant in the room. Everyone knew it was there, but it was typically only discussed by neuroscientists in their old age, usually without the rigor of their younger years. What *is* consciousness? The eminent philosopher John Searle offered a definition:

> Consciousness refers to those states of sentience and awareness that typically begin when we awake from a dreamless sleep and continue until we go to sleep again, or fall into a coma or die or otherwise become "unconscious." Dreams are a form of consciousness though of course quite different from full waking states. Consciousness so defined switches on and off. By this definition a system is either conscious or it isn't, but within the field of consciousness there are states of intensity ranging from drowsiness to full awareness. Consciousness so defined is an inner, first-person, qualitative phenomenon. Humans and higher animals are obviously conscious, but we do not know how far down the phylogenetic scale consciousness exists.[1]

It is a basic postulate of neural science that everything that we think, see, or feel is related to the actions of the brain. And so neuroscientists (and philosophers) have speculated on the way in which consciousness might be related to the structure and function of the brain. There are several levels of consciousness. Our brains work when we are asleep, but they work in a different way than when we are awake. How does the functioning of our brain differ when we are awake and when we are asleep? Sleep cannot be simply a turning-off of the brain because a dream can be as vivid as a true experience. Thus, there are at least two states of consciousness. But even while we are fully awake, there is an obvious difference between sitting quietly with our eyes closed and actively solving a mathematical puzzle. Indirect measurements allow us to glimpse something about brain activity in different states of consciousness, but there is no agreement on how it all works. The literature on consciousness is vast, and it is far beyond the scope of this book to do more than address the basic questions and look at the historical and current techniques used to study the phenomenon of consciousness.

The Electroencephalogram and Sleep

An obvious place to begin is with sleep. The electroencephalogram (EEG) has been a major tool in the study of sleep. From time to time since the end of the nineteenth century physiologists had noted that a resting electrical oscillation can be recorded from the brains of experimental animals. These potentials occur without any obvious stimulus; they seemed to reflect ongoing activity of the brain.

In the late 1920s Hans Berger, a German psychiatrist, found that such potentials can be recorded from the human scalp, and he described the characteristic rhythms that can be observed. Such a recording is called an *electroencephalogram*, meaning electrical activity recorded from the brain—EEG for short. The EEG can serve as an indicator of the state of consciousness, and it can help in the diagnosis of certain diseases. Electrodes are placed on the scalp and the forehead. The electrode may be a small cup-like silver disk, and the attached wires are led to a device that amplifies the voltages so they can drive a pen that writes on a moving chart. Although the potentials from the scalp recordings are only a few microvolts in amplitude, amplification allows them to be recorded.

In recordings of the normal human subject, there are characteristic differences in the frequencies; they vary, depending on the state of arousal. In a relaxed but wakeful state, with the eyes closed, it is common to record an alternating voltage frequency of ten per second. Berger termed this the *alpha rhythm*. If a subject is brought out of this relaxed state by a bright light or by asking him or her to do mental arithmetic, the alpha wave is replaced by much faster lower-voltage activity called *beta rhythm*.

The alpha rhythm typically appears intermittently from electrodes placed near the back of the head. The alpha rhythm is broken up, or *desynchronized*, by sensory input or mental activity to be replaced by the low-voltage activity characteristic of the more alert state.

As a person falls asleep the EEG becomes dominated progressively by slower activity interrupted by brief bursts of *sleep spindles*. Deeper and deeper sleep is characterized by increasingly high-voltage slow activity called *delta waves*. In a typical night's sleep there is a gradual transition from drowsiness through successive stages to the very deepest levels of sleep, all in about an hour (figure 20.1). What happens then?

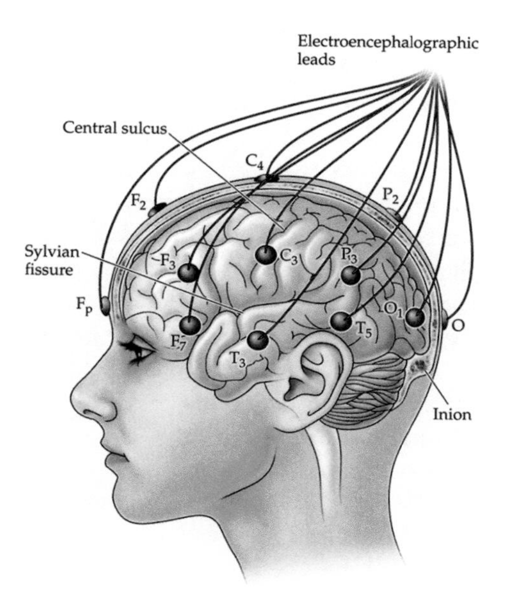

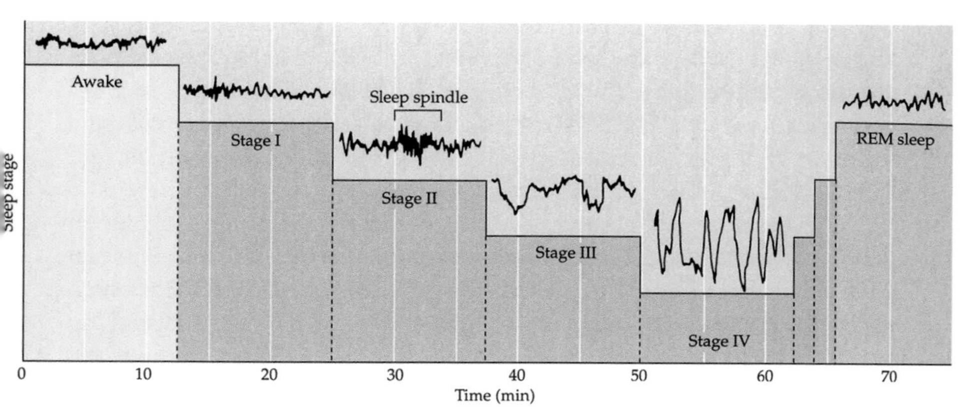

Figure 20.1
Subject with EEG wires attached and six traces representing stages of sleep and arousal. Note the unique, low-voltage character of REM sleep. Courtesy Dale Purves and Sinauer Associates.

In the early 1950s a graduate student, Eugene Aserinsky, working in the laboratory of Nathaniel Kleitman at the University of Chicago, made an important observation. He noted that although subjects remained asleep, they had occasional periods about 20 min in length in which their eyes would move rapidly under their lids (figure 20.2). The periods of rapid eye movements occurred repeatedly through the night, alternating in a regular way with ordinary sleep. The periods of rapid eye movement were always accompanied by an activated pattern of the EEG in which fast activity is recorded from all leads. If subjects are awakened during periods of rapid eye movements, they almost always report a vivid dream.

The apparent duration of the action in the dream is highly correlated with the amount of time that elapsed since the period of rapid eye movements began. Dreams may also be reported when a subject is awakened during periods when the eyes are not moving, but then the dreams are typically more fragmentary and less vivid.

Rapid eye movement sleep, abbreviated REM sleep, is very different from that other form of sleep—so-called slow-wave sleep. Not only is the EEG in a more aroused state characteristic of very light sleep but, in humans, the body remains remarkably still. Animals show a similar sequence of alternating slow-wave and REM-like sleep, often in their case associated with the whisker twitching of a sleeping cat or the paw movements of a sleeping dog. Surprisingly, sleepwalking, bedwetting, and talking do not occur in REM sleep.

These modern sleep studies have begun to clarify some age-old human problems. Some people suffer from insomnia—the inability to fall asleep or to stay asleep. Physicians sometimes prescribe "sleeping pills" for such people; these pills are often barbiturates. But people who depend on barbiturates to fall asleep usually have far less REM sleep because barbiturates interfere with the normal cycling of wakefulness/slow-wave sleep/REM sleep. In those cases slow and systematic withdrawal of barbiturates over time may benefit the insomniacs. Another unexpected finding of modern sleep researchers is that frequently people who believe that they are insomniacs often have normal patterns of sleep.

Clinical Use of EEG

In 1940 the Second World War was being fought in Europe, and the United States instituted the first peacetime draft in its history. Draftees were given a series of mental and physical examinations in order to assess

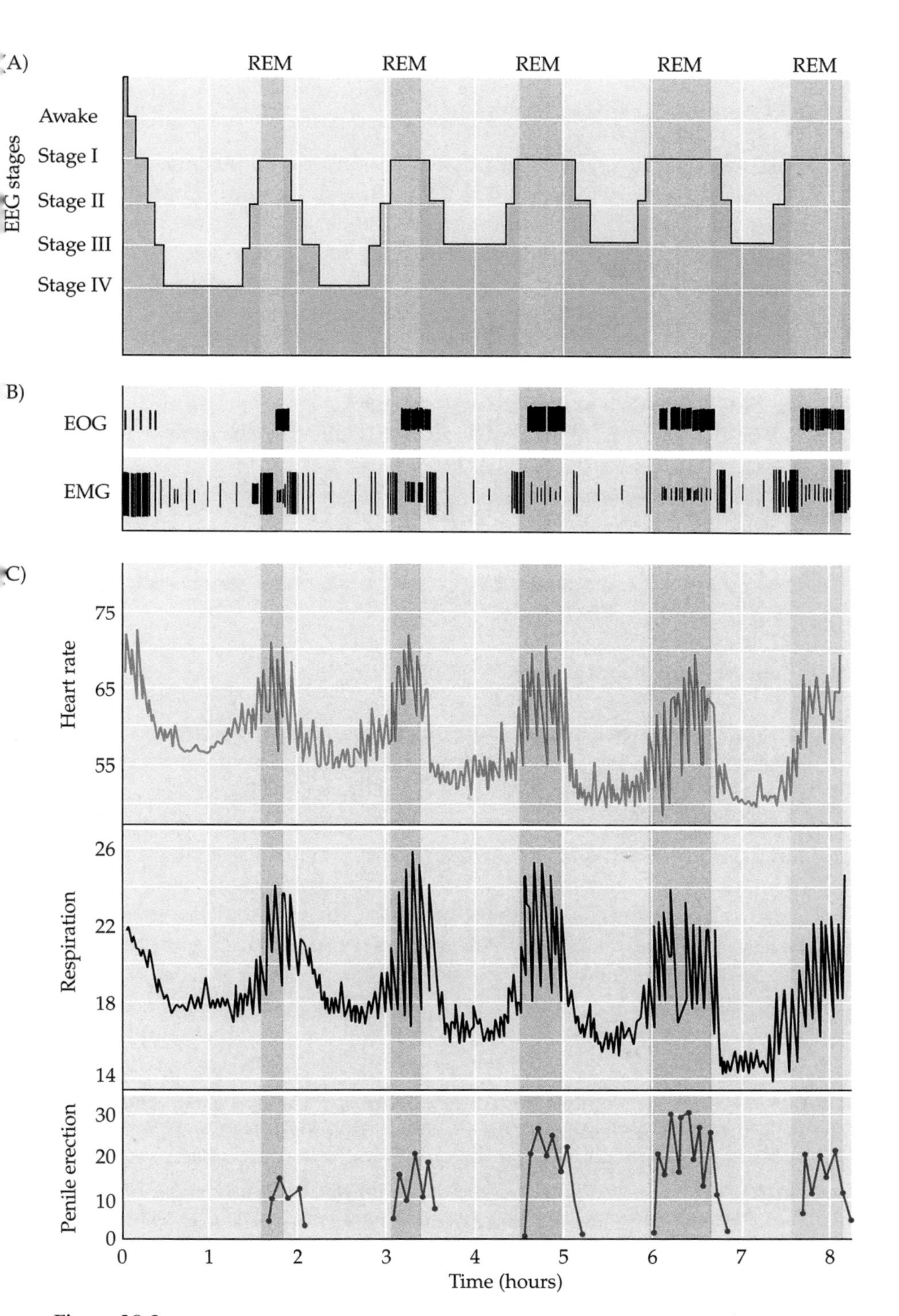

Figure 20.2
The figure shows recordings from a typical night's sleep in a male subject. Periods of alternating deep and REM sleep with associated changes in heart rate, respiration, and penile erection. Courtesy Dale Purves and Sinauer Associates.

their suitability for service. Some potential draftees claimed unsuitability because of a history of epilepsy. In a famous case a group of draft evaders claimed a history of epilepsy and were on their way by bus to have their EEGs recorded. On the way there roughly half of the men in the bus, fearing exposure, gave up and withdrew their objections to being drafted. They need not have done so. The EEG is not infallible. There are people with an aberrant-appearing EEG who have no history of neurological disease and people with a history of seizures who have a normal-appearing EEG when they are not having an attack.

The EEG, however, remains a valuable adjunct to diagnosis of neurological disease, such as epilepsy. During one type of seizure—the *grand mal*—high-voltage and irregular spike-like activity is recorded from all leads of the electroencephalogram. During another type of seizure—the *petit mal*—there is a characteristic pattern of activity that helps to diagnose the disorder. Brain tumors also may produce a characteristic slow-wave focus in the EEG.

Brainstem and Arousal

In the 1930s the Belgian physiologist Frédéric Bremer studied the effects of making transverse cuts through the brainstem of cats at various levels. Cuts through the midbrain at the level of the colliculi produced a state resembling coma in cats. Transverse cuts at a deeper level, although they paralyzed movement, left the cerebral cortex in a waking state.

In addition to identifiable nuclei, the brainstem contains a system of nerve cells and their interconnections called the reticular formation. The Italian physiologist Giuseppe Moruzzi and the American Horace Magoun showed that stimulation of the reticular formation would change the EEG from a sleeping to a waking state. They named this system the *reticular activating system* and argued that it provided the physiological basis for normal sleep and wakefulness. Although the original concept is somewhat oversimplified, the interaction of brainstem and cortex is probably an important component of the normal conscious state.

These experiments bear on the question of interpreting an apparently comatose state. Some people may look conscious but are not; they are in a *persistent vegetative state*. Some people may look unconscious but are not; they are in the *"locked in" syndrome*. These conditions and the difference between them raise fundamental moral and legal questions. Suppose that a person had left an advance directive such as a living will stating that he had no wish to be maintained on artificial life support if he were no longer capable of recovery. Should the person be kept alive when there seems to be no chance of recovery?

Imaging of Brain Activity: The Brain and Its Blood Supply

At the end of the eighteenth century the Scottish anatomist Alexander Monro[2] observed that because the brain is in a confined and rigid space, it must contain a relatively constant amount of blood at any one time. Deprived of blood for only a few seconds, people faint. An excess amount of blood would cause high pressure in the cerebral vessels and an attendant danger of stroke.

In the late nineteennth century it became apparent that the brain receives a high share of the total amount of blood pumped through the body. The brain weighs only about 1.5 kg—about 2 percent of the total body weight—yet it uses about one-quarter of the available energy provided by the blood supply. Although the net amount of blood within the brain varies over a very narrow range, the blood pressure, as recorded from the brain, reflects the state of arousal.

Angelo Mosso (figure 20.3), professor of physiology in Torino, studied a patient named Bertino (figure 20.4), who had sustained damage to the entire top of the skull. Bertone's skull defect was replaced by a cap made of *gutta percha*, a rubber-like substance. The flexible cap allowed Mosso to record pulsations of the brain associated with calculating and emotional arousal. The flexible cap allowed Mosso to record pulsations of the brain arising from different rates of blood pressure. The pulsations were correlated with activity such as mathematical calculation and emotional arousal. These same pulsations can be observed in newborn infants, whose skull bones are not yet fully closed until the second month. Pulsations of the brain can be seen and felt until the skull bones fuse.

Looking Inside the Skull

The development of x-rays in the 1890s allowed physicians (and shoe salesmen) to image the body's bones. X-rays quickly became standard for looking at fractures or dental decay or even the fit of a shoe. Although x-rays are poor at making soft tissue visible, they can help in looking at gross distortions within the brain. In the early twentieth century physicians injected air into the cerebrospinal fluid. The resultant x-rays could make the cerebral ventricles visible, which allowed physicians to see distortions of the brain caused by tumors or cysts.

Figure 20.3
Angelo Mosso (1846–1910) recorded the pulsation of the brain in a man who had sustained a skull defect. He recorded increased blood flow during mental activity.

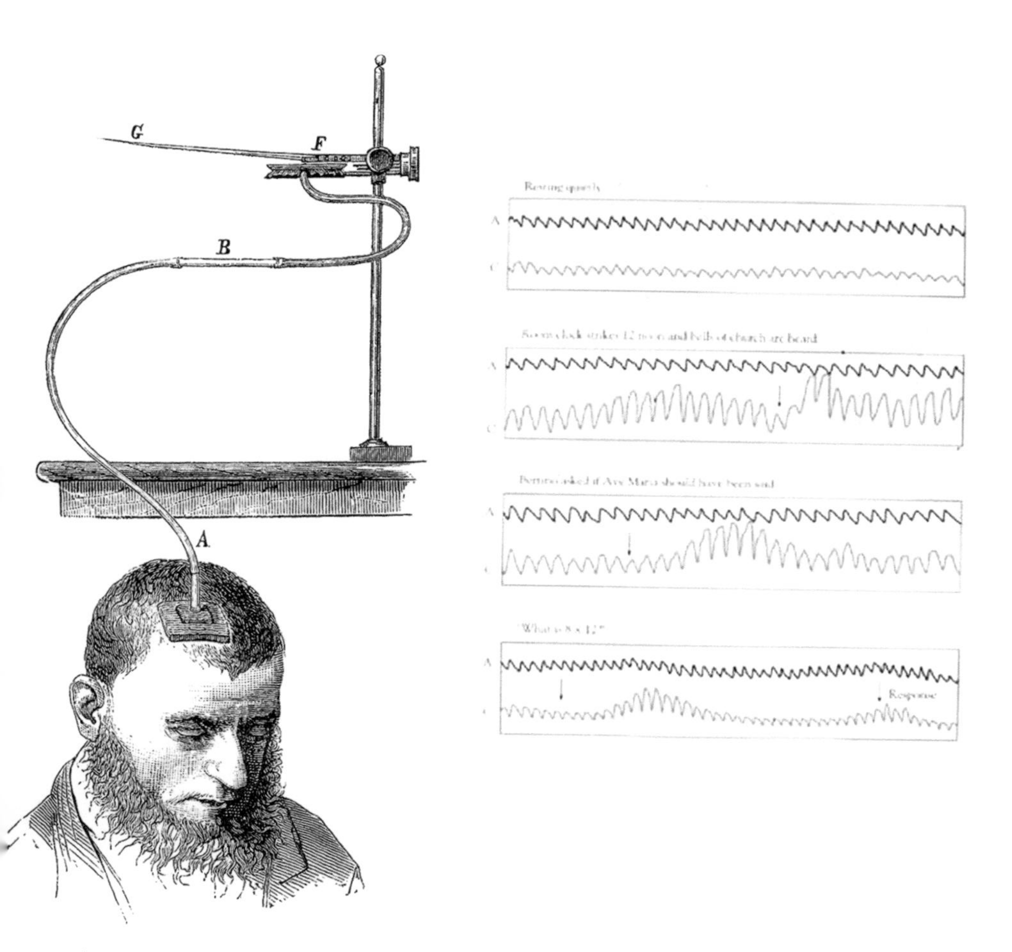

Figure 20.4
Bertino, a patient of Mosso's with the record of brain pulsations.

The availability of high-speed computing in the twentieth century reopened the possibility of using x-rays for analysis of structure. In *computerized axial tomography*—CAT scans—a beam of high-energy x-rays is aimed at the head with detectors on the side opposite the source of the x-rays. Computer storage and analysis allow a useful view of the brain in CAT scans. But CAT scans are relatively coarse in their resolution and expose the patient to radiation, so a great and welcome advance was the introduction in the 1980s of magnetic resonance imaging.

Magnetic resonance imaging (MRI) reveals the water content of brain tissue, which differs in gray and white matter. A most remarkably accurate

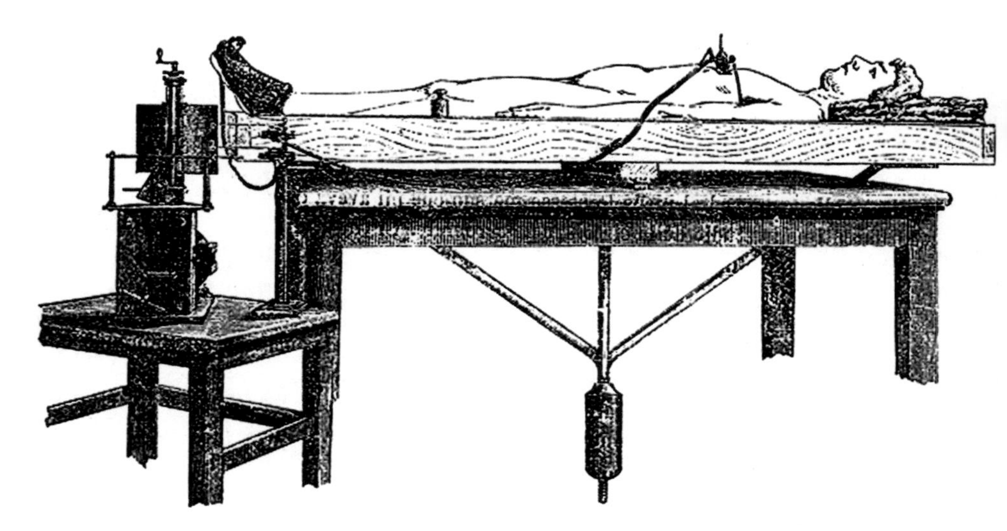

Figure 20.5
Mosso's table "La Bilancia." The subject lay on the table, which was balanced in the middle. Mosso hoped to record an increase of blood flow to the head during mental activity.

image of the brain is constructed by the computer. The image is stored as a set of voxels, each representing the density of water in a tiny volume of brain tissue at a specific location. Hence, the brain can be imaged in slices apparently cut in any desired plane, although the horizontal plane is conventional. Figure 20.6 shows an example of a normal MRI.

The MRI is a most useful tool for clinical investigations and as an aid in diagnosis. Tumors, demyelinization, and vascular accidents are among the conditions that it can identify.

Brain Metabolism and fMRI

The success of MRI suggested that similar physical measurement might be used to detect active sites in the brain. When a given region of brain increases its activity, there is an increase in the use of glucose by its cells, and a small increase in regional blood flow to that part of the brain follows. Because there is a difference in oxygen content of blood going to and returning from a given structure, that effect can serve as the basis for identifying active sites.

Functional magnetic resonance imaging (fMRI) detects changes in brain oxygen levels, yielding what is called the *blood-oxygen-level-*

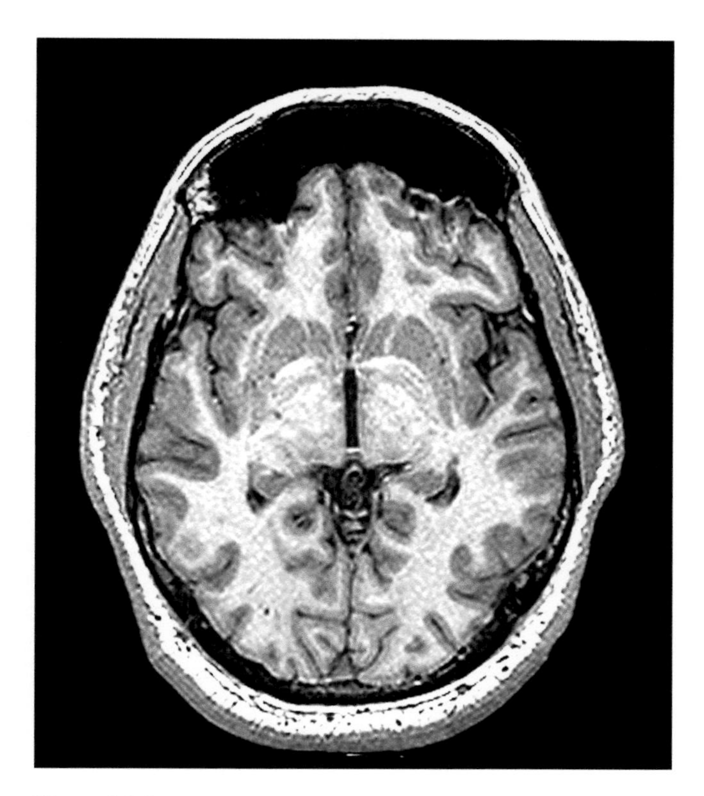

Figure 20.6
MRI structural image of brain. Courtesy University of Washington.

dependent or BOLD signal. The BOLD signal can identify a region that is active, when the brain is involved in control of movement, for instance, or thinking about that control. It also allows researchers to follow brain visual activity beyond the primary visual cortex. To many researchers fMRI seems to hold out the possibility of actually watching a conscious brain in action after centuries of only being able to speculate about brain activity in living creatures. But fMRI remains a limited tool. Not only are the spatial and temporal resolution poor, but the results can be open to poor statistical analysis and misinterpretation.

Conclusion

Myself when young did eagerly frequent

Doctor and Saint, and heard great argument

About it and about: but evermore

Came out by the same door where in I went.[3]

In 1959 a symposium was held at the New York University Institute of Philosophy. The conference to discuss consciousness was attended by many of the leading philosophers and scientists of the day. Then and now, every philosopher and neuroscientist seems to know how consciousness works, but no two of them seem to agree. There are good ideas and better apparatus now, but consciousness remains a challenge for neuroscience.

21

Are We There Yet?

Neuroscience tries to understand the structure and function of the brain and spinal cord—how they are made up of small elements and how those elements are put together so that we can see and hear, that we can walk and dance, that we can speak and understand each other when we listen. It tries to understand what has gone wrong when a crippling disease strikes. Neuroscience also extends beyond us to other animals. How can a rattlesnake hunt in the dark? How can a mouse judge whether it can fit through a tiny hole in the wall of your kitchen?

We know a great deal about all of these questions, although much remains to be discovered. This book is about how we know rather than what we know. Others have written large textbooks that summarize our current knowledge. I have tried to reveal how we got there. Neuroscience is complex, and there are often barriers to our understanding. Some are serious—modern physics can require sophisticated mathematical understanding. But other barriers are easily toppled. The great proliferation of funny names for brain subdivisions can be demystified if you translate them. The lateral geniculate nucleus looked like a little knee to some observer hundreds of years ago.

Some principles should emerge from the book. Neuroscience is a product of individual humans no less than the painting of a picture or the writing of a moving aria. It differs from art in that, like other sciences, it must ultimately be based on evidence.

Sometimes, a simple observation no one had ever noted before can lead to a deeper understanding; Eugene Aserinsky was a graduate student at the University of Chicago when he noticed that people had bursts of eye movements that seemed to come in phases during sleep. Sometimes the observation may extend over years, as with Huntington observing the tragic course of the disease that bears his name.

Technology has kept pace with the field and has been essential in allowing us to see in new ways. In the 1820s the engineers designing microscopes discovered a way to minimize chromatic aberration, one of the major limitations to seeing small objects. Within a few years a Czech genius, Jan Evangelista Purkinje, described a class of cells in the cerebellum that still bears his name. One hundred years later the oscilloscope allowed Gasser and Erlanger to see a true picture of the activity in a compound nerve.

But technology can be seductive. When the EEG was discovered, it seemed to some the ultimate tool for unlocking the secrets of the human brain. Functional scanners have taken over that role. Both are important techniques. Both tend to have been, or may now be, overrated.

Science, like the tango, might take two. The careful and systematic development by Camillo Golgi of a stain that allowed complete visualization of the structure of a single neuron gave Santiago Ramon y Cajal the method he needed to study and describe in exquisite detail the structure of the entire nervous system in vertebrates. Unlike most people who dance the tango together, the participants may not like each other. But although there is often antagonism, there is, more frequently, appreciation. The Internet has allowed people of similar interest to get to know each other, to work together, and to like each other as much as the friend in the house across the street. There are no international boundaries. You know your Dutch colleague, and you know his children and his grandchildren.

There are different sorts of discovery. Some are the product of pure intuition. Thomas Young reasoned that since there could not be an *infinite* number of color receptors, there must be some *definite* number: he suggested three. In this case confirmation came many years later, but sometimes the reasoning led the reasoner himself to test the idea. Wheatstone thought that pictures of distant scenes looked more realistic than pictures of nearby objects. He thought that maybe the disparity in the two eyes was the cause. He built an apparatus with which he could test that idea—the stereoscope—which gave him a 3-D image when each eye looked at a scene that was slightly displaced from the other.

Are we there yet? The child in the back seat has a right to ask. Maybe half-way. Fundamental questions, like those regarding consciousness and free will, remain to be understood. Beware the neuroscientist who tells you he or she knows the answer. Neuroscientists are often willing to tell you how "it" all works and derive from their knowledge rules for running your life or raising your child. Be cautious.

Neuroscience, like other sciences, and like art and like music, is a human creation. It is fallible, but it is no less a valuable part of our heritage.

Notes

Chapter 1

1. Adams, F., trans. (1972). *The genuine works of Hippocrates*. Malabar, FL: Krieger Publishing Company. Republished in 1972 from 1946 reprint of the 1849 original.

Chapter 2

1. Cotugno, D. F. A. C. (1764). *De ishiade nervosa commentarius*. Naples: Fratres Simonii. Translated in Clarke & O'Malley (1968).

Chapter 3

1. Sherrington's original suggestion for the point of contact had been *syndesm*. Michael Foster, Professor of Physiology and Fellow of Trinity College, Cambridge, discussed it with his friend Arthur Woollgar Verrall, a noted classics don at Trinity College. Verrall suggested *synapse*, from the ancient Greek *to clasp*. Foster prevailed. Sherrington wrote to his former student John Fulton about it forty years later:

> You enquire about the introduction of the term synapse; it happened thus. M. Foster had asked me to get on with the Nervous System part, . . . and [I] had not got far with it before I felt the need of some name to call the junction between nerve-cell & nerve-cell (because that place of junction now entered physiology as carrying functional importance). I wrote him of my difficulty, & my wish to introduce a specific name. I suggested using "syndesm." . . . He consulted his Trinity friend Verrall, the Euripidean scholar about it, & Verrall suggested "synapse," & as that yields a better adjectival form, it was adopted for the book.

History clarified by Dr. Martyn Bracewell. Quote from Sherrington's letter cited in Fulton (1938).

Chapter 6

1. Müller, J. (1838). *Handbuch der Physiologie des Menschen für Vorlesungen*. 2 Vols. Coblenz: Verlag von J. Hölscher. [Author's translation.]

2. Pacini, F. (1840). *Nuovi organi scoperti nel corpo umano*. Pistoia: Tipografia Cino. [Author's translation.]

3. Bianchi, A. (1889). *Relazione e catalogo dei manoscritti di Filippo Pacini*. Florence and Rome: Tipografia Bencini. Cited in Bentivoglio, M. & Pacini, P. (1995). Fillipo Pacini: A determined observer. *Brain Research Bulletin*, *38* (2), 161–165.

Chapter 7

1. Da Vinci, Leonardo. (1989). 16th century diagram from the Windsor Collection, Plate 97 in the catalog of an exhibition at the Hayward Gallery London. London and New Haven: South Bank Centre and Yale University Press.

Chapter 8

1. Newton, I. (1704). *Opticks or, a treatise of the reflexions, refractions, inflexions and colors of light: also two treatises of the species and magnitude of curvilinear figures* (4th ed.). London.

2. Wollaston, W. H. (1824). On the semi-decussation of optic nerves. *Philosophical Transactions of the Royal Society of London, 114*, 222–231.

3. Bastian, H. C. (1880). *The brain as an organ of mind*. London: Kegan.

4. Gowers, W. R. (1888). *A manual of diseases of the nervous system*. American edition. Philadelphia: Blakiston, Son & Co.

5. Doty, R., Glickstein, M., & Calvin, W. (1966). Laminar structure of the lateral geniculate nucleus in the squirrel monkey, *Saimiri sciureus*. *Journal of Comparative Neurology, 127*, 335–340.

6. David Whitteridge was professor of physiology in Edinburgh at the time. He later was appointed Waynflete professor of physiology at Oxford. His friend and colleague Peter Daniel joined him summers for their joint research on the projection of the visual fields. In keeping with British custom at the time, their authorship was presented alphabetically.

Chapter 10

1. Sherrington, C. (1900). Cutaneous sensations. In E. A. Schäfer (Ed.), *Textbook of physiology, Vol. II*, (920–1001). Edinburgh & London: Young J. Pentland.

2. Nomenclature can be confusing. Technically a fiber that conducts *toward* the cell body is a dendrite. But sensory fibers from the body look and function like axons in the rest of the nervous system, so the convention is to call them sensory axons.

Chapter 12

1. Magendie, F. (1822). Expériences sure les fonctions des racines des nerfs rachidiens. *Journal of Physiology and Experimental Pathology, Paris, 2,* 276–279. Cited and translated in Clarke & O'Malley (1968).

2. Goddard, J. (1669). Experiment performed before the Royal Society on April 1, 1669.

Chapter 13

1. Fritsch, G., & Hitzig, E. (1870). Uber die elektrische Erregbarkeit des Grosshirns. *Archiv für Anatomie und Physiologie, 37,* 300–32. Translation from von Bonin (1960).

2. Betz, V. A. (1874). Anatomischer Nachweis zweier Gehirncetnra. *Zentralblad für die Medizinische Wissenschaft, 12,* 578–580, 595–599. Translation from Clarke & O'Malley (1968).

Chapter 14

1. Yakovlev, V. M., & Yakovlev, P. I. (1955). S. S. Korsakoff's psychic disorder in conjunction with peripheral neuritis: A translation of Korsakoff's original article with brief comments on the author and his contribution to clinical medicine. *Neurology, 5,* 394–406.

2. Morgan, C. T., & Stellar, E. (1950). *Physiological psychology* (2nd ed.). New York: McGraw-Hill.

3. Tanzi, E. (1893). I fatti e le induzioni dell'odierna istologia del sistema nervoso. *Rivista Sperimentale di Freniatria e Medicina Legale, 19,* 419–472. Translation by Giovanni Berlucchi in Berlucchi, G., & Buchtel, H. A. (2009). Neuronal plasticity: Historical roots and evolution of meaning. *Experimental Brain Research, 192,* 307–319.

4. Hebb, D. O. (1949). *Organization of behavior: A neuropsychological theory.* New York: John Wiley & Sons.

Chapter 15

1. Nothnagel, H. (1881). Durst und Polydipsie. *Archiviv Für pathologische Anatomie und Physiologie, 86,* 435–447. Translation by author.

Chapter 16

1. Premack, D. (2004). Is language the key to human intelligence? *Science, 303* (5656), 318–320.

2. Gall, F. J., & J. C. Spurzheim. (1819). *Anatomie et physiologie du système nerveux en géneral*, . . . Vol. IV. Paris: Schoell et al. Translated in Clarke & O'Malley (1968).

3. Flourens, P. (1823). Recherches physique sur les propriétés et les fonctions du système nerveux dans les animaux vertébrés. *Archives Générale de Médecine (Paris)*, 2, 321–370. Translated in Clarke & O'Malley (1968).

4. Boiullaud, J.-B. (1825). Recherches cliniques propres à dèmontrer que la perte de la parole correspond à la lesion de lobules antérieurs de cerveau, ... *Archives Générale de. Médecine (Paris)*, *8*, 25–45. Translated in Clarke and O'Malley (1968).

5. Aubertin, S. A. E. (1861). Reprise de la discussion sur la forme et le volume de cerveau. *Bulletins de la Société d'Anthropologie (Paris)*, 2, 209–220.

6. Broca, P. (1861). Remarques sure le siège de la faculté du language articulé. *Bulletins de la Société d'Anthropologie (Paris), 6,* 330–357, 398–407. Translation by Gerhard von Bonin, in von Bonin, G. (1951). *Some papers on the cerebral cortex*. Springfield, IL: Charles C. Thomas.

7. Wernicke, C. (1874). Der aphasische Symptomcomplex; eine psychologische Studie auf Anatomischeer Basis. Breslau: Cohn and Weigert. Translation by Dalman and Eling, in Koehler, P. J., Bruyn, G. W., & Pearce, J. M. S. (2000). *Neurological eponyms*. New York: Oxford University Press.

8. Dandy, W. E. (1936). Operative experience in cases of pineal tumor. *Archives of Surgery, 33, 19–46.* As quoted in Finger, S. (1994). *Origins of neuroscience: A history of explorations into brain function*. New York: Oxford University Press.

9. Sperry, R. W. (1977). Forebrain commissurotomy and conscious awareness. *Journal of Medicine and Philosophy, 2* (2), 101–126.

Chapter 17

1. Lloyd, G. E. R. (Ed.). (1950). *Hippocratic writings*. London: Penguin.

2. Gowers, W. R. (1888). *A manual of diseases of the nervous system*. Phildelphia: Blakiston, Son & Co.

3. Hughlings-Jackson, J. (1867). Clinical remarks on the disorderly movements of chorea and convulsion. *Medical Times and Gazette, 2*, 669–670. Reprinted in York, G., & Koehler, P. (2000). Jacksonian epilepsy. As cited in Koehler, P., Bruyn, G., & Pearce, J. (Eds.). *Neurological eponyms*. New York: Oxford University Press.

4. Frith, D. (1948). *The case of Augustus D'Este*. New York: Cambridge University Press.

5. Parkinson, J. (1817). *An essay on the shaking palsy*. London: Whittingham and Rowland.

6. Huntington, G. (1872). On chorea. *Medical and Surgical Reporter, 26*, 317–321.

7. Alzheimer, A. (1907). Über eine eigenartige Erkrankung der Hirnrinde. *Allgemeine Zeitschrift für Psychiatrie und Psychisch-Gerichtliche Medizin, 64*, 146–148. Cited and translated in Koehler, P., Bryce, G., & Pearce, M. (2000). *Neurological eponyms*. New York: Oxford University Press.

8. Charcot, J.-M. (1874). Sclerose laterale amyotrophique. In Charcot, J.-M. Oeuvres Complete. *Bureaux du Progres Medical, 2*, 267–300. Cited in Koehler, P., Bryce, G., & Pearce, M. (2000). *Neurological eponyms*. New York: Oxford University Press.

9. Cited by Sir Geoffrey Keynes, The Grey Turner Memorial Lecture, University of Durham, delivered at the Royal Infirmary, Newcastle-upon-Tyne, February 22, 1961.

Chapter 18

1. Brown, S., & Schafer, E. A. (1888). An investigation into the functions of the occipital and temporal lobes of the monkey's brain. *Philosophical Transactions of the Royal Society of London, B, 179*, 303–327.

2. Papez, J. (1937). A proposed mechanism of emotion. *Archives of Neurology and Psychiatry, 38*, 725–743.

3. Feinstein, J. S. R., Adolphus, A. E., Damasio A. R., & Tranel, D. (2011). The human amygdala and the induction and experience of fear. *Current Biology, 21*, 34–38.

Chapter 19

1. Willis, Thomas. (1683). *Two discourses concerning the soul of brutes*. London: John Dring.

2. Kraepelin, E. (1913). *Lectures on clinical psychiatry* (3rd English ed.). T. Johnstone (Ed.). New York: William Wood & Co. Originally published in 1905 as *Einführung in die Psychiatrische Klinik. Zweiunddreißig Vorlesungen.*

3. Ibid.

Chapter 20

1. Searle, J. R. (1993). The problem of consciousness. In G. R. Bock & J. Marsh (Eds.), *Ciba Foundation Symposium 174: Experimental and theoretical studies of consciousness*. New York: Wiley, 61–62.

2. Alexander Monro, called "Secundus," was the son of the first professor of Anatomy at Edinburgh, Alexander Monro Primus. Monro Primus and Secundus

were both good anatomists. It was Secundus who clearly identified the passage between the lateral ventricles and the third ventricle, still known as the *Foramen of Monro*. Secundus's son Alexander Monro Tertius became the third Monro to occupy the chair. By all accounts he was a poor anatomist and a bad lecturer. However, it was probably Monro Tertius who made the greatest contribution of the three. Charles Darwin was enrolled as a medical student at Edinburgh but left after two years without completing the medical degree. It was Monro Tertius's lectures that were the final straw. Darwin said, "Professor Monro made anatomy as dull as he was."

3. Fitzgerald, E., trans. (1954). *Rubaiyat of Omar Khayyam*. London & Glasgow: Collins.

Further Readings

There are two directions for further readings. One is for advanced reference works and textbooks of neuroscience. Four comprehensive works stand out:

Gazzaniga, M., ed. (2009). *The cognitive neurosciences* (4th ed.). Cambridge, MA: MIT Press.

Kandel, E., Schwartz, J., Jessell, T., Siegelbaum, S., & Hudspeth, A. J., eds. (2012). *Principles of neural science* (5th ed). New York: McGraw-Hill Professional.

Purves, D., Augustine, G. J., Fitzpatrick, D., Hall, W. C., LaMantia, A.-S., & White, L. E., eds. (2012). *Neuroscience* (5th ed.). Sunderland, MA: Sinauer Associates.

Squire, L., Berg, D., Bloom, F. E., du Lac, S., Ghosh, A., & Spitzer, N., eds. (2012). *Fundamental neuroscience* (4th ed.). New York: Academic Press.

The other direction is in the history of neuroscience. Readings here fall into three broad categories: books that provide a broad narrative history of the field, classic textbooks, and biographies.

Overviews

Clarke, E., & O'Malley, C. D. (1968). *The human brain and spinal cord: A historical study illustrated by writings from antiquity to the twentieth century.* Berkeley: University of California Press. Reprinted 1996 by Jeremy Norman, San Francisco.

An excellent source book for the study of the history of neuroscience. Major contributors from ancient writers up to the 1960s are represented in chronological order, with translations into English of key works from several different languages.

Finger, S. (1994). *Origins of neuroscience: A history of explorations into brain function.* New York: Oxford University Press.

A scholarly work tracing the history of brain function from its earliest roots to the present.

Classic Textbooks:

Anatomy

Ramon y Cajal, S. (1995). *Histology of the nervous system of man and vertebrates*. (N. Swanson & L. W. Swanson, Trans.). New York: Oxford University Press. Original work published 1899.

Cajal's monumental work *Textura del Sistema Nervioso del Hombre y de los Vertebrados* was published in three volumes, the first of which appeared in Spanish in 1899. The work was translated into French and published in two volumes in 1953 as *Histologie du Systeme Nerveux*. An English translation was published in 1995. Over one hundred years since its first publication Cajal's work is still consulted for its remarkable breadth and accuracy in its descriptions and drawings of the structure of the vertebrate brain and spinal cord.

Physiology

Schäfer, E. A., ed. (1898, 1900). *Text-book of physiology*. Edinburgh & London: Young J. Pentland.

This massive tome, published in two volumes, was the standard textbook of its day. It is a multiauthored work, and the chapters on the nervous system were written by Charles Sherrington. Some topics were controversial at the time, but Sherrington's intuition usually proved correct.

Hodgkin, A. L. (1971). *The conduction of the nervous impulse*. Liverpool: Liverpool University Press.

Alan Hodgkin and Andrew Huxley collaborated to study the ionic basis of nerve conduction. Their work formed the basis for most subsequent analysis of mechanism. In this slim volume, Hodgkin describes those experiments simply and clearly.

Neurology

Gowers, W. R. (1888). *A manual of diseases of the nervous system* (American edition). Philadelphia: Blakiston, Son & Co.

Gowers was one of the foremost neurologists of his day, and this scholarly work is a most valuable source for the understanding of neurological disease and its relation to basic neuroscience.

Psychology

James, W. (1890). *The principles of psychology, in two volumes*. New York: Henry Holt & Co. Reprinted by Dover, New York, 1950.

This very well-written work by William James, the brother of the novelist Henry James, explores a broad set of questions on the mechanism of sensation perception and voluntary action. James's work represents a transition from an earlier speculative philosophical approach to these questions to a more modern experimentally based approach.

Murchison, C., ed. (1934). *A handbook of general experimental psychology*. Worcester MA: Clark University Press.

This volume was a milestone in the field. The editor assembled papers by contemporary experts on sensation, motivation, and learning.

Biography

Two great anatomists provided the structural basis for modern neuroscience. Cajal's autobiography is an excellent way to understand his contributions. Mazzarello's biography of Golgi is helpful in understanding his life and work.

Koehler, P. J., Bruyn, G. W., & Pearce, J. M. S. (2000). *Neurological eponyms*. New York: Oxford University Press.

Structures in the brain and diseases that affect it are often named for the person whose work is most associated with that structure or disease. This pleasant volume tells you a bit about the life and work of Purkinje, Babinski, Broca, and fifty-three others whose names are attached to a structure, process, or disease of the brain.

Mazzarello, P. (2010). *Golgi: A biography of the founder of modern neuroscience* (A. Badiani & H. A. Buchtel, Trans.). New York: Oxford University Press.

Ramon y Cajal, S. (1989). *Recollections of my life*. (E. Horne Craigie with J. Cano, Trans.) Cambridge, MA: MIT Press.

Squire, L. R. (1996–2011). *The history of neuroscience in autobiography*. Six volumes. New York: Elsevier and Oxford University Press.

These autobiographical essays describe the scientific careers and the influences on the work of nearly one hundred neuroscientists, most of whom are alive today.

References

General

Clarke, E., & O'Malley, C. D. (1968). *The human brain and spinal cord: A historical study illustrated by writings from antiquity to the twentieth century.* Berkeley: University of California Press.

Finger, S. (1994). *Origins of neuroscience: A history of explorations into brain function.* New York: Oxford University Press.

Koehler, P. J., Bruyn, G. W., & Pearce, J. M. S. (2000). *Neurological eponyms.* New York: Oxford University Press.

von Bonin, G. (1960). *Some papers on the cerebral cortex: Translated from the French and German.* Springfield, IL: Charles C. Thomas.

Chapter 1

Adams, F. (Ed.). (1972). *The genuine works of Hippocrates.* Malabar, FL: Krieger Publishing Company. Republished in 1972 from 1946 reprint of the 1849 original.

Chapter 2

Cotugno, D. F. A. C. (1764). *De ishiade nervosa commentarius.* Naples: Fratres Simonii.

Chapter 3

Ben Geren, B. (1954). The formation from the Schwann cell surface of myelin in the peripheral nerves of chick embryos. *Experimental Cell Research*, *7*(2), 558–562.

Ben Geren, B., & Raskind, J. (1953). Development of the fine structure of the myelin sheath in sciatic nerves of chick embryos. *Proceedings of the National Academy of Sciences of the United States of America*, *39*(8), 880–884.

Deiters, O. (1865). *Untersuchungen über Gehirn und Rückenmark des Menschen und der Säugethiere*. Braunschweig: Vieweg und Sohn.

Fulton, J. (1938). *Physiology of the nervous system*. London: Oxford University Press.

Golgi, C. (1883). Recherches sur l'histologie des centres nerveaux. *Archives Italiennes de Biologie, 3*, 285–317; (1884) *4*, 92–123; (1959) *97*, 279–299.

Golgi, C. (1903). *Opera Omnia. Collection of all Golgi's work in 3 volumes*. Milano: Hoepli.

Purkinje, J. E. (1837). Bericht über die Versammlung deutscher Naturforscher und Aerzte in Prague [Report on the conference of German scientists and doctors in Prague]. *Anatomisch-physiologische Verhandlungen*, *3*, 177–180.

Ramon y Cajal, S. (1972). *The structure of the retina* (A. S. Thorpe & M. Glickstein, Trans.). Springfield, IL: Charles C. Thomas.

Chapter 4

Galvani, L. (1791). *De viribus electriciatis in motu musculari commentarius. Pars prima*. Bologna: Accademia delle Scienze.

Gasser, J., & Erlanger, H. (1968). *Electrical signs of nervous activity* (2nd ed.). Philadelphia: University of Pennsylvania Press. Original work published 1937.

Hartline, H. K. (1934). Intensity and duration in the excitation of single photoreceptor units. *Journal of Cellular and Comparative Physiology*, *5*, 229–247.

Hartline, H. K., & Graham, C. H. (1932). Nerve impulses from single receptors in the eye. *Journal of Cellular and Comparative Physiology*, *1*, 277–295.

Helmholtz, H. von. (1852). Messungen über Fortpflanzungsgeschwindigkeit der Reizung in den Nerven. *Archiv für Anatomie, Physiologie und wissenschaftliche Medicin*, 199–216.

Hodgkin, A. L. (1958). The Croonian lecture: Ionic movements and electrical activity in giant nerve fibres. *Proceedings of the Royal Society of London. Series B, Biological Sciences*, *148*(930), 1–37.

Hodgkin, A. L. (1971). *The conduction of the nervous impulse*. Liverpool: Liverpool University Press.

Hodgkin, A. L., & Huxley, A. F. (1939). Action potentials recorded from inside a nerve fibre. *Nature*, *144*, 710–711.

Chapter 5

Ahlquist, R. P. (1948). A study of the adrenotropic receptors. *American Journal of Physiology*, *153*(3), 586–600.

Bain, W. A. (1932). A method of demonstrating the humoral transmission of the effects of cardiac vagus stimulation in the frog. *Experimental Physiology*, *22*(3), 269–274.

Dale, H. H. (1914–15). The action of certain esters and ethers of choline, and their relation to muscarine. *Journal of Pharmacology and Experimental Therapeutics*, 6, 188.

Dale, H. H., Feldberg, W. S., & Vogt, M. (1936). Release of acetylcholine at voluntary motor nerve endings. *Journal of Physiology*, *86*, 353–380.

Elliott, T. R. (1904). On the action of adrenaline. *Journal of Physiology*, *31*, xx–xxi.

Langley, J. N. (1905). On the reaction of cells and of nerve-endings to certain poisons, chiefly as regards the reaction of striated muscle to nicotine and to curari. *Journal of Physiology*, *33*, 374–413.

Löwi, O. (1921). Über humorale Übertragkeit der Herznervenwirkung. *Pflügers Archiv für die gesamte Physiologie*, *189*, 201–213.

Pert, C. B., & Snyder, S. H. (1973). Opiate receptor: Demonstration in nervous tissue. *Science*, *179*, 1011–1014.

Chapter 6

Bentivoglio, M., & Pacini, P. (1995). Fillipo Pacini: A determined observer. *Brain Research Bulletin*, *38*(2), 161–165.

Blough, D. S. (1958). A method for obtaining psychophysical thresholds from the pigeon. *Journal of the Experimental Analysis of Behavior*, *1*, 31–43.

Fechner, G. (1860). *Elemente der Psychophysik*. Leipzig: Breitkopf & Härtel.

Loewenstein, W. R. (1960). Biological transducers. *Scientific American*, *203*, 98–108.

Müller, J. (1838). *Handbuch der Physiologie des Menschen für Vorlesungen*. 2 Vols. Coblenz: Verlag von J. Hölscher.

Pacini, F. (1840). *Nuovi organi scoperti nel corpo umano*. Pistoia: Tipografia Cino.

Weber, E. H. (1846). Tastsinn und Gemeingefühl. In R. Wagner (Ed.), *Handwörterbuch der Physiologie* (Vol. III, pp. 481–588). Brunswick: Vieweg.

Chapter 7

Da Vinci, L. (1989). *16th century diagram from the Windsor collection, Plate 97 in the catalog of an exhibition at the Hayward Gallery London*. London, New Haven: South Bank Centre and Yale University Press.

Davson, H. (1962). *The eye: The visual process*. New York: Academic Press.

Descartes, R. (1637). *Discours de la méthode pour bien conduire sa raison, et chercher la vérité dans les sciences*. Leiden: Maire.

Kepler, J. (1604). *Ad vitellionem paralipomena, quibus Astronomiae pars optica traditur*. Frankfurt: Claudius Marnius. *English translation: Kepler, J. (2000). Optics* (W. H. Donahue, Trans.). Santa Fe, NM: Green Lion Press.

Polyak, S. (1957). *The vertebrate visual system*. Chicago: University of Chicago Press.

Ramon y Cajal, S. (1972). *The structure of the retina* (A. S. Thorpe & M. Glickstein, Trans.). Springfield, IL: Charles C. Thomas.

Scheiner, C. (1620). *Oculus, hoc est: Fundamentum opticum*. Innsbruck: Agricola.

Schultze, M. (1867). Ueber Stäbchen und Zapfen der Retina. *Archiv für mikroskopische Anatomie, 3*, 215–247.

Walls, G. L. (1942). *The vertebrate eye and its adaptive radiation*. New York: Hafner Publishing. Copyright Cranbrook Institute of Science.

Young, T. (1802). On the theory of lights and colours. *Philosophical Transactions of the Royal Society*, *92*, 12–48.

Chapter 8

Allman, J., & Kaas, J. (1971). Representation of the visual field in striate and adjoining cortex of the owl monkey (*Aotus trivirgatus*). *Brain Research*, *35*(1), 89–106.

Balint, R. (1909). Seelenlähmung des "Schauens," optische Ataxie, räumliche Störung der Aufmerksamkeit. *Monatsschrift für Psychiatrie und Neurologie*, *25*, 51–81.

Bard, P. (1938). Studies on the cortical representation of somatic sensibility. Harvey Lecture, February 17, 1938. *Bulletin of the New York Academy of Medicine*, *14*(10), 585–607.

Bastian, H. C. (1880). *The brain as an organ of mind*. London: Kegan.

Brouwer, B., & Zeeman, W. (1926). The projection of the retina in the primary optic neuron in monkeys. *Brain*, *49*, 1–35.

Brown, S., & Schäfer, E. (1888). An investigation into the functions of the occipital and temporal lobes in the monkey's brain. *Philosophical Transactions of the Royal Society of London. Series B, Biological Sciences*, *179*, 303–327.

Chow, K. L. (1951). Effects of partial extirpation of posterior association cortex on visually mediated behavior. *Comparative Psychology Monographs*, *20*, 187–218.

Daniel, P., & Whitteridge, D. (1961). The representation of the visual field on the cerebral cortex in monkeys. *Journal of Physiology*, *159*, 203–221.

Doty, R., Glickstein, M., & Calvin, W. (1966). Laminar structure of the lateral geniculate nucleus in the squirrel monkey, *Saimiri sciureus*. *Journal of Comparative Neurology*, *127*, 335–340.

Ferrier, D. (1876). *The functions of the brain*. London: Smith-Elder.

Flourens, P. (1824). *Recherches experimentales sur le propriétés et les fonctions du système nerveux dans les animaux vertébrés*. Paris: Cheviot.

Foerster, O. (1929). Beitrag zur Pathophysiologie der Sehbahn und der Sehsphäre. *Journal für Psychologie und Neurologie (Leipzig)*, *39*, 463–485.

Gennari, F. (1782). *De peculiari structura cerebri nonnulisque ejus morbis*. Parma: Regio Typographeo.

Glickstein, M., King, R. A., Miller, J., & Berkley, M. (1967). Cortical projections from the dorsal lateral geniculate nucleus of cats. *Journal of Comparative Neurology*, *130*, 55–76.

Glickstein, M., & May, J. G. (1982). Visual control of movement: The visual input to the pons and cerebellum. In W. D. Neff (Ed.), *Progress in sensory physiology* (Vol. 7). New York: Academic Press.

Glickstein, M., May, J., & Mercier, B. (1985). Corticopontine projection in the macaque: The distribution of labelled cortical cells after large injections of horseradish peroxidase in the pontine nuclei. *Journal of Comparative Neurology*, *235*, 343–359.

Glickstein, M., & Rizzolatti, G. (1984). Francesco Gennari and the structure of the cerebral cortex. *Trends in Neurosciences*, *17*, 464–467.

Glickstein, M., & Whitteridge, D. (1987). Tatsuji Inouye and the mapping of the visual field on the human cerebral cortex. *Trends in Neurosciences*, *10*(9), 350–353.

Gowers, W. R. (1888). *A manual of diseases of the nervous system* (American edition). Philadelphia: Blakiston, Son & Co.

Henschen, S. (1890). *Klinische und anatomische Beitträge zur Pathologie des Gehirns. Part I*. Uppsala: Almquist and Wiksell.

Holmes, G. (1918a). Disturbances of vision by cerebral lesions. *British Journal of Ophthalmology*, *2*, 253–384.

Holmes, G. (1918b). Disturbances of visual orientation. *British Journal of Ophthalmology*, *2*, 449–468, 506–516.

Hubel, D., & Wiesel, T. (1965). Receptive fields and functional architecture in two non-striate areas (18 and 19) of the cat. *Journal of Neurophysiology*, *28*, 229–289.

Inouye, T. (2000). Visual disturbances following gunshot wounds of the cortical visual area. Based on observations of the wounded in the recent Japanese wars. (M. Glickstein & M. Fahle, Trans.) [Originally published in 1909 as *Die Sehstörung bei Schussverletzungen der kortikalen Sehsphäre.]*. *Brain*, *123*(suppl), 1–100.

Klüver, H., & Bucy, P. (1938). An analysis of certain effects of bilateral temporal lobectomy in the rhesus monkey with special reference to psychic blindness. *Journal of Psychology*, *5*, 33–54.

Minkowski, M. (1920). Über den Verlauf die Endigung und die zentrale Repräsentation von gekreutzten und ungekreutzten Sehnervfasern bei einigen Säugetieren und beim Menschen Schweiz. *Archives of Neurology and Psychiatry*, *6*, 201–252, 268–303.

Mishkin, M. (1966). Visual mechanisms beyond the striate cortex. In R. Russell (Ed.), *Frontiers in Physiological Psychology*. New York: Academic Press.

Munk, H. (1881). *Über die Funktionen der Grosshirndinde*. Berlin: Hirschwald.

Newton, I. (1704). *Opticks or, a treatise of the reflexions, refractions, inflexions and colours of light: also two treatises of the species and magnitude of curvilinear figures* 4th ed. London.

Obersteiner, H. (1888). *Anleitung beim Studium des baues der Nervösen Centralorgane im gesunde und kranken Zustände*. Leipzig, Vienna: Toeplitz und Deuticker.

Penfield, W., & Rasmussen, T. (1952). *The cerebral cortex of man*. New York: Macmillan.

Talbot, S. (1942). A lateral localization in cat's visual cortex. *Federation Proceedings*, *1*, 84.

Talbot, S. A., & Marshall, W. H. (1941). Physiological studies on neural mechanisms of visual localization and discrimination. *American Journal of Ophthalmology*, *24*, 1255–1263.

Ungerleider, L., & Mishkin, M. (1982). Two cortical visual systems. In D. F. Ingle & M. Goodale (Eds.), *Analysis of visual behavior* (459–486). Cambridge, MA: MIT Press.

Van Buren, J. (1963). Trans-synaptic retrograde degeneration in the visual system. *Journal of Neurology, Neurosurgery, and Psychiatry*, *26*, 402–409.

Van Essen, D., Anderson, C., & Felleman, D. (1992). Information processing in the visual system: An integrated systems perspective. *Science*, *225*, 419–423.

Wollaston, W. H. (1824). On the semi-decussation of optic nerves. *Philosophical Transactions of the Royal Society of London*, *114*, 222–231.

Zeki, S. (1978). Uniformity and diversity of structure and function in the rhesus monkey prestriate visual cortex. *Journal of Physiology*, *277*, 273–290.

Chapter 9

Griffin, D. (1958). *Listening in the dark*. New Haven, CT: Yale University Press.

Gourevitch, G., & Hack, M. (1966). Audibility in the rat. *Journal of Comparative and Physiological Psychology*, *62*(2), 288–291.

Helmholtz, H. (1954). *On the sensation of tone*. New York: Dover. Original work published 1896.

Rayleigh. (1896). *Theory of sound* (Vol. I). London: MacMillan.

Stebbins, W., Green, S., & Miller, J. (1966). Auditory sensitivity in the monkey. *Science*, *153*, 1646–1647.

Von Bekesy, G. (1961). Concerning the pleasures of observing and the mechanics of the inner ear. Nobel lecture, 1961.

Weaver, E., & Bray, C. (1930). Auditory nerve impulses. *Science*, *71*, 215.

Willis, T. (1683). *Two discourses concerning the soul of brutes*. London: John Dring.

Chapter 10

Blix, M. (1884). Experimentelle Beiträge zur Lösung der Frage über die specifische Energie der Hautnervern. *Zeitscrift für Biologie*. *20*, 141–156. Cited in E. A. Schäfer (Ed.), *Text-book of physiology*, (Vol. II, pp. 920–1001). Edinburgh & London: Young J. Pentland.

Breuer, J. (1873). Über Bogengänge des Labyrinths. *Allg. Wien. med. Ztg.*, *18*, (S. 598, 606).

Brown-Sequard, C. E. (1860). *Course of lectures on the physiology and pathology of the central nervous system*. Philadelphia: Collins.

Foerster, O. (1933). The dermatomes in man. (Schorstein Lecture, London, 1932.). *Brain*, *56*, 1–39.

Ramon y Cajal, S. (1989). *Recollections of my life* (E. Horne Craigie with J. Cano, Trans.). Cambridge, MA: MIT Press. Original work published in Madrid, 1901–1917.

Sherrington, C. (1900). Cutaneous sensations. In E. A. Schäfer (Ed.), *Text-book of physiology* (Vol. II, pp. 920–1001). Edinburgh: Young J. Pentland.

Vallbo, A. B., Hagbarth, K. E., & Wallin, B. G. (2004). Microneurography: How the technique developed and its role in the investigation of the sympathetic nervous system. *Journal of Applied Physiology*, *96*(4), 1262–1269.

Chapter 11

Bullock, T. H., & Cowles, R. B. (1952). Physiology of an infrared receptor: The facial pit of pit vipers. *Science*, *115*(2994), 541–543.

Butenandt, A. (1955). Über Wirkstoffes Insektenreiches. *Naturwissenschaftliche Rundschau*, *8*, 457–464.

Dethier, V. G. (1962). *To know a fly*. San Francisco: Holden Day.

Fabre, J.-H. (1921). *Le Monde Merveilleux des Insectes*. Paris: Delagrave.

Henning, H. (1916). Die Qualitätsreihe des Geschmacks. *Zeitschrift fur Psychologie mit Zeitschrift fur Angewandte Psychologie*, *74*, 203–219.

Linneaus, K. (1752). Odores medicamentorum. *Amoenitates Academicae*, *3*, 183–201.

Lissman, H. W. (1951). Continuous electrical signals from the tail of a fish (*Grymnarchus niloticus*). *Nature*, *167*(4240), 201–202.

Noble, G. K., & Schmidt, A. (1937). The structure and function of the facial and labial scales of snakes. *Proceedings of the American Philosophical Society*, *B*, *77*, 263.

Chapter 12

Babinski, J. (1903). De l'abduction des orteils. *Revue Neurologique, 11*, 728–729.

Bell, C. (1870). *Letters of Sir Charles Bell selected from his correspondence with his brother George Joseph Bell.* London: Murray.

Descartes, R. (1637). *Discours de la méthode pour bien conduire sa raison, et chercher la vérité dans les sciences.* Leiden: Maire.

Erb, W. H. (1875). Über Sehnenereflexe bei gesunden und Rückenmarkskranken. *Archiv für Psychiatrie und Nervenkrankheiten, 5*, 792–802.

Goddard, J. (1669). Experiment performed before the Royal Society, London, on 1 April, 1669.

Hall, M. (1833). On the reflex function of the medulla oblongata and medulla spinalis. *Philosophical Transactions of the Royal Society, 123*, 635–655.

Liddell, E. G. T., & Sherrington, C. (1924). Reflexes in response to stretch (myotatic reflexes). *Proceedings of the Royal Society of London. Series B, Containing Papers of a Biological Character, 96*, 212–242.

Lloyd, D. (1960). Spinal mechanisms involved in somatic activities. In J. Field, W. Magoun, & V. E. Hall (Eds.), *Handbook of Physiology* (pp. 929–950). Washington, DC: American Physiological Society.

Magendie, F. (1822). Expériences sure les fonctions des racines des nerfs rachidiens. *Journal de Physiologie Expérimentale et Pathologie, Paris.*, 2, 276–279.

Ramon y Cajal, S. (1909–1911). *Histologie du systeme nerveux* (L. Azouolay, Trans.). Madrid: Instituto Cajal (C.S.I.C.) English translation: Ramon y Cajal, S. (1995). Histology of the nervous system of man and vertebrates (N. Swanson & L. W. Swanson, Trans.). New York: Oxford University Press.

Renshaw, B. (1940). Activity in the simplest spinal reflex pathways. *Journal of Neurophysiology, 3*(5), 373–387.

Westphal, K. (1875). Über einige Bewegungs-erscheinungen an gelähmten Gliedern. *Archiv für Psychiatrie und Nervenkrankheiten, 5*, 803–834.

Whytt, R. (1763). *An essay on the vital and other involuntary motions of animals.* Edinburgh: Balfour.

Chapter 13

Betz, V. A. (1874). Anatomischer Nachweis zweier Gehirncetnra. *Zentralblad für die Medizinische Wissenschaft. 12*, 578–580, 595–599.

Broca, P. (1861). Remarques sur le siège de la faculté du langage articulé, suivies d'une observation d'aphémie (perte de la parole). *Bulletin de la Société d'Anthropologie, Paris, 36*, 330–357.

Fritsch, G., & Hitzig, E. (1870). Über die elektrische Erregbarkeit des Grosshirns. *Archiv für Anatomie und Physiologie, 37*, 300–332.

Lawrence, D. G., & Kuypers, H. G. J. M. (1968). The functional organization of the motor system in the monkey. I. The effects of bilateral pyramidal lesions. *Brain*, *91*(1), 1–14.

Sultan, F., & Braitenberg, V. (1993). Shapes and sizes of different mammalian cerebella: A study in quantitative comparative neuroanatomy. *Journal fur Hirnforschung*, *34*, 79–92.

Chapter 14

Chow, K. L. (1951). Effects of partial extirpations of the posterior association cortex on visually mediated behavior. *Comparative Psychology Monograph*, *20*, 187–217.

Fuster, J. M. (1980). *The prefrontal cortex*. New York: Raven Press.

Goldman-Rakic, P. S. (1987). Circuitry of primate prefrontal cortex and regulation of behavior by representational memory. In F. Plum & V. Mountcastle (Eds.), *Handbook of physiology* (Vol. 5, pp. 373–517). Washington, DC: The American Physiological Society.

Gross, C. G., Rocha-Miranda, C. E., & Bender, D. B. (1972). Visual properties of neurons in inferotemporal cortex of the macaque. *Journal of Neurophysiology*, *35*, 96–111.

Hebb, D. O. (1949). *Organization of behavior: A neuropsychological theory*. New York: John Wiley & Sons.

Hunter, W. S. (1913). The delayed reaction in animals and children. *Behavior Monographs*, *2*, 1–86.

Hunter, W. S. (1930). A consideration of Lashley's theory of the equipotentiality of cerebral action. *Journal of General Psychology*, *3*(4), 455–468.

Jacobsen, C. F. (1936). Studies of cerebral function in primates: I. The functions of the frontal association areas in monkeys. *Comparative Psychology Monographs*, *13*, 1–68.

Kebabian, J. W., Petzold, G. L., & Greengard, P. (1972). Dopamine-sensitive adenylate cyclase in caudate nucleus of rat brain, and its similarity to the "dopamine receptor." *Proceedings of the National Academy of Sciences of the United States of America*, *69*, 2145–2149.

Klüver, H., & Bucy, P. C. (1938). An analysis of certain effects of bilateral temporal lobectomy in the rhesus monkey, with special reference to "psychic blindness." *Journal of Psychology*, *5*, 33–54.

Lashley, K. S. (1929). *Brain mechanisms and intelligence: A quantitative study of injuries to the brain*. Chicago: University of Chicago Press.

McCormick, D. A., & Thompson, R. F. (1984). Cerebellum: Essential involvement in the classically conditioned eyelid response. *Science*, *223*, 296–299.

Milner, B., Corkin, S., & Teuber, H. L. (1968). Further analysis of the hippocampal amnesiac syndrome: 14-year follow-up study of H. M. *Neuropsychologia*, *6*, 215–234.

Mishkin, M., & Pribram, K. H. (1954). Visual discrimination performance following partial ablations of the temporal lobe. I. Ventral vs. lateral. *Journal of Comparative and Physiological Psychology*, *47*, 14–20.

Morgan, C. T., & Stellar, E. (1950). *Physiological psychology* (2nd ed.). New York: McGraw-Hill.

Myers, R. E. (1956). Function of corpus callosum in interocular transfer. *Brain*, *79*(2), 358–363.

Pinsker, H., Kupfermann, I., Castellucci, V., & Kandel, E. R. (1970). Habituation and dishabituation of the gill-withdrawal reflex in *Aplysia*. *Science*, *167*, 1740–1742.

Scoville, W. B., & Milner, B. (1957). Loss of recent memory after bilateral hippocampal lesions. *Journal of Neurology, Neurosurgery, and Psychiatry*, *20*(1), 11–21.

Tanzi, E. (1893). I fatti e le induzioni dell'odierna istologia del sistema nervoso. *Rivista Sperimentale di Freniatria e Medicina Legale*, *19*, 419–472. Translation by Giovanni Berlucchi in Berlucchi, G., & Buchtel, H. A. (2009). Neuronal plasticity: Historical roots and evolution of meaning. *Experimental Brain Research*, *192*, 307–319.

Yakovlev, V. M., & Yakovlev, P. I. (1955). S. S. Korsakoff's psychic disorder in conjunction with peripheral neuritis: A translation of Korsakoff's original article with brief comments on the author and his contribution to clinical medicine. *Neurology*, *5*, 394–406.

Yeo, C. H., Hardiman, M. J., & Glickstein, M. (1984). Discrete lesions of the cerebellar cortex abolish the classically conditioned nictitating membrane response of the rabbit. *Behavioural Brain Research*, *13*, 261–266.

Chapter 15

Cannon, W. B. (1918). The Croonian lecture: The physiological basis of thirst. *Proceedings of the Royal Society of London. Series B, Containing Papers of a Biological Character*, *90*, 283–301.

Cannon, W., & Washburn, A. (1912). An explanation of hunger. *American Journal of Physiology*, *29*, 441–454.

Carlisle, H. J. (1966). Heat intake and hypothalamic temperature during behavioral temperature regulation. *Journal of Comparative and Physiological Psychology*, *61*(3), 388–397.

Epstein, A. N., & Milestone, R. (1968). Showering as a coolant for rats exposed to heat. *Science*, *160*(3830), 895–896.

Garcia, J., Kimeldorf, D. J., & Koelling, R. A. (1955). Conditioned aversion to saccharin resulting from exposure to gamma radiation. *Science*, *122*(3160), 157–158.

Nothnagel, H. (1881). Durst und Polydipsie. *Archiv für pathologiosche Anatomie und Physiologie*, *86*, 435–447.

Raisman, G., & Field, P. M. (1971). Sexual dimorphism in the preoptic area of the rat. *Science*, *173*(3998), 731–733.

Ranson, S. W. (1934). The hypothalamus: Its significance for visceral innervation and emotional expression. *Transactions of the College of Physicians of Philadelphia*, *2*, 222–242.

Chapter 16

Aubertin, S. A. E. (1861). Reprise de la discussion sur la forme et le volume de cerveau. *Bulletins de la Société d'Anthropologie (Paris)*, *2*, 209–220.

Bogen, J. E., & Vogel, P. J. (1963). Treatment of generalized seizures by cerebral commissurotomy. *Surgical Forum*, *14*, 431–433.

Boiullaud, J.-B. (1825). Recherches cliniques propres à dèmontrer que la perte de la parole correspond à la lesion de lobules antérieurs de cerveau. *Archives Générale de Médecine (Paris)*, *8*, 25–45.

Broca, P. (1861). Remarques sure le siège de la faculté du language articulé. *Bulletins de la Société d'Anthropologie*, *6*, 330–357, 398–407.

Dandy, W. E. (1936). Operative experience in cases of pineal tumor. *Archives of Surgery*, *33*, 19–46.

Flourens, P. (1823). Recherches physique sur les propriétés et les fonctions du système nerveux dans les animaux vertébrés. *Archives Générale de Médecine (Paris)*, *2*, 321–370.

Gall, F. J., & Spurzheim, J. C. (1819). *Anatomie et physiologie du système nerveux en géneral* (Vol. IV). Paris: Schoell et al.

Gardner, R., & Beatrice, T. (1969). Teaching sign language to a chimpanzee. *Science*, *165*(3894), 664–672.

Gazzaniga, M., Bogen, J., & Sperry, R. W. (1962). Some functional effects of sectioning the cerebral commissures in man. *Proceedings of the National Academy of Sciences of the United States of America*, *48*, 1765–1769.

Hayes, C. (1951). *The ape in our house*. New York: Harper.

Kellog, L., & Kellog, A. (1933). *The ape and the child*. New York: McGraw-Hill.

Lenneberg, E. (1967). *Biological foundations of language*. New York: Wiley.

Lorch, M. (2012). Speaking for yourself: The medico-legal aspects of aphasia in 19th century Britain. In S. Casper & S. Jacyna (Eds.), *The neurological patient in history* (pp. 63–80). New York: Rochester University Press.

Pilley, J. W. & Reid, A. K. (2011). Border collie comprehends object names as verbal referents. *Behavioural Processes*, *86*(2), 184–195.

Premack, D. (2004). Is language the key to human intelligence? *Science*, *303*(5656), 318–320.

Sperry, R. W. (1977). Forebrain commissurotomy and conscious awareness. *Journal of Medicine and Philosophy*, *2*(2), 101–126.

Sperry, R. W., Gazzaniga, M. S., & Bogen, J. E. (1969). Interhemispheric relationships: The neocortical commissures; syndromes of hemisphere disconnection. In P. M. Vinken & G. W. Bruyn (Eds.), *Handbook of clinical neurology, vol. IV: Disorders of speech, perception and symbolic behaviour* (pp. 273–290). Amsterdam: North-Holland.

Trescher, J. H., & Ford, F. R. (1937). Colloid cyst of the third ventricle. Report of a case: Operative removal with section of posterior half of corpus callosum. *Archives of Neurology and Psychiatry*, *37*, 973.

Wernicke, C. (1874). Der aphasische Symptomcomplex; eine psychologische Studie auf Anatomischer Basis. Breslau: Cohn and Weigert.

Chapter 17

Alzheimer, A. (1907). Über eine eigenartige Erkrankung der Hirnrinde. *Allgemeine Zeitschrift für Psychiatrie und Psychisch-Gerichtliche Medizin*, *64*, 146–148.

Carlsson, A. (1993). Thirty years of dopamine research. *Advances in Neurology*, *60*, 1–10.

Charcot, J.-M. (1874). Sclerose laterale amyptrophoque. In J.-M. Charcot, Oeuvres complete. *Bureaux du Progres Medical, 2,* 267–300.

Cotzias, G. C., Papavasiliou, P. S., & Gellene, R. (1969). Modification of Parkinsonism—chronic treatment with L-dopa. *New England Journal of Medicine*, *280*(7), 337–345.

Creutzfeld, H.-G. (1920). Über eine eigenartige herdförmige Erkrankung des Zentralnervensystems. *Zeitschrift für die Gesamte Neurologie und Psychiatrie*, *57*, 1–18.

Frith, D. (1948). *The case of Augustus D'este*. New York: Cambridge University Press.

Gajdusek, D. C., Gibbs, C. J. J., Jr., & Alpers, M. (1967). Transmission and passage of experimenal "kuru" to chimpanzees. *Science*, *155*(3759), 212–214.

Gowers, W. R. (1888). *A manual of diseases of the nervous system*. Phildelphia: Blakiston Son & Co.

Hamill, O. P., Marty, A., Neher, E., Sakmann, B., & Sigworth, F. J. (1981). Improved patch-clamp techniques for high-resolution current recording from cells and cell-free membrane patches. *Pflügers Archiv European Journal of Physiology*, *391*(2), 85.

Hornykiewicz, O. (1962). Die topische Lokalisation und das Verhalten von Noradrenalin und Dopamin (3-Hydroxytyramin) in der Substantia nigra des normalen und Parkinsonkranken Menschen. *Wiener Klinische Wochenschrift*, *75*, 309–312.

Hughlings-Jackson, J. (1867). Clinical remarks on the disorderly movements of chorea and convulsion. *Medical Times and Gazette*, 2, 669–670.

Huntington, G. (1872). On chorea. *Medical and Surgical Reporter*, *26*, 317–321.

Jakob, A. M. (1921). Über eigenartige Erkrankungen des Zentralnervensystems mit bemerkenswerten anatomischen Befunden (spastische Pseudosclerose-encephalomyelopathie mit disseminierten Degenerationsherden). *Zeitschrift für die Gesamte Neurologie und Psychiatrie*, *64*, 147–228.

Lloyd, G. E. R. (Ed.). (1950). *Hippocratic writings*. London: Penguin.

Parkinson, J. (1817). *An essay on the shaking palsy*. London: Whittingham and Rowland.

Prusiner, S. B. (1982). Novel proteinaceous infectious particles cause scrapie. *Science*, *216*(4542), 136–144.

Tretiakoff, K. (1919). *Contribution à l'etude de l'anatomie pathologique du Locus Niger*. Thése de Paris, No. 293.

Willis, T. (1683). *Two discourses concerning the soul of brutes*. London: John Dring.

Chapter 18

Bigelow, H. J. (1850). Dr. Harlow's case of recovery from the passage of an iron bar through the head. *American Journal of the Medical Sciences*, *20*, 13–22.

Brown, S., & Schafer, E. A. (1888). An investigation into the functions of the occipital and temporal lobes of the monkey's brain. *Philosophical Transactions of the Royal Society of London. Series B, Biological Sciences*, *179*, 303–327.

Feinstein, J. S., Adolphs, R., Damasio, A., & Tranel, D. (2011). The human amygdala and the induction and experience of fear. *Current Biology*, *21*(1), 34–38.

Harlow, J. M. (1848). Passage of an iron rod through the head. *Boston Medical and Surgical Journal*, *39*, 389–393.

Jacobsen, C. F. (1936). Studies of cerebral function in primates: I. The functions of the frontal association areas in monkeys. *Comparative Psychology Monographs*, *13*, 1–68.

Klüver, H., & Bucy, P. (1939). Preliminary analysis of functioning of the temporal lobes in monkeys. *Archives of Neurology and Psychiatry*, *42*, 979–1000.

Lombroso, C. (1876). *L'uomo deliquente: In rapporto all'antropologia, alla qiurisprudenza ed alle discipline carcerarie*. Rome: Bocca.

Macmillan, M. (2000). *An odd kind of fame: Stories of Phineas Gage*. Cambridge, MA: The MIT Press.

Papez, J. (1937). A proposed mechanism of emotion. *Archives of Neurology and Psychiatry*, *38*, 725–743.

Polyak, S. (1957). *The vertebrate visual system*. Chicago: University of Chicago Press.

Chapter 19

American Psychiatric Association. (2000). *Diagnostic and statistical manual of mental disorders* (4th ed., text rev.). Washington, DC: Author.

Blueler, E. (1911). *Dementia praecox oder die Gruppe der Schizophrenien*. Leipzig: Deuticke.

Jacobsen, C. F. (1935). Functions of frontal association area in primates. *Archives of Neurology and Psychiatry*, *33*, 558–569.

Jacobsen, C. F., Wolf, J. B., & Jackson, T. A. (1935). An experimental analysis of the functions of the frontal association areas in primates. *Journal of Nervous and Mental Disease*, *82*, 1–14.

Kraepelin, E. (1913). *Lectures on clinical psychiatry* (3rd English ed.). T. Johnstone (Ed.). New York: William Wood & Co. Original work published in 1905 as *Einführung in die Psychiatrische Klinik. Zweiunddreißig Vorlesungen.*

López-Muñoz, F., Alamo, C., Cuenca, E., Shen, W. W., Clervoy, P., & Rubio, G. (2005). History of the discovery and clinical introduction of chlorpromazine. *Annals of Clinical Psychiatry*, *17*(3), 113–135.

Meduna, L. (1935). Versuche über die biologische Beeinflussung des Ablaufes de Schizophrenia. I. Camphour und Cardiazol Krampfe. *Zeitschrift für die Gesamte Neurologie und Psychiatrie*, *1935*, 235–262.

Moniz, E. (1936). *Tentatives opératoires dans le traitement de certaines psychoses*. Paris: Masson.

Willis, T. (1683). *Two discourses concerning the soul of brutes*. London: John Dring.

Chapter 20

Aserinsky, E., & Kleitman, N. (1953). Regularly occurring periods of eye motility, and concomitant phenomena, during sleep. *Science*, *118*(3062), 273–274.

Bremer, F. (1935). Cerveau "isole" et physiologie du sommeil. *Comptes Rendus des Séances et Mémoires de la Société de Biologie (Paris)*, *118*, 1235–1241.

Bremer, F. (1938). L'activité cérébrale et le problème physiologique du sommeil. *Bollettino della Societa Italiana di Biologia Sperimentale*, *13*, 271–290.

Fitzgerald, E. (1954). *Rubaiyat of Omar Khayyam*. London: Collins.

Moruzzi, G., & Magoun, H. W. (1949). Brain stem reticular formation and activation of the EEG. *Electroencephalography and Clinical Neurophysiology*, *1*, 455–473.

Mosso, A. (1881). *Über den Kreislauf des Blutes in menschlichen Gehirn*. Leipzig: Viet.

Searle, J. R. (1993). The problem of consciousness. In G. R. Bock & J. Marsh (Eds.), *Ciba Foundation Symposium 174: Experimental and Theoretical Studies of Consciousness*. New York: Wiley.

Index

Printed in the United States
by Baker & Taylor Publisher Services